Also available in the Inverse and Ill-Posed Problems Series:

Ill-Posed Boundary-Value Problems
S.E. Temirbolat

Linear Sobolev Type Equations and Degenerate Semigroups of Operators
G.A. Sviridyuk and V.E. Fedorov

Ill-Posed and Non-Classical Problems of Mathematical Physics and Analysis
Editors: M.M. Lavrent'ev and S.I. Kabanikhin

Forward and Inverse Problems for Hyperbolic, Elliptic and Mixed Type Equations
A.G. Megrabov

Nonclassical Linear Volterra Equations of the First Kind
A.S. Apartsyn

Poorly Visible Media in X-ray Tomography
D.S. Anikonov, V.G. Nazarov, and I.V. Prokhorov

Dynamical Inverse Problems of Distributed Systems
V.I. Maksimov

Theory of Linear Ill-Posed Problems and its Applications
V.K. Ivanov, V.V. Vasin and V.P. Tanana

Ill-Posed Internal Boundary Value Problems for the Biharmonic Equation
M.A. Atakhodzhaev

Investigation Methods for Inverse Problems
V.G. Romanov

Operator Theory. Nonclassical Problems
S.G. Pyatkov

Inverse Problems for Partial Differential Equations
Yu. Ya. Belov

Method of Spectral Mappings in the Inverse Problem Theory
V. Yurko

Theory of Linear Optimization
I.I. Eremin

Integral Geometry and Inverse Problems for Kinetic Equations
A.Kh. Amirov

Computer Modelling in Tomography and Ill-Posed Problems
M.M. Lavrent'ev, S.M. Zerkal and O.E. Trofimov

An Introduction to Identification Problems via Functional Analysis
A. Lorenzi

Coefficient Inverse Problems for Parabolic Type Equations and Their Application
P.G. Danilaev

Inverse Problems for Kinetic and Other Evolution Equations
Yu.E. Anikonov

Inverse Problems of Wave Processes
A.S. Blagoveshchenskii

Uniqueness Problems for Degenerating Equations and Nonclassical Problems
S.P. Shishatskii, A. Asanov and E.R. Atamanov

Uniqueness Questions in Reconstruction of Multidimensional Tomography-Type Projection Data
V.P. Golubyatnikov

Monte Carlo Method for Solving Inverse Problems of Radiation Transfer
V.S. Antyufeev

Introduction to the Theory of Inverse Problems
A.L. Bukhgeim

Identification Problems of Wave Phenomena - Theory and Numerics
S.I. Kabanikhin and A. Lorenzi

Inverse Problems of Electromagnetic Geophysical Fields
P.S. Martyshko

Composite Type Equations and Inverse Problems
A.I. Kozhanov

Inverse Problems of Vibrational Spectroscopy
A.G. Yagola, I.V. Kochikov, G.M. Kuramshina and Yu.A. Pentin

Elements of the Theory of Inverse Problems
A.M. Denisov

Volterra Equations and Inverse Problems
A.L. Bughgeim

Small Parameter Method in Multidimensional Inverse Problems
A.S. Barashkov

Regularization, Uniqueness and Existence of Volterra Equations of the First Kind
A. Asanov

Methods for Solution of Nonlinear Operator Equations
V.P. Tanana

Inverse and Ill-Posed Sources Problems
Yu.E. Anikonov, B.A. Bubnov and G.N. Erokhin

Methods for Solving Operator Equations
V.P. Tanana

Nonclassical and Inverse Problems for Pseudoparabolic Equations
A. Asanov and E.R. Atamanov

Formulas in Inverse and Ill-Posed Problems
Yu.E. Anikonov

Inverse Logarithmic Potential Problem
V.G. Cherednichenko

Multidimensional Inverse and Ill-Posed Problems for Differential Equations
Yu.E. Anikonov

Ill-Posed Problems with A Priori Information
V.V. Vasin and A.L. Ageev

Integral Geometry of Tensor Fields
V.A. Sharafutdinov

Inverse Problems for Maxwell's Equations
V.G. Romanov and S.I. Kabanikhin

INVERSE AND ILL-POSED PROBLEMS SERIES

Inverse Problems of Mathematical Physics

INVERSE AND ILL-POSED PROBLEMS SERIES

Inverse Problems of Mathematical Physics

M.M. Lavrentiev, A.V. Avdeev,
M.M. Lavrentiev, Jr. and V.I. Priimenko

///VSP///

UTRECHT · BOSTON
2003

VSP
an imprint of Brill Academic Publishers
P.O. Box 346
3700 AH Zeist
The Netherlands

Tel: +31 30 692 5790
Fax: +31 30 693 2081
vsppub@compuserve.com
www.brill.nl
www.vsppub.com

© Koninklijke Brill NV, 2003

First published in 2003

ISBN 90-6764-396-3

All rights reserved. No part of this publication may be reproduced, stored in a retrieval system, or transmitted in any form or by any means, electronic, mechanical, photocopying, recording or otherwise, without the prior permission of the copyright owner.

Printed in The Netherlands by Ridderprint bv, Ridderkerk.

Preface

The present book describes a part of the theory of the so-called *Inverse Problems of Mathematical Physics* and some applications of such problems. Mostly the theoretical aspects of Inverse Problems are discussed. Besides, we also consider some applications and numerical methods of solving the problems under study. Descriptions of particular numerical experiments are also included.

The theory of *Inverse Problems of Mathematical Physics* is a vast and intensively developing field of modern mathematics. Plenty of publications appear, and even a number of specialized journals are published. Because of extended area of applications, many various statements of problems are considered, and diverse methods are used for their solution.

We stress attention at providing a concept of versatility and complexity of inverse problems arising in applications. We did not pursue the aim of giving the complete review of literature, instead we pointed out the most popular textbooks and characteristic statements of the problems. Also, we often pointed out the connections of such problems with various applications of methods of mathematical simulation. The references cited are mainly of illustrative character. At the same time, we gave references to the most frequently cited monographs which contain further references and a more complete account of the history of this field. Meaning to provide an introduction intended for specialists in other fields, we tried to emphasize the basic general principles and approaches to solution of various problems, supplying them with concrete examples of results obtained.

The monograph is arranged as follows. In Introduction we explain our understanding of the concept of Mathematical Modeling, outline the general differences between direct and inverse problems, and give strict mathematical definitions of correct and ill-posed problems.

In Chapter 1, we show up a list of applied areas, where inverse problems have been successfully used for years. Of course, it is impossible to list all such applications, so we took a liberty to mention somehow the *basic fields*, some of which historically developed along with the theory of inverse problems.

In practice, almost all problems are solved approximately, in mathematical sense. So, in Chapter 2, we give some of the basic definitions related to various regularizations of inverse problems. We hope that the reader who does not need strict definitions can get all the general ideas just browsing the *text part* of the chapter.

Chapter 3 is dedicated to the problems of Integral Geometry, which are both classical (accounting Radon transform, e.g.) and non standard. Some recent results, previously available only through specialized journals, are included.

Chapter 4 is arranged in non traditional way providing the reader with a sort of overview of one-dimensional inverse problems. In spite of usual separation of model equations of hyperbolic and parabolic type, the chapter is compiled as follows. First, Lamé system is described from physical statement to model simplifications and uniqueness and stability results. Second, the so-called "quasi-stationary approximation" of Maxwell system is concerned. The point is that the complete proof through analytic relations between solutions to equations of hyperbolic and parabolic types is given. Then, a concept of relations among inverse problems for different type governing equations are discussed. Such concept is not that new, but perhaps is not well known. The next sections is dedicated to brief descriptions of such fundamental methods of investigation as the separation of singularities and the reduction of one-dimensional inverse problem with a focused source of disturbances to a linear integral equation. Finally, the determination of the piece-wise constant coefficient for wave equation is considered. The example of inverse problem (arising in applications) is given, in which case recurrent algorithm for exact determination of equation coefficients is available.

In Chapter 5 we consider some inverse problems for the coupled Maxwell and Lamé systems. First of all, the solution of the one-dimensional inverse problems for the equations of electromagnetoelasticity in the case of seismomagnetic interaction is studied. Then we give some results of the solution of inverse problems for the system of electromagnetoelasticity in the case of piezoelectric effect. In the next section, the linear process of interaction of electromagnetic and elastic waves in a weakly conducting elastic medium is considered. Finally, we give some results of solution of direct and inverse problems for the system of electromagnetoelasticity in the case of nonlinear interactions between electromagnetic and elastic fields.

Chapter 6 contains some examples of numerical solution of the different type inverse problems arising in applications. The first section is a small survey in numerical methods for inverse problems. The next section represents the numerical solution of a 3D inverse kinematic problem of seismics. Then we describes how the proposed algorithm of numerical solution could be used for the determination of the structure of the Earth's upper mantle. The numerical solutions of two inverse problems of electromagnetoelasticity are discussed in the next section. We would like to draw attention on the last section, which represents the results of numerical modeling of coastal profile evolution. This application of inverse problems is new and not well developed at the moment.

Acknowledgement

The authors are grateful to Academician of RAS, Professor A. S. Alekseev, Corresponding Members of RAS, Professors V. G. Romanov, V. V. Vasin, and B. G. Mikhailenko, and to Professors Yu. E. Anikonov, S. I. Kabanikhin, V. G. Yakhno, and V. A. Cheverda for useful discussions and kind permission to use some their results.

Also, the authors are grateful to A. Nazarov and D. Nechaev for their kind and helpful association in preparation of the camera-ready.

Finally, we would like to mention that this book was written under financial support, in part, of the Russian Foundation for Basic Research (projects Nos. 01-05-64704, 02-05-64939) and the U.S. Civil Research Development Foundantion (project No. RG1-2415-NO-2).

Mikhail M. Lavrentiev,
Alexander V. Avdeev,
Mikhail M. Lavrentiev, Jr.,
Viatcheslav I. Priimenko

Contents

Introduction   **1**
0.1. The concept of mathematical simulation 1
0.2. Direct and inverse problems 3
0.3. On correctness of direct and inverse problems of mathematical
       physics 5

Chapter 1. Some physical motivations of inverse problems   **13**
1.1. Inverse problems of geophysics 13
1.2. Inverse tomography problems 22

**Chapter 2. Approximate methods of solution of ill-posed
       problems**   **29**
2.1. On some aspects of statement and solution of ill-posed problems . 29
2.2. Solutions on compact sets. The concept of a quasi-solution 30
2.3. The method of quasi-inversion 41
2.4. Regularization methods 47

Chapter 3. Integral geometry problems   **53**
3.1. Statement of integral geometry problems 53
3.2. The Radon problem 53
3.3. The problem of general form on the plane 57
3.4. Problems of general form in the space 65
3.5. Problems of Volterra type with manifolds invariant under
       the motion group 77
3.6. Integral geometry problems with perturbation on the plane 81
3.7. Mathematical problems of tomography and hyperbolic mappings . 88

Chapter 4. One dimensional inverse problems — 103
4.1. Some inverse problems for Lamé system 106
4.2. Inverse problems for quasi-stationary Maxwell's equations 112
4.3. Connections among inverse problems of hyperbolic, elliptic,
and parabolic type 118
4.4. Problems with a focused source of disturbances 125
4.5. Reducing the problem with a focused source of disturbances to a
linear integral equation: necessary and sufficient conditions
for solvability of the inverse problems 135
4.6. Determination of the piece-wise constant coefficient for wave
equation 140

Chapter 5. Inverse problems for the coupled Maxwell and Lamé systems — 147
5.1. One-dimensional inverse problem of electromagnetoelasticity
in the case of the seismomagnetic effect 148
5.2. Inverse problems of electromagnetoelasticity in the case of
piezoelectric interaction 156
5.3. Inverse problems of electromagnetoelasticity for weakly
conducting media 165
5.4. An inverse problem of electromagnetoelasticity in the case of
nonlinear interaction 179

Chapter 6. Numerical solution of inverse problems: some examples — 191
6.1. Short review of numerical approaches to solving inverse problems 191
6.2. Numerical solution of a 3D inverse kinematic problem of seismics 196
6.3. Determination of the structure of the Earth's upper mantle 209
6.4. Numerical solution of inverse problems of
electromagnetoelasticity 229
6.5. Simulation of the long-term coastal profile evolution 245

Bibliography — 259

Introduction

0.1. THE CONCEPT OF MATHEMATICAL SIMULATION

The statement of direct and inverse problems of mathematical physics implies preliminary schematic representation of the real process in a certain mathematical form. At present, modern technology is inconceivable without mathematical modeling, which is understood as replacement of the original object under investigation by its *model*. The aim of such a replacement is to study the properties of the object with the help of its model.

There are *experimental* (physical) modeling and *theoretical* (mathematical) simulation. In the former case the process is studied on the basis of real experiments on mock-ups, physical models, laboratory installations, etc. Mathematical simulation consists in construction and investigation of quantitative values for physical parameters through a simplified model of the process under study, which is formulated in mathematical terms. As a rule, a mathematical model should be *adequate* to the physical process.

Mathematical models contain unknown characteristics. The choice and evaluation of these characteristics is a difficult problem which is solved on the basis of accumulated experience, available experimental data, physical laws, etc. To develop a mathematical model it is necessary to pass through two main stages: *identification*, i.e., the choice of the type (structure) of the model, and *determination* of the numerical values of the model parameters and characteristics.

The structure of a model is conceived as a qualitative character of the mathematical description of the processes under investigation. Thus, physical laws are most frequently represented in the form of differential equations. One can distinguish models with *concentrated parameters*, in which case physical parameters are independent of space and only their evolution in time is the subject of study. Such models are described by systems of

Ordinary Differential Equations. On the contrary, models with *distributed parameters* involve spatial distribution of physical fields in addition to temporal evolution. Such models are based on *Partial Differential Equations* (equations in partial derivatives).

Such classification is very rough. Indeed, one can distinguish stationary (steady-state) and non-stationary (dynamic) models, linear and nonlinear models, one-dimensional and multidimensional (in space variables) models, etc.

Henceforth we shall deal with models represented in the form of differential equations in partial derivatives (PDE's). According to the general theory of such equations, a number of additional conditions should be attached to the equation itself to single out a unique element from the whole of the set of solutions to this equation.

We shall mostly speak about *linear governing equations of second order.* It means that the changes of physical variables (displacement, temperature, concentration, field intensity, etc.) are described by linear differential equations, whose coefficients represent the physical properties of the medium where the process being modeled takes place.

Example 0.1.1. The equation

$$\frac{\partial^2 u}{\partial t^2} = c^2 \frac{\partial^2 u}{\partial z^2}, \qquad (0.1.1)$$

describes small oscillations of a string. In this case the function $u(z,t)$ is the displacement of the string, at a point z and at a moment t, from the equilibrium state, which is supposed to coincide with the z-axis. The coefficient c characterizes the speed of disturbance propagation along the string. If the string is bounded, $z \in [0, l]$, and fixed at the endpoints $z = 0$ and $z = l$, then the boundary conditions for (0.1.1) have the form

$$u(0, t) = u(l, t) = 0, \qquad t \in \overline{\mathbb{R}_+}, \qquad (0.1.2)$$

where $\overline{\mathbb{R}_+} = [0, \infty)$. To set the boundary conditions is not sufficient to single out a unique solution to (0.1.1). We arrive at a unique solution only if some additional initial conditions are used. In this case they are the initial string displacement and the initial velocities of its points:

$$u(z, 0) = \varphi(z), \qquad \frac{\partial u}{\partial t}(z, 0) = \psi(z), \qquad z \in [0, l]. \qquad (0.1.3)$$

One can prove that conditions (0.1.2) and (0.1.3) define a unique solution to (0.1.1).

0.2. DIRECT AND INVERSE PROBLEMS

In some statements, the coefficients of the governing equations are considered as given, thus, the problem is to study the properties of solutions to the model equations. In the case where the initial and boundary conditions are set "properly" (there exists a unique solution to the problem under study and this solution depends continuously on the parameters of the problem), we shall refer to such statements as *Direct Problems*.

The statement of each direct problem implies prescribing some set of functions, i.e., the coefficients of the equations, sources (right-hand sides of the equations), external actions (nonhomogeneities and the coefficients of the boundary conditions), etc. In the result of solving a direct problem, a new set of functions — the solutions to the direct problem — is placed into correspondence to that original set. Thus, the operator of a given direct problem is defined; i.e., the operator which maps the data of the problem into its solution.

Assume now that some functions among the data of a "properly stated" direct problem are unknown; and, instead, some additional information on the solution of the problem is given. Such problems will be referred to as *Inverse Problems*.

In particular, inverse problems of mathematical physics are often understood as problems of determining the internal characteristics of a medium (as a rule, they cannot be measured directly) from a certain information on the values of various physical fields (parameters) at the boundary of a certain domain.

Example 0.2.1. The wave propagation in a vertically-inhomogeneous medium can be described by the following problem:

$$\frac{\partial^2 u}{\partial t^2} = c^2(x_3)\Delta u, \qquad x \in \mathbb{R}^3, \quad t \in \mathbb{R}, \qquad (0.2.1)$$

$$u\big|_{t<0} \equiv 0, \qquad x \in \mathbb{R}^3, \qquad (0.2.2)$$

$$\frac{\partial u}{\partial x_3}\bigg|_{x_3=0} = f(t)g(x_1, x_2), \qquad (x_1, x_2, t) \in \mathbb{R}^2 \times \overline{\mathbb{R}}_+, \qquad (0.2.3)$$

where $c(x_3)$ is a function characterizing the velocity of wave propagation in the medium, and $f(t)$ and $g(x_1, x_2)$ describe the duration of action and the space distribution of sources on the free surface, respectively.

Inverse Problem 0.2.1 (IP 0.2.1). *Let the wave propagation be described by the system (0.2.1)–(0.2.3). It is necessary to reconstruct the* velocity distribution (to find the function $c(x_3)$) in the medium and/or the characteristics of the sources, $f(t)$ and $g(x_1, x_2)$, using the additional information about the vibration regime of the observation surface $x_3 = 0$

$$u|_{x_3=0} = u_0(x_1, x_2, t),$$
$$t \in [0, T], \quad (x_1, x_2) \in S \subset \{x \in \mathbb{R}^3 \mid x_3 = 0\}. \qquad (0.2.4)$$

To clarify better the relations and distinctions between Direct and Inverse Problems of Mathematical Physics, we give their "cause-and-effect" interpretation. Consider the given physical parameters of a medium (e. g., density, conductivity, etc.) along with the boundary and initial conditions, the geometry of the domain, etc. as *causal* characteristics. As *effects* we obtain the states of physical fields (temperature or concentration distributions, velocity fields, etc.), which are determined by solving the corresponding direct problem. So, to solve a *Direct Problem* means to describe the *effect* of given *causal* factors. On the contrary, solution of an *Inverse Problem* is interpreted as reconstruction of causal characteristics from their effect.

Therefore, in contrast to Direct Problems, the statements of some Inverse Problems do not correspond to any physically realizable events. Indeed, one cannot invert the direction of time-flow (in order to reconstruct the initial distribution of a physical field from its state at a given moment); it is also impossible to reverse the process of reagent diffusion or heat propagation. In this sense, one can say that a number of inverse problems are *"physically incorrect"*. In mathematical statements, naturally, this difficulty displays itself as *mathematical incorrectness*, which results in such complications as instability of a solution, multiple solutions, even absence of solutions, etc. These natural causes give rise to difficulties in development of reliable methods and algorithms to solve inverse problems.

That is why, in spite of existence of many general methods for solution of inverse problems, each concrete statement requires special theoretical treatment. Note that without such preliminary "analytic" investigation, it is practically impossible to create cost-effective and efficient numerical algorithms.

A natural approach to solving complex problems consists in constructing a series of models with increasing complexity that describe the initial statement more and more comprehensively. Consecutive study of these models allows us to determine, at initial stages, the most general qualitative properties of solutions. Later these general properties are determined more exactly in the course of study of more complex models.

0.3. ON CORRECTNESS OF DIRECT AND INVERSE PROBLEMS OF MATHEMATICAL PHYSICS

The notion of correctness is usually considered in the theory of Direct Problems of Mathematical Physics. When dealing with Inverse Problems of Mathematical Physics it is convenient to alter this notion a little. Below we give a short overview.

0.3.1. General notes about correct problems

The theory of differential equations states that a differential equation defines a whole set of its solutions which depend on a certain number of arbitrary constants or arbitrary functions. For the problem to have definite physical sense, we need to single out a unique solution. Usually it is achieved by setting initial and boundary conditions. This is illustrated by examples below.

Example 0.3.1.

Consider the so-called *heat equation*, which describes, e.g., the temperature evolution in a cooling body,

$$\frac{\partial u}{\partial t} = \Delta u \qquad (0.3.1)$$

in a cylindrical domain $\mathcal{G} = \Omega \times \overline{\mathbb{R}}_+$, where $\Omega \subset \mathbb{R}^3$ is a domain bounded by a closed surface S. To single out a unique solution in this case it is sufficient to set the heat regime on the surface S, for instance,

$$u(x,t) = 0, \qquad x \in S, \qquad t \in \overline{\mathbb{R}}_+, \qquad (0.3.2)$$

and the initial distribution of temperature inside Ω,

$$u(x,0) = g(x), \qquad x \in \Omega. \qquad (0.3.3)$$

Setting of the initial and boundary conditions is aimed at singling out a unique solution from the whole class of solutions to a differential equation. But the number of these conditions should be minimal, for otherwise they may contradict one another, in which case a solution to such problem does not exist.

As it is known (see, e.g., Ladyženskaya, Solonnikov, and Ural'tseva, 1968), there exists a unique solution to the problem (0.3.1)–(0.3.3). Moreover, small enough perturbations of the initial profile $g(x)$ cause *arbitrarily small* deviations of this solution over any finite time interval $t \in [0, T]$. One should remember that the main goal of solving mathematical problems is to describe certain physical processes in mathematical terms. In this case the initial data are obtained experimentally; and since measurements cannot be absolutely precise, the data contain measurement errors. For a mathematical model to describe a real physical process, the problem should be supplemented with some additional requirements reflecting, in a physical sense, the fact that the solution should have only small variations under slight changes in initial data or, to put it conventionally, the *stability* of the solution under small perturbations in the initial data.

Generally speaking, in such case the problem is said to be *correct*, while in alternative cases, *ill-posed* or *incorrect*.

0.3.2. Mathematical definitions

Now we put the above general idea on the strict theoretical (mathematical) basis. Given a differential equation with concrete initial and boundary conditions, we can pose the problem of finding its solutions belonging to various functional spaces. The choice of a concrete function class depends on the physical interpretation of the problem. For example, we can consider the problem of finding a solution to (0.1.1)–(0.1.3) in the class of functions $C^2(D)$, where $D = \{(z,t) \mid z \in [0,l],\ t \in \overline{\mathbb{R}_+}\}$, or in other classes. In other words, one can choose a functional space of solutions to a differential equation in quite an arbitrary way.

The functions involved in the boundary and initial conditions of the problems for differential equations cannot be chosen arbitrarily; they should ensure that the solution belongs to the chosen functional space. For this, they should belong to the certain special functional space corresponding to the space of solutions. This becomes clearer if one considers problems for differential equations from the viewpoint of functional analysis. Choose a space U for the solutions of a differential equation. The differential equation together with some additional conditions defines the operator A that relates any solution $u \in U$ to the set of functions involved in the additional (initial and/or boundary) conditions. For (0.1.1) these are the functions φ and ψ, for (0.3.1) it is the function g. Considering this set of functions as an element f of a functional space F, one comes to a conclusion that solving a problem for a differential equation is equivalent to solving the formal operator equation

$$Au = f \qquad (0.3.4)$$

under the condition that $u \in U$.

The solution to this equation exists if and only if the element f is the image of a certain element $u \in U$, i.e., it should belong to the set of values of the operator A. Thus, the set of data of the problem is defined by specifying the space of solutions U.

If there exists a unique solution to (0.3.4), then the inverse operator A^{-1} exists, too,

$$u = A^{-1}f, \qquad (0.3.5)$$

that solves the problem by fitting the solution $u \in U$ to the initial data of the problem, i.e., to the element f.

Let u be a solution to (0.3.4) wherein the operator A acts from the normed space U into the normed space F. A solution to (0.3.4) is said to be *stable* under small variations in the right-hand side $f \in F$ if for every $\varepsilon > 0$ there exists $\delta > 0$ such that for every element $\tilde{f} \in F$ satisfying

$$\|f - \tilde{f}\|_F \leq \delta,$$

the inequality

$$\|u - \tilde{u}\|_U \leq \varepsilon$$

holds true. Here $Au = f$ and $A\tilde{u} = \tilde{f}$.

The stability of a nonlinear operator A depends, generally speaking, on the element f; the operator may be stable for one set of elements and unstable for another set. In the case of a linear operator A, either stability or instability takes place for all elements $f \in F$ at once.

A mathematical problem of solving (0.3.4) which obeys the requirements of existence, uniqueness, and stability of the solution under small variations of the initial data is called a *correct problem*. The concept of well-posedness (correctness) was developed by a prominent French mathematician Hadamard. Below we give a detailed discussion. Let us abstract ourselves from the concrete nature of the operator A. Let A be an operator acting from a normed space U to a normed space F.

Definition 0.3.1. The problem of solving (0.3.4) is said to be formulated *correctly* in the spaces F and U if the solution to the problem

1) exists for every $f \in F$ and belongs to the space U;
2) is unique in the space U;
3) is stable at every element $f \in F$.

Problems of solving (0.3.4) that do not satisfy one or more of the above requirements will be called *ill-posed*.

Condition 1 means that the problem should not have redundant data which make it overdetermined. Condition 2 means that the data should be sufficient for singling out the unique solution. Condition 3 is associated with the following circumstance. If the problem is related to a physical phenomenon, its data cannot be considered as known exactly. We can only assume that the exact values are approximated arbitrarily close. Consequently, if the solution is not stable under small variations of data, it is not actually determined at all.

Note that a problem may be correct in one pair of spaces and ill-posed in another pair. It is clear, for instance, that when the space F is extended, the requirement that a solution should exist for every $f \in F$ may be violated. In the case of a linear operator A, the problem (0.3.4) is correct in a pair of Banach spaces U and F if and only if for the operator A there exists a bounded inverse operator A^{-1} acting from F into U, the domain of definition of the inverse operator coinciding with the space F.

Below we give an example of a correct problem.

Example 0.3.2. Consider equation (0.1.1) (with $c \equiv 1$) in the half-plane $(z, t) \in \mathbb{R} \times \mathbb{R}_+$. In this case, it is sufficient to set the Cauchy data

$$u(z, 0) = \varphi(z), \qquad \frac{\partial u}{\partial t}(z, 0) = \psi(z), \qquad z \in \mathbb{R}. \qquad (0.3.6)$$

The solution to the problem (0.1.1), (0.3.6) is unique and is given by the d'Alembert formula

$$u(z, t) = \frac{1}{2}[\varphi(z - t) + \varphi(z + t)] + \frac{1}{2}\int_{z-t}^{z+t} \psi(\xi)\, d\xi. \qquad (0.3.7)$$

For arbitrary $T \in \mathbb{R}_+$ consider the characteristic triangle

$$D_T = \{(z, t) \mid 0 \leq t \leq T - |z|\}.$$

Formula (0.3.7) yields that for $\varphi \in C^2[-T, T]$ and $\psi \in C^1[-T, T]$ the solution $u(z, t)$ belongs to the function class $C^2(D_T)$.

The uniqueness of the solution to the problem (0.1.1), (0.3.6) in the class $C^2(D_T)$ is proved easily. Moreover, it can be shown that the solution given by (0.3.7) is stable:

$$\|u\|_{C^2(D_T)} \leq K\left(\|\varphi\|_{C^2[-T,T]} + \|\psi\|_{C^1[-T,T]}\right). \qquad (0.3.8)$$

In other words, for arbitrary $\varepsilon > 0$ we have $\|u\|_{C^2(D_T)} < \varepsilon$ in case the initial functions φ and ψ are small enough (small variations of the functions φ and ψ result in small changes of the solution in the norm $\|\cdot\|_{C^2(D_T)}$).

Thus, the problem is correct for $\varphi \in C^2[-T,T]$, $\psi \in C^1[-T,T]$, and $u \in C^2(D_T)$ for every $T \in \mathbb{R}_+$.

Note that the same problem becomes ill-posed if one assumes that both functions φ and ψ are only continuous, still considering the same space $C^2(D_T)$ for solutions. Indeed, the first and third conditions of correctness are not satisfied in this case.

Now consider an example of an ill-posed problem which remains ill-posed for arbitrarily smooth initial information. Such a problem was first formulated by Hadamard to emphasize the importance of the third condition of correctness.

Example 0.3.3. Let $u(x_1, x_2)$ be a solution to the Laplace equation

$$\Delta u = 0 \qquad (0.3.9)$$

in the domain $\mathcal{G} = \{(x_1, x_2) \mid x_1 \in (-\pi, \pi); \, x_2 \in \mathbb{R}_+\}$ satisfying the conditions

$$u(-\pi, x_2) = u(\pi, x_2) = 0, \quad u(x_1, 0) = 0,$$
$$\frac{\partial u}{\partial x_2}(x_1, 0) = e^{-\sqrt{n}} \sin n x_1. \qquad (0.3.10)$$

It is easy to check that the solution to this problem is

$$u(x_1, x_2) = \frac{1}{n} e^{-\sqrt{n}} \sin n x_1 \, \sinh n x_2. \qquad (0.3.11)$$

One can prove that the solution to this problem is unique. As $n \to \infty$ the function $e^{-\sqrt{n}} \sin n x_1$, which represents the data of the problem (0.3.9), (0.3.10), tends uniformly to zero together with all its derivatives. As is seen from (0.3.11), for every fixed $x_2 \in \mathbb{R}_+$ the solution to the problem has the shape of a harmonic curve with arbitrarily big amplitude if n is sufficiently large. Hence, arbitrarily small changes in the problem data in $C^k[-\pi, \pi]$ at every finite k result in substantial variations in the solution. Thus, the problem (0.3.9), (0.3.10) is ill-posed due to its instability.

The above examples show that, generally, there exist two types of ill-posed problems. There are problems that are ill-posed in one set of spaces but can be made correct in another set. There also exist ill-posed problems (such as, for instance, (0.3.9), (0.3.10)) which are ill-posed in any normed spaces whose norms involve a finite number of derivatives.

For problems which are not correct in the classical sense, Tikhonov (1943) suggested a new notion of correctness which was physically justified for many applied problems. Within his approach, Tikhonov considered a very restricted set $M \subset U$ that is essentially narrower than the whole of the normed space U. Usually M is a compact set. We shall denote by R the image AM of the set M in the normed space F under the mapping realized by the operator A; $R = AM$.

Definition 0.3.2. The problem of solving (0.3.4) is called *conditionally correct* (or *well-posed in the Tikhonov sense*) if

1) it is known *a priori* that the solution u exists and belongs to a certain given set M of the normed space U, i.e., $u \in M \subset U$;

2) the solution is unique in the set M;

3) for every $\varepsilon > 0$ there exists $\delta > 0$ such that for any $f, \tilde{f} \in R = AM$ satisfying the condition $\|f - \tilde{f}\|_F \leq \delta$ the inequality $\|u - \tilde{u}\|_U \leq \varepsilon$ holds.

The set M is called *a set of correctness* (*set of well-posedness*) of the problem.

Below we explain the difference between problems well-posed in the Tikhonov sense and problems well-posed in the classical sense. In the classical definition of correctness (Definition 0.3.1) the problem data f are assumed to belong to a certain normed space F. The first condition demands to prove that for every $f \in F$ the solution to (0.3.4) exists and belongs to U. The corresponding condition of correctness in the Tikhonov sense postulates that the solution to (0.3.4) belongs to the given set M without discussing the necessary properties of the set $R = AM$. This is associated with the fact that for many ill-posed problems the conditions of the belonging $f \in AM$ cannot be verified practically.

There is no difference between studying the questions of uniqueness for problems well-posed in the Tikhonov sense on the set M and for problems well-posed in the classical sense on the space U.

Now consider the third condition of correctness. Within the classical approach the solution to the problem exists for every $f \in F$ and belongs to the space U. It is important that any small change in the element f in the norm of the space F leads to a small change in the solution in the norm of the space U. Tikhonov stated that correctness does not require numerous initial data at all and postulated, instead, that the solution belongs to a certain set M. In this case any small change in the initial data in the norm of F can result either in the nonexistence of a solution to the problem or, even if the solution still exists, its non-belonging to M. Therefore it would be reasonable to modify the requirement of continuous dependence of the solution on the problem data by demanding the continuous dependence for such variations in f that retain the solution within the set M; this is reflected in the third condition of correctness.

In connection with the definition of conditional correctness note also that two of the three conditions of correctness are modified in it. Of course, if in the case of an unstable problem one can constructively describe the set of data, i. e., the set $R = AM$, then it is not reasonable to abandon the first requirement of correctness in its classical variant.

Many problems that are ill-posed in the classical sense are conditionally correct. For instance, the problem (0.3.9), (0.3.10) is conditionally correct in the set of functions $u(x_1, x_2)$, $x_1 \in [-\pi, \pi]$, $x_2 \in [0, 1]$, belonging at every fixed $x_2 \in [0, 1]$ to the space $L_2[-\pi, \pi]$ and satisfying the additional condition

$$\int_{-\pi}^{\pi} u^2(x_1, 1)\, dx_1 \leq C^2,$$

where C is a given constant.

When the operator A is continuous and M is a compact set, general theorems of functional analysis yield that if the first and second conditions of correctness are satisfied, then the third condition is satisfied, too. This result serves as the basis for defining the conditional correctness of the problem. A problem unstable throughout the whole space F may be stable on the set $R = AM$; the correct type of stability estimation depends crucially on the set M. Within the new approach to the definition of correctness the focus point is the uniqueness of the problem solution, i. e., establishing the uniqueness theorem.

Let us explain why the concept of the conditional well-posedness is natural in studying many applied and theoretical problems and especially in studying the inverse problems for differential equations. Consider, e. g., the inverse problem connected with oscillations of a string. When the density of the string changes from one point to another, the process of small oscillations of the string is described by equation (0.1.1) with nonconstant parameter $c = c(z)$. Consider the inverse problem of determining the function $c(z)$ with the help of some additional information on the solution to the direct problem (0.1.1)–(0.1.3). It follows from the physical statement of the problem that the solution to the problem should be sought in the class of positive functions that are bounded from above and below by some constants. One can assign the values of these constants if one knows the possible set of materials of which the string can be made. As a result, we can determine the set M to which the function $c(z)$ belongs.

If the data of the problem are taken from a concrete physical experiment with a string such that $c(z) \in M$, then, obviously, there exists the actual function $c(z)$ corresponding to this experiment. Now the question is how to find this function and whether this information is sufficient for its unique determination.

It can be easily understood that the theory of ill-posed problems consists not only of inverse problems (though they constitute the widest field of application of the theory of ill-posed problems). To understand this it suffices to recall such classical problems as differentiation of a function which is given approximately, or the problem of numerical summation of the Fourier series. Consider, for example, the problem of differentiating a function $f(z)$ which is given approximately. Let $f_1'(z) = u_1(z)$, $z \in [0, \pi]$. The function $f_2(z) = f_1(z) + \sin(nz)$ differs from the function $f_1(z)$ in the norm of the space $C[0, \pi]$ by the value $\|f_1 - f_2\|_{C[0,\pi]} = 1$ for every $n \in \mathbb{N}$. But the derivative $f_2'(z) = u_2(z)$ differs from $u_1(z)$ in the norm of $C[0, \pi]$ by the value $\|u_1 - u_2\|_{C[0,\pi]} = n$ which can be arbitrarily large for sufficiently large n. Therefore, the problem under consideration is ill-posed in this pair of spaces, since the third condition of Definition 0.3.2 is violated.

However, if the distance between the functions f_1 and f_2 is calculated in the norm of the space $C^1[0, \pi]$:

$$\|f_1 - f_2\|_{C^1[0,\pi]} = \max_{z \in [0,\pi]} \{|f_1(z) - f_2(z)| + |f_1'(z) - f_2'(z)|\},$$

and the distance between the functions u_1 and u_2 is estimated in the norm of the space $C[0, \pi]$, then the problem under study is well-posed in this pair of spaces, since the smallness of the quantity $\|f_1 - f_2\|_{C^1[0,\pi]}$ implies the smallness of $\|u_1 - u_2\|_{C[0,\pi]}$.

In the sequel, we shall sometimes transcend the frames of the study of inverse problems for differential equations and shall consider such ill-posed problems that are not inverse problems for differential equations.

Ill-posed problems are called weakly ill-posed if there exits a pair of functional spaces whose norm is defined by a finite number of derivatives and the solution depends by a finite number of derivatives and the solution depends continuously on data in these spaces. An example of such problem is the noted problem of differentiation. Ill-posed problems for which such a pair cannot be found are called strongly ill-posed (Examples 0.1.1, 0.3.3).

Chapter 1.

Some physical motivations of inverse problems

1.1. INVERSE PROBLEMS OF GEOPHYSICS

One of the widest and most developed application area of the theory of inverse problems is connected with geophysics. It includes the study of the Earth's structure, search for natural resources, study of the ocean, etc.

The geophysical methods are based on observations of physical fields whose distributions in space (and in time, if the fields are nonstationary) depend on the structure and composition of the Earth and, in particular, the Earth's crust.

Various methods can be classified according to the physical nature of the fields involved. In that view we distinguish *seismics*, which involves the study of elastic waves; *geoelectrics*, which involves the study of electric and electromagnetic fields; *gravimetry*, which deals with gravitational fields; *magnetometry*, which deals with stationary magnetic fields; and *radiometry*, which involves the study of the natural or induced radioactivity of rock and soil.

Geophysical prospecting involves two major types of problems:

- structural problems which deal with the structure of sedimentary deposits and the Earth's crust, in particular, the structures that are probably associated with oil and gas resources;

– location problems connected with determining local inhomogeneities in sedimentary deposits and crust that are due to mineral deposits such as unstructured oil and gas deposits, ore bodies, etc.

Without going into details, here we just mention that we refer to *inverse problems of geophysics* as the problems to recover the inner structure of the Earth (Earth's crust) through the surface measurements of physical fields.

Depending on the prospecting goals, different types of "model structure" hypotheses are used to solve inverse problems. For structural prospecting, the basic model is that of a stratified medium in which the medium's parameter $\alpha = \alpha(x_3)$ is supposed to be a piecewise continuous function of depth. Usually this parameter is taken to be piecewise constant to describe a layer-homogeneous medium. A more complicated model is that of a quasi-stratified medium in which seams with constant α_k have variable thickness $h_k(x_1, x_2)$.

For location problems, two classes of models are usually considered:

– a homogeneous medium containing a system of local bodies with constant parameters α_k;

– a homogeneous medium with a zone with variable parameter $\alpha(x)$, $x = (x_1, x_2, x_3) \in \mathbb{R}^3$.

1.1.1. Inverse problems of seismics

The problems of seismics are understood as the study of elastic vibrations in the Earth's crust using measurements made at its surface. Elastic fields carry information about, for instance:

– properties of the field sources;

– the structure of areas of the Earth's interior, where seismic waves travel through.

So, as was mentioned above, such problems are referred to as inverse problems.

These problems can be mathematically formulated in several different ways depending on the nature of the seismic waves (volume, surface, etc.), on relative dimensions of the irregularities sounded (beam and wave approximations), and so on. In any case, a description of the Earth's interior structure results from certain additional information about seismic fields. The corresponding mathematical problems are ill-posed (see Introduction).

The basic problem of geophysics consists in studying the internal Earth's structure on the basis of observations on its surface or in a near-surface layer. Consider some examples of the inverse problems of seismics.

At the end of the 19th century the following question was posed in geophysics. Is it possible to determine the velocity of propagation of seismic waves inside the Earth if we know the data on the seismic wave fronts on the Earth's surface from different earthquakes? It should be noted that this velocity depends on the elastic properties of the matter. For example, the velocities of longitudinal waves v_p and transverse (shear) waves v_s are connected with the elastic properties of the medium λ, $\varkappa$, and ρ by the formulas

$$v_p = \sqrt{\frac{\lambda + 2\varkappa}{\rho}}, \qquad v_s = \sqrt{\frac{\varkappa}{\rho}},$$

where λ and $\varkappa$ stand for the Lamé parameters and ρ is the density of the material.

From the point of view of mechanics, the Lamé parameters λ and $\varkappa$ and the density ρ characterize completely an elastic substance. Therefore, one of the most important problems of seismology consists in finding the propagation velocities v_p and v_s of longitudinal and transverse seismic waves, respectively, as functions of spatial location.

The physical statement of the above problem is as follows. In the domain $D \subset \mathbb{R}^3$ bounded by a surface S we consider the process of propagation of seismic waves generated by sources of perturbations concentrated at points $x^0 \in S$; the waves from the sources of perturbations propagate in the domain D with finite velocity $v(x)$; at the points of the boundary we measure the "travelling time" $\tau(x, x^0)$ that it takes the wave to travel from the *source* $x^0 \in S$ to the *receiver* $x \in S$.

In the case of an isotropic medium the function $\tau(x, x^0)$ inside the domain D satisfies the differential equation of first order (the well-known *eikonal equation*)

$$|\mathrm{grad}_x \tau(x, x^0)| = \frac{1}{v(x)}, \qquad x \in D, \tag{1.1.1}$$

under the condition that

$$\tau(x, x^0) = O(|x - x^0|), \qquad x \to x^0,$$

where grad_x is the gradient with respect to x. The equation $\tau(x, x^0) = t$ defines the wave front from the source x^0 at the time t; and $\tau(x^0, x^0) = 0$.

The mathematical formulation of our problem is as follows.

Inverse Problem 1.1.1 (IP 1.1.1). *Given the times $\tau(x, x^0)$, where x is an arbitrary point of the surface S and x^0 runs over some set $M \subset D \cup S$, it is required to find the wave propagation velocity $v(x)$ inside the domain D satisfying equation (1.1.1).*

This problem is known as the *inverse kinematic problem of seismics*. In 1905–1907 Herglotz and Wiechert (see (Herglotz and Wiechert, 1905)) considered the first inverse kinematic problem assuming a spherically symmetric model of the Earth and, moreover, that the velocity $v(x)$ grows monotonically with depth. They showed the unique solvability of the problem and obtained explicit formulas for finding the function $r/v(r)$, where $r = |x|$. In alternative cases the inverse kinematic problem, as was shown, may have multiple solutions. On this basis the first conclusions on the internal structure of the Earth were made.

The above "pure kinematic approach" has certain limitations due to a possible ambiguity and instability of solutions. Besides, to extract kinematic data from real seismograms observed, it is necessary, as a first step, to solve a very complicated problem of wave correlation (to determine the nature of waves whose travel times are used in solving the inverse problem). In complex subsurface geometry this problem of *wavefield decomposition* to kinematically isolated waves leads to the loss of its physical meaning.

These limitations of the kinematic approach compel us to use the wave field on the Earth's surface. It is possible to register, on the Earth's surface, the displacement vector $\boldsymbol{U}(x, t, x^0)$ of surface points as a function of t, too. The function $\boldsymbol{U}(x, t, x^0)$, as a function of the variables x and t, satisfies the dynamical system of elasticity

$$\rho \frac{\partial^2 \boldsymbol{U}}{\partial t^2} = \text{Div}\, T + \boldsymbol{F},$$

where

$$\text{Div}\, T = \Big(\sum_{j=1}^{3} \frac{\partial}{\partial x_j} T_{ij} \Big)_{i=1}^{3},$$

and $\boldsymbol{F}(x, t, x^0)$ represents external forces. For the stress tensor T we have the following defining relation in the case of an isotropic medium:

$$T = \lambda \, \text{tr}\, S \cdot I + 2\varkappa S,$$

where S is the strain tensor defined by the formula

$$S_{ij} = \frac{1}{2}\Big(\frac{\partial U_i}{\partial x_j} + \frac{\partial U_j}{\partial x_i} \Big), \qquad i,j = 1,2,3,$$

and I is the unit matrix of order 3×3.

Chapter 1. Some physical motivations

It is necessary to add to this equation some boundary condition which, for example, in the case of a free surface S has the form

$$T \cdot n\big|_{x \in S} = 0,$$

where n is the outward unit normal vector to S. In some cases it is reasonable to simulate the earthquake by a function $\boldsymbol{F}(x, t, x^0)$ of the type

$$\boldsymbol{F}(x, t, x^0) = \delta(x - x^0) f(t, x^0),$$

where $\delta(x - x^0)$ is Dirac's mass and $f(t, x^0) \equiv 0$ for $t < 0$. The function $f(t, x^0)$ may be regarded as an unknown function together with the Lamé parameters λ, $\varkappa$, and the density of the medium ρ which describe the medium. In this case the inverse problem can be formulated as follows:

Inverse Problem 1.1.2 (IP 1.1.2). *Find the density of the medium ρ, the elastic Lamé parameters λ and $\varkappa$, and the vector-function of sources $f(t, x^0)$ in the domain D if the vector $u(x, t, x^0)$ is known for $x \in S$, $t \in \overline{\mathbb{R}}_+$ and $x^0 \in M \subset \overline{D}$ for a certain M belonging to the boundary S or "near-boundary" layers of D.*

This problem is called the *inverse dynamic problem of seismics*.

1.1.2. Inverse problems of geoelectrics

The electromagnetic methods in geophysical prospecting use the fact that alternating electromagnetic fields penetrate into the Earth according to their electromagnetic properties and so it is possible to determine the rock structure. For example, the electrical conductivity of granite and basalt is comparatively low whereas that of sediments is relatively high. Interaction of the electromagnetic field with a medium is described by Maxwell's system of equations

$$\operatorname{rot} \boldsymbol{H} = \varepsilon \frac{\partial}{\partial t} \boldsymbol{E} + \sigma \boldsymbol{E} + \boldsymbol{j}, \qquad \operatorname{div} \varepsilon \boldsymbol{E} = \rho_e,$$

$$\operatorname{rot} \boldsymbol{E} = -\mu \frac{\partial}{\partial t} \boldsymbol{H}, \qquad \operatorname{div} \mu \boldsymbol{H} = 0,$$

where $\boldsymbol{E}$ and $\boldsymbol{H}$ are the intensities of the electric and magnetic fields, respectively; ε and μ are the dielectric and magnetic permeabilities, respectively; σ stands for the electric conductivity; ρ_e is the density of free charges; and $\boldsymbol{j}$ is the source of electric oscillations.

Thus, the task of electromagnetic prospecting is to determine the distribution of electromagnetic parameters ε, μ, and σ within the Earth by measuring the electromagnetic field at the Earth's surface.

Before the 50's, all electrical prospecting problems were based on steady electric currents. Dynamic statements involving nonstationary electromagnetic (EM) fields became available after the well-known paper of Tikhonov (Tikhonov, 1950).

Electromagnetic methods of geophysical prospecting use both artificial (generated) EM fields and natural ones (magnetotelluric fields), the sources of the latter being located in the Earth's magnetosphere and ionosphere. Natural fields vary slowly in time (with periods from one second to 24 hours) and, therefore, penetrate deeply into the Earth. Also these fields change slowly over the Earth's surface since the sources of the magnetotelluric fields are located high above the Earth. This make it possible to simplify the mathematical model of the magnetotelluric field substantially and to consider the field as depending only on the depth $x_3 = z$ in a stratified medium. Thus, a one-dimensional problem is reasonable in view of various applications. Such model looks like:

$$e''(z) + i\omega\mu_0\sigma(z)e(z) = 0, \qquad z \in [0, H],$$

$$e\big|_{z=0} = 1, \qquad (e' - i\sqrt{i\omega\mu_0\sigma_H}\, e)\big|_{z=H} = 0,$$

where $e(z) = E_1(z)/E_1(0)$ is the relative electric field in the Earth, ω is the frequency of the field variations over time, μ_0 is the magnetic permeability of vacuum, and σ_H is the electric conductivity of the homogeneous half-space ($z > H$).

In the above approximation, the inverse problem of magnetotelluric sounding (MTS) is stated as follows.

Inverse Problem 1.1.3 (IP 1.1.3). *Determine the electric conductivity $\sigma(z)$ from the known frequency dependence of the admittance of the electromagnetic field*

$$Y(\omega) = \frac{H_2(0)}{E_1(0)} = -\frac{i}{\omega\mu_0 E_1(0)}\frac{\partial E_1}{\partial z}\bigg|_{z=0} = -\frac{i}{\omega\mu_0}\frac{\partial e}{\partial z}\bigg|_{z=0}$$

measured at the Earth's surface $z = 0$.

The MTS method was proposed by Tikhonov (1946) along with a proper theoretical background. A uniqueness theorem for the MTS inverse problem in the class of piecewise analytic functions $\sigma(z)$ was proved. We would also like to refer to the papers by Cagniard (1953), Kato and Kikuchi (1950), Rikitake (1950) which were very influential for the further development of the MTS method.

1.1.3. Inverse problems of gravimetry and magnetometry

The gravitational field at the Earth's surface has a complicated structure, which is determined by the interior mass distribution. This field is conventionally subdivided into a standard component (the field of a "geoid" with an averaged density), a regional component which varies slowly across geological regions (in a large spatial scale, in other words), and an anomalous component due to local inhomogeneities (sources).

The purpose of gravitational prospecting is to study local sources of anomaly. For this, the standard field is excluded by conventional methods, while the regional component is considered as a background over which the anomalous field can be discerned using an initial treatment.

It is natural that the inverse problems of determining the characteristics of the field sources from an observed surface field are classified according to the components of the total field. Hence the inverse problems of Gravimetry and Magnetometry can be divided into:

– problems of determination to single out ore bodies (or groups of bodies), which result in anomalies;

– problems involving multilayered media or their fragments.

In addition, there is the third important class of problems, namely, to determine how a potential field could be extended towards its source. This is also an inverse problem since the procedure of stable extension is used to obtain approximate information about the depth of occurrence of the sources in accordance with the location of some characteristic points, such as the center of mass, or some surface points, etc., of the gravitating bodies.

The problems of magnetometry in their scalar statements are formally equivalent to those of gravimetry. Many problems of gravitational and magnetic prospecting are connected with the so-called *inverse potential problem*. Its physical formulation is as follows: the potential created by a body lying inside some domain bounded by a surface S is known outside this domain; it is necessary to find the shape and density of the body.

Consider the simplest mathematical problem connected with the inverse problem of Newton's potential. Suppose that we have a body with the density $\rho(x)$ which is located inside a finite domain in the space $\mathbb{R}^3$ bounded by a surface S. Denote the support of $\rho(x)$ (i.e., the set of points x such that $\rho(x) \neq 0$) as $\mathcal{D}$, and the characteristic function of $\mathcal{D}$ as $\chi_\mathcal{D}(x)$. Newton's potential $u(x)$ created by this body satisfies the Poisson equation

$$\Delta u = -4\pi \rho(x) \chi_\mathcal{D}(x). \tag{1.1.2}$$

Inverse Problem 1.1.4 (IP 1.1.4). *Determine the functions $\rho(x)$ and $\chi_\mathcal{D}(x)$ if we know the solution to equation (1.1.2) outside S.*

Since the solution to (1.1.2) has the form

$$u(x) = \int_\mathcal{D} \frac{\rho(\xi)\,d\xi}{|x-\xi|}, \qquad \xi = (\xi_1, \xi_2, \xi_3),$$

the above IP 1.1.4 is reduced to investigation of integral equations of first kind with respect to the functions $\rho(x)$ and $\chi_\mathcal{D}(x)$. Poincaré (1899) has shown that the geometrical characteristics of a body and its density cannot be determined simultaneously and uniquely from the gravitational field. Therefore, it is necessary to impose some simplifying assumptions about the density, for example, that the density is known. Under such assumption the first uniqueness theorem was proved. Later the inverse potential problem in different statements was studied by Lavrentiev (1962), Lavrentiev, Romanov, and Shishatskii (1986), Strakhov (1969), Tikhonov and Arsenin (1977) and others.

Nowadays the theory of inverse potential problems still attracts much attention, both in views of the theory and numerical methods of their solution.

1.1.4. Inverse problems in combined statements

Investigation of inverse problems in combined statements began in the middle of the fifties of the last century. The interest of geophysicists to this theme was caused by the fact that there appeared a possibility to analyze simultaneously different geophysical fields in the study of the Earth's structure both at the qualitative and quantitative levels by using computers.

The work by Lines, Schultz, and Treitel (1988) should be mentioned here, in which the numerical study of a possibility of reconstructing the medium structure was carried out using the data of the wave and gravity fields given at the surface. In this case it was assumed that there exists a functional relation between the velocity and density. Two approaches were compared: minimization of a complex functional taking into account the relation between the parameters of the medium and sequential minimization of data misfit functionals for each of the problems with recalculation of the parameters.

Numerical study of a combined inverse problem of seismics and geoelectrics was considered in Avdeev, Goryunov, and Priimenko (1996a), Avdeev, Goryunov, and Skazka (1995) and others.

The problem of multidisciplinary earthquake prediction study deserves special mention. The main concept of prediction is based on the maximally reliable prediction of large earthquakes. This leads to the use of the largest possible number of precursors, which provides a comprehensive control of the process of preparation of large earthquakes.

However, the use of multifactor criteria makes more complicated the conditions of reliable application of statistical methods and techniques of pattern recognition because in this case it is necessary to have a larger volume of *a priori* information. Moreover, "logical deadlocks" appear in cases where the multifactor and multidisciplinary criteria in different combinations give contradictory estimates of the fact itself of earthquake preparation and its tempo.

The direct problems of geophysics give only a description of transformation of the variations of the medium properties into the changes of the fields-precursors. The reverse procedure is necessary for the earthquake prediction: passing from the fields-precursors to the mechanical parameters of the medium.

However, in the geophysical monitoring of a seismic zone, each method of observation can give only insufficient reliability and few details in the result of solution of the corresponding inverse problem when the time and coordinates of the expected earthquakes are not definite. Therefore, the monitoring based on the solution of combined inverse problems using all the observed geophysical anomalies as initial data is necessary.

Omitting here the geophysical aspect of the problem, we would like to point out the paper by Alekseev (1992) which is, in our opinion, of substantial mathematical and methodological importance. It turned out that, in essence, the inverse problems in combined statements provide a possibility of a more successful solution than the study of each of the inverse problems separately, taking into account the data obtained in order to get the general idea of the medium in study.

The property of individual inverse problems to "bring" more information and "of better quality" into the combined inverse problem than it might appear at first glance was called by Alekseev *the property of complementability* (Alekseev, 1992). He gave a mathematical definition of the combined inverse problem and showed that it is not equivalent to a simple set of individual problems. Here it should be pointed out that the widely used term "joint inversion" is not the same as the solving of the combined inverse problem. The first term denotes a technical method whereas the second one denotes a new mathematical problem. There exist many theoretical questions which are connected with the uniqueness, existence, and stability of solutions of combined inverse problems. These questions are more difficult as compared with the same questions concerning individual inverse problems. Here we have no possibility to discuss in detail their distinctions and would like to refer the readers to Priimenko and Vishnevskii (2003a).

1.2. INVERSE TOMOGRAPHY PROBLEMS

Tomography is a scientific-technical direction that develops intensively in recent years. The word originates from the Greek words "tomos" (layer) and "graphy" (image). This word appeared in the scientific-technical literature in the twenties of the last century in connection with creation of medical X-ray tomographs. Intensive development of tomography began in the 70s of the last century. There are two basic aspects in a modern tomograph. First, this is a technical device that allows one to obtain the data of X-raying of an object under examination. Second, this is a transformation of the data obtained into an image of a layer of the object. The second aspect is based on mathematical theory.

Though until recently the term tomography was used mainly in connection with the problems of medicine, virtually the same principles have already been used in geophysics for a long time in reconstruction of sections of regions of the Earth's crust from the data of geophysical measurements along a profile.

1.2.1. Mathematical problems of tomography and hyperbolic mappings

Below, in Chapter 3, we consider a new class of mathematical problems related to interpretation of tomography data. The main assumption is that the sought distribution of absorption is an identically one function in the domain to be determined. These problems are connected with three known directions of mathematical physics: the Dirichlet problems for hyperbolic equations, the problems of small oscillations of a rotating fluid, and the problems of supersonic flows of an ideal gas.

By inverse tomography problems we mean the mathematical problems that are related to interpretation of tomography data. The first high-quality tomogram of a human brain was obtained in 1972. Since then tomography has found rather broad applications in medicine. Progress was made in industrial tomography, tomography in gas dynamics and plasma physics, and seismic tomography (see Natterer, 1986; Nikolaev, 1988; Pikalov and Preobrazhenskii, 1987).

It is X-ray tomography that is used most widely in medicine and industry. Interpretation of tomography data in medical X-ray tomography is connected with the Radon transform. We state the corresponding mathematical problem.

Suppose that $u(x,y)$ is a continuous function compactly supported in a bounded domain D (we can assume that D is the disk $D = \{(x,y) \mid x^2 + y^2 < 1\}$). The *Radon transform* of $u(x,y)$ is the function

$$f(x, y, \alpha) = \int_{-\infty}^{\infty} u(x + s \cos \alpha, y + s \sin \alpha) \, ds. \qquad (1.2.1)$$

We can consider (1.2.1) as a linear operator equation in $u(x,y)$. It is required to determine $u(x,y)$ if the function $f(x,y,\alpha)$ is given.

In the mathematical model of X-ray tomography $u(x,y)$ relates to absorption of X-rays. In medical tomography it is reasonable to consider the sought function to be an arbitrary continuous or piecewise continuous function, since different patients possess rather different distributions of absorption. Moreover, the distribution of absorption of the same patient may vary depending on whether he or she stands or lies.

In industrial tomography another situation also arouses interest. The distribution of absorption has usually a standard shape. The goal of the tomography exploration is to find deviations from this standard shape, flaws of a product. If an industrial product is obtained by casting, then there may be some cavities and cracks inside. Thus, we arrive at the following mathematical problem: the function $u(x,y)$ in (1.2.1) equals unity in some unknown domain $D_0 \subset D$ and vanishes outside D_0:

$$u(x,y) = 1, \quad (x,y) \in D_0 \subset D; \qquad u(x,y) = 0, \quad (x,y) \notin D_0.$$

Moreover, the boundary of D_0 is a piecewise smooth curve. Since the function $u(x,y)$ in this statement is determined by two functions of one variable, it is natural to suppose that a solution to (1.2.1) can be obtained from an

The reduction of the seismic tomography problem to an inverse problem for a differential equation of the Hamilton–Jacobi type makes it possible to develop the general numerical method aimed at solving the problems of integral geometry, both linear (Romanov, M. E., 1975, 1983) and nonlinear (Romanov, M. E., 1972, 1988). This method employs the same common algorithm in which the reconstruction of a medium is accomplished by combination of downward continuation of data along the rays and local projection techniques.

it along a given set of curves determined by the locations of sources and receivers.

However, in practice, the methods and algorithms developed for medical tomography have been observed to be unsuitable for the seismic tomography problems. In contrast to the problems of medical tomography, the problems of seismic tomography are characterized, on the whole, by the nonlinearity of governing equations (the ray traces depend on the reconstructed velocity), which results in "curved rays" (refraction) which, in addition, form a non-regular set. It should also be noted that unlimited scanning of the medium is not always possible. Besides, practical observational systems do not always provide total covering of the reconstruction domain with rays. Finally, seismic tomography problems belong to the class of ill-posed problems and this is their principal peculiarity.

All these complications require development of new efficient numerical algorithms and methods based on the principles of geometrical optics.

The inverse kinematic problem appears first in geophysics (see IP 1.1.1). It arose in connection with the attempt to study the Earth's internal structure from observations, on the Earth's surface, of the propagation of seismic wave fronts generated by earthquakes.

From the mathematical viewpoint the inverse kinematic problem can be formulated as the problem of finding the right-hand side of equation (1.1.1) such that the solution to this equation under the condition $\tau(x^0, x^0) = 0$ takes on the given values at $x \in S$, $x^0 \in M$. Here the point x^0 is the parameter of the direct problem for equation (1.1.1).

It is important to extend consideration from one-dimensional (1D) models, where the function v is dependent on one variable, to multidimensional models, where v is essentially dependent on all three coordinates.

Along these lines we would like to mention the linear variant of the inverse kinematic problem of seismics. The function $v(x)$ is assumed to be represented in the form

$$v(x) = v_0(x) + v_1(x), \qquad (1.2.2)$$

where $v_0(x)$ is supposed to be known and $v_1(x)$ is small as compared with $v_0(x)$. In this case, the problem of determining the small addition $v_1(x)$ to the function $v_0(x)$ is reduced to the problem of solving an integral equation of first kind, i.e., the problem of integral geometry.

The main theoretical results concerning the questions of uniqueness and stability of solutions to such problems were obtained by Bukhgeim (1975,

1983a, 1999, 2000), Lavrentiev and Romanov (1966), Lavrentiev, Romanov, and Shishatskii (1986), Lavrentiev, Romanov, and Vasiliev (1970), Mukhometov (1977, 1978, 1981), Mukhometov and Romanov (1978), Romanov (1967a, 1969, 1978a, 1978b).

In the nonlinear problem, the first results related to the uniqueness of solution to the multidimensional inverse kinematic problem were obtained by Anikonov (1978), Romanov (1973a, 1973b, 1974a, 1974b, 1974c, 1975). It is also necessary to point out the results of Bernstein and Gerver (1978, 1980), Goldin (1986).

In view of theoretical and applied reasons it is important to specify the necessary and sufficient conditions upon which the solution of the inverse kinematic problem exists and belongs to the given functional space. For the general case this problem is still unsolved.

1.2.3. Numerical methods and algorithms of seismic tomography

Numerical techniques for solving multidimensional seismic tomography problems can be conventionally divided into three groups:

- computer trial-and-error techniques;
- methods of projecting into a finite-dimensional subspace;
- general numerical methods.

Trial-and-error techniques are based on the possibility of effective solution of the direct problem for a certain selected class of models of the medium. The first results see, e.g., Alekseev (1967), Matveeva (1972), Matveeva and Alekseev (1964).

If the available information is not sufficient, trial-and-error techniques can give a general idea of the medium structure. State-of-the-art in the methods of solution of direct problems increases the possibilities of trial-and-error techniques. Note, however, that this requires high computational costs and even the best agreement between the calculated and observed data does not guarantee that the profile obtained is the right one.

The inverse kinematic problem is nonlinear since the geometry of seismic rays is unknown. Such geometry must be determined in order to compute the travel times. In seismic investigations an approximate mathematical model of the inhomogeneous medium under study, i.e., a certain velocity model corresponding to the refraction coefficient, can be pre-selected on the basis of "*a priori*" data or can be obtained by approximate methods based on selected data from the wave field observed.

Then a nonlinear seismic tomography problem is reduced by linearization to a linear one, i. e., to the solution of a linear integral equation of first kind. This is the usual way, which is used in most works on seismic tomography. The advantage of this approach is that for the linear inverse problems, computational techniques are developed on the basis of efficient algorithms.

One of such methods is the projecting of the infinite-dimensional model space into a suitably selected finite-dimensional subspace. Combinations of different forms of v_0, which is given, and v_1, which is to be reconstructed, see (1.2.2), constitute that variety of methods and algorithms which are used in seismic tomography.

If v_0 is constant, the rays are straight lines and v_1 is determined from the Radon inverse transformation. To implement this transformation, algorithms of back projection and algebraic reconstruction techniques are normally used which are most commonly employed in medical tomography. However, under conditions of limited scanning geometry, which are specific for seismic problems, their application is rather difficult.

These methods are used in seismics when studying the medium between the boreholes (Dines and Lytle, 1979; Ivansson, 1985; McMechan, 1983), in investigation of an inhomogeneous layer by reflected or refracted waves (Bishop *et al.*, 1985; Neumann-Denzau and Behrens, 1984; Nolet, 1985, 1987), and in determining global horizontal inhomogeneities along the Earth's surface from the observations of surface waves.

Modifications of the algebraic reconstruction technique and back projection algorithms are also used in the case where v_0 is a one-dimensional function of depth and in the inverse problem in terms of the generalized Radon transform (Fawcett and Clayton, 1984; Humphreys, Clayton, and Hager, 1984). In the papers Anderson (1984), Anderson and Dziewonski (1984), Dziewonski (1984), Nakanishi and Anderson (1982), Nataf, Nakanishi, and Anderson (1984), Woodhouse and Dziewonski (1984), a spherically symmetric model corresponds to the initial profile $v_0(x)$ and the velocity variation $v_1(x)$ is represented as an expansion in a series with respect to spherical harmonics.

Under the same assumptions, the first results on the Earth's seismic tomography were obtained as early as in 1969–1971; see Chapter 6, where the problem is considered to determine the P-wave propagation velocity in the upper layers of the Earth's mantle from the earthquake travel-time curves. In that problem, it is assumed that the Earth without the crust is a sphere and that we know the travel-times along seismic rays between two arbitrary points of a certain connected domain on the surface of the sphere.

The reduction of the seismic tomography problem to an inverse problem for a differential equation of the Hamilton–Jacobi type makes it possible to develop the general numerical method aimed at solving the problems of integral geometry, both linear (Romanov, M. E., 1975, 1983) and nonlinear (Romanov, M. E., 1972, 1988). This method employs the same common algorithm in which the reconstruction of a medium is accomplished by combination of downward continuation of data along the rays and local projection techniques.

Chapter 2.

Approximate methods of solution of ill-posed problems

2.1. ON SOME ASPECTS OF STATEMENT AND SOLUTION OF ILL-POSED PROBLEMS

The study of any ill-posed problem connected with physical phenomenon is carried out on the basis of a certain mathematical model. It means that the operator A and the normed space U (see Introduction) containing the sought characteristic $\bar{u}$ of the process are chosen. The normed space F for the right-hand side f of the equation

$$A\bar{u} = f \qquad (2.1.1)$$

is defined by the type of the experimental information which is used to solve the ill-posed problem (2.1.1). Here we assume that in the normed space F there exists an element $\bar{f} = A\bar{u}$ which is given as an experimental measurement. But these measurements are made with some error, therefore, the element $\bar{f}$ is unknown. Instead, an approximate element f_δ such that $\|f_\delta - \bar{f}\|_F \leq \delta$ is given along with the value of the error δ, which characterizes the distance between f_δ and $\bar{f}$ in the norm of the space F.

Thus, of special importance in the problem concerned are the questions of uniqueness and stability of the solution. Indeed, the study of the uniqueness of the solution to an ill-posed problem gives an answer to the question whether the experimental information is sufficient for unique determination of the sought characteristic of the process under study. The problem of stability is connected with construction of such methods that would allow us to determine the approximate solutions u_δ, which solve the equations $Au_\delta = f_\delta$. Such solutions must be close to the sought function $\bar{u}$. It is natural to look for these approximations on the basis of the information f_δ, which is given approximately, and the number δ.

The basic idea of how to construct the approximate solutions u_δ to ill-posed problems was proposed in the 50's (see Tikhonov and Arsenin, 1977). The point is to use the *a priori* information on the exact solution to the ill-posed problem (2.1.1) in construction of the approximate solution u_δ. As a rule, the *a priori* information is connected with the fact that the unknown characteristic stands for some physical quantity. This quantity has certain properties (positiveness, boundedness, the monotonicity of the derivative, etc.). The *a priori* information of such kind makes it possible to solve the desired inverse problem on a certain subspace (or subset) M of the original space, $M \subset U$, on which the solution to the problem is stable. The idea of reducing the class of possible solutions of an ill-posed problem to some set on which its solution is stable lies in the foundation of the concept of correctness by Tikhonov (see Introduction) which was introduced by Lavrentiev (1962), Lavrentiev, Romanov, and Shishatskii (1986).

As a rule, the set M on which the solution is stable is compact.

However, one often encounters the situation where the *a priori* information on the belonging of the solution to the set M on which the solution to the problem is stable is absent. To construct approximate solutions to the ill-posed problem in this case, the fundamental concept of a regularizing operator is used (see, e.g., Tikhonov, 1963; Tikhonov and Arsenin, 1977). On the basis of this concept approximate methods of solution of ill-posed problems are developed, these methods being stable to small perturbations of initial data. Concrete examples realizing this approach are given in this chapter.

2.2. SOLUTIONS ON COMPACT SETS. THE CONCEPT OF A QUASI-SOLUTION

The following theorem (Tikhonov and Arsenin, 1977) is the substantiation of the approach to solution of ill-posed problems based on singling out a compact set $M \subset U$.

Chapter 2. Approximate methods

Theorem 2.2.1. *Let a set M of a normed space U be mapped by an operator A onto a set R $(R = AM)$ of a normed space F in a one-to-one manner. Let the operator A be continuous on the set $M \subset U$, and let M be a compactum. Then the inverse operator A^{-1} is continuous on the set R.*

Proof. Since the set $R = AM$ is the one-to-one image of the set M, therefore, there exists the inverse operator A^{-1} on the set R. Let us show that this operator is continuous. Assume that it is not so. Then there exist an element $f_0 \in R$, a constant $\varepsilon > 0$, and a sequence $\{f_n\}_{n \in \mathbb{N}}$, $f_n \in R$, such that $\lim_{n \to \infty} \|f_n - f_0\|_F = 0$, while

$$\|u_n - u_0\|_U \geq \varepsilon > 0, \tag{2.2.1}$$

where $u_n = A^{-1} f_n$ and $u_0 = A^{-1} f_0$. Since the sequence $\{u_n\}_{n \in \mathbb{N}}$ belongs to the compact set M, therefore, one can select a subsequence $\{u_{n_m}\}_{m \in \mathbb{N}}$ from the sequence $\{u_n\}_{n \in \mathbb{N}}$ such that $\lim_{m \to \infty} \|u_{n_m} - u\|_U = 0$, where $u \in M$. Then it follows from inequality (2.2.1) that

$$\|u - u_0\|_U \geq \varepsilon.$$

But A is continuous on the compact set M, and $\lim_{m \to \infty} \|u_{n_m} - u\|_U = 0$. Then we have

$$\lim_{m \to \infty} \|f_{n_m} - f_0\|_F = \lim_{m \to \infty} \|A u_{n_m} - f_0\|_F = 0.$$

Hence,

$$\|Au - f_0\|_F = 0,$$

which implies $Au = Au_0 = f_0$.

Since the mapping A is a one-to-one mapping on the set M, therefore, $u = u_0$. But this equality contradicts the inequality $\|u - u_0\|_U \geq \varepsilon$. Consequently, our assumption at the beginning of the proof was wrong, and the operator A^{-1} is continuous on R. □

In case the problem (2.1.1) is conditionally well-posed (see Introduction) on a pair of normed spaces U and F and the set $M \subset U$ is a compactum, one can construct an estimate of the conditional stability.

Theorem 2.2.2. *Let the operator A be a continuous one-to-one mapping of the compact set M of the normed space U on the set $R = AM$ of the normed space F. Then there exists a function $\omega(\tau) : \omega(0) = 0$ continuous at zero and such that for any elements $u_1, u_2 \in M$ the following estimate takes place:*

$$\|u_1 - u_2\|_U \leq \omega(\|Au_1 - Au_2\|_F). \tag{2.2.2}$$

Proof. Assume that such a function $\omega(\tau)$ does not exist. Then for every number $n \in \mathbb{N}$ there exist elements $u_n^1, u_n^2 \in M$, a number τ_n, $\lim_{n \to \infty} \tau_n = 0$, and a number $\varepsilon > 0$ such that

$$\|Au_n^1 - Au_n^2\|_F \leq \tau_n, \qquad (2.2.3)$$
$$\|u_n^1 - u_n^2\|_U \geq \varepsilon. \qquad (2.2.4)$$

Since the sequences $\{u_n^k\}_{n \in \mathbb{N}}$, $k = 1, 2$, belong to the compact set M, therefore, one can select subsequences $\{u_{n_m}^k\}_{m \in \mathbb{N}}$, $k = 1, 2$, such that

$$\lim_{m \to \infty} \|u_{n_m}^k - u^k\|_U = 0, \qquad k = 1, 2,$$

where $u^k \in M$, $k = 1, 2$. Then it follows from the estimate (2.2.3) that $\|Au^1 - Au^2\|_U = 0$. Since the operator A is a one-to-one map on the compact set M, we have $u^1 = u^2$. But inequality (2.2.4) implies that $\|u^1 - u^2\|_U \geq \varepsilon > 0$. We have arrived at a contradiction, hence, the initial assumption was wrong, and the function $\omega(\tau)$ possessing the necessary properties does exist. □

Stability estimates of the type (2.2.2) obtained under the assumption that the solution belongs to some set M are called *the conditional stability estimates*. Examples of estimates of this type will be given below.

Consider the case where A is a continuous operator acting from a normed space U into a normed space F such that the problem of solving equation (2.1.1) is ill-posed.

Assume that for the exact right-hand side $\bar{f}$ of equation (2.1.1) there exists the unique solution $\bar{u}$ from some compact set M. But the element $\bar{f}$ is unknown, and instead of $\bar{f}$ an approximate right-hand side f_δ is given, along with the value of the error δ, such that $\|\bar{f} - f_\delta\|_F \leq \delta$. With f_δ and δ known, it is required to propose a method of construction of approximate solutions u_δ which would tend to the exact solution $\bar{u}$ as $\delta \to 0$ in the norm of the space U. Consider the set

$$U_\delta = \{u \in U \mid \|Au - f_\delta\|_F \leq \delta\}.$$

The problem to solve equation (2.1.1) is ill-posed, therefore, we cannot consider an arbitrary element of this set as an approximate solution. But we have the *a priori* information that the exact solution belongs to the compact set M. Therefore, let us narrow down the set U_δ, taking its intersection with the compactum M, namely, consider the set $U_\delta^M = U_\delta \cap M$. It is clear that the set U_δ^M is not empty since it contains the exact solution $\bar{u}$. Let us show that the elements of the set U_δ^M can be treated as approximate solutions to equation (2.1.1).

Theorem 2.2.3. *Under the above notations, we have*
$$\lim_{\delta \to 0} \sup_{u \in U_\delta^M} \|\bar{u} - u\|_U = 0.$$

Proof. Assume that the theorem is not true. Then there exist a sequence of numbers $\{\delta_n\}_{n \in \mathbb{N}}$, $\lim_{n \to \infty} \delta_n = 0$, and a sequence of elements $\{u_{\delta_n}\}_{n \in \mathbb{N}}$, $u_{\delta_n} \in U_{\delta_n}^M$, such that $\|\bar{u} - u_{\delta_n}\|_U \geq \varepsilon > 0$. But the sequence $\{u_{\delta_n}\}_{n \in \mathbb{N}}$ belongs to the compact set M, therefore, it contains a converging subsequence $\{u_{\delta_{n_k}}\}_{k \in \mathbb{N}}$ (for simplicity, we shall denote it as $\{u_{\delta_k}\}_{k \in \mathbb{N}}$), $\lim_{k \to \infty} \|u_{\delta_k} - u_0\|_U = 0$, where $u_0 \in M$. Then we have
$$0 < \varepsilon \leq \lim_{k \to \infty} \|\bar{u} - u_{\delta_k}\|_U = \|\bar{u} - u_0\|_U. \qquad (2.2.5)$$

Since $u_{\delta_k} \in U_{\delta_k}^M$, therefore, $\|Au_{\delta_k} - f_{\delta_k}\|_F \leq \delta_k$. Consequently, we have
$$\|Au_0 - \bar{f}\|_F = \lim_{k \to \infty} \|Au_{\delta_k} - f_{\delta_k}\|_F \leq \lim_{k \to \infty} \delta_k = 0.$$

This is possible if $\|Au_0 - \bar{f}\|_F = 0$. Thus, $Au_0 = \bar{f} = A\bar{u}$, and from the uniqueness of the solution to equation (2.1.1) with the right-hand side $\bar{f}$ we have $u_0 = \bar{u}$. But this equality contradicts inequality (2.2.5). Thus, the initial assumption was wrong, and the theorem is proved. $\square$

Remark 2.2.1. In fact, the proof of Theorem 2.2.3 coincides with the proof of Theorem 2.2.1 of the continuity of the inverse operator on the compact set M. But in contrast to Theorem 2.2.1, it was assumed in Theorem 2.2.3 that the solution to equation (2.1.1) is unique not on the whole of the compact set M but only for the exact right-hand side $\bar{f}$.

Let us present an example showing the absence of the convergence of all elements of the set U_δ to the exact solution $\bar{u}$.

Example 2.2.1. Consider the operator $A : \mathbb{R} \to \mathbb{R}$ which acts as follows: $Au = u/(u^2 + 1)$. Then the equation $Au = \bar{f}$ with the exact right-hand side $\bar{f} = 0$ has the unique solution $\bar{u} = 0$. Let the approximate right-hand side be $f_\delta = \delta/2$ and the value of the error δ be given instead of $\bar{f}$. The set U_δ in this case is the set of real numbers satisfying the inequality
$$\left| \frac{u}{u^2 + 1} - \frac{\delta}{2} \right| \leq \delta.$$

Obviously, for every $\delta > 0$ the number $u_\delta = 2/\delta$ will be a solution to this inequality. This number does not tend to the exact solution $\bar{u} = 0$ as $\delta \to 0$. On the other hand, if as the compact set M we consider the segment $[-1, 1]$, then for a sufficiently small δ the set U_δ^M is the set of real numbers satisfying the inequalities

$$\frac{-1 + \sqrt{1 - \delta^2}}{\delta} \leq u \leq \frac{1 - \sqrt{1 - 9\delta^2}}{3\delta},$$

and for it Theorem 2.2.3 holds true. Indeed, in this case

$$\lim_{\delta \to 0} \sup_{u \in U_\delta^M} \|u - \bar{u}\|_U = \lim_{\delta \to 0} \max_{u \in U_\delta^M} |u| = 0$$

since

$$\lim_{\delta \to 0} \frac{-1 + \sqrt{1 - \delta^2}}{\delta} = \lim_{\delta \to 0} \frac{\delta^2}{\delta \left[1 + \sqrt{1 - \delta^2}\right]} = 0$$

and, similarly,

$$\lim_{\delta \to 0} \frac{1 - \sqrt{1 - 9\delta^2}}{3\delta} = 0.$$

Prior to proceeding to the concept of quasi-solution, we would like to stress attention on some difficulties connected with solution of ill-posed problems on an example of the simplest variant of a classical inverse problem: the problem with inverse time for the heat equation.

As the direct problem, consider the problem of determining the function $u(z, t)$ which satisfies the heat equation

$$\frac{\partial u}{\partial t} = \frac{\partial^2 u}{\partial z^2}, \qquad z \in (0, l), \qquad t \in (0, T), \tag{2.2.6}$$

the boundary conditions

$$u(0, t) = u(1, t) = 0, \qquad t \in [0, T], \tag{2.2.7}$$

and the initial condition

$$u(z, 0) = \varphi(z), \qquad z \in [0, l]. \tag{2.2.8}$$

Inverse Problem 2.2.1 (IP 2.2.1). The inverse problem will be understood as the problem of finding the function $\varphi(z)$ if we know the values of the solution to the direct problem (2.2.6)–(2.2.8) at the moment $t = T$

$$u(z, T) = f(z), \qquad z \in [0, l]. \tag{2.2.9}$$

Such problem is ill-posed.

Let us assume that the additional information (2.2.9) is known from an experiment not exactly but approximately with some error δ

$$f_\delta(z) = f(z) + \delta \sin(n\pi z), \quad n \in \mathbb{N},$$

where δ is a small positive number. The exact solution to this problem has the form

$$u_\delta(z,t) = u(z,t) + \delta e^{(n\pi)^2(T-t)} \sin(n\pi z), \quad n \in \mathbb{N}.$$

The point is that for any given δ and $t \in [0,T)$, the multipliers $e^{(n\pi)^2(T-t)}$ can be arbitrarily large if n is chosen properly. Let us show how this problem can be made correct by proper choice of norms. The idea is almost evident: one needs to define the norm $\|\cdot\|_F$ in the space of data of the inverse problem in such a way that the quantity $\delta\|\sin(n\pi z)\|_F$ grows rapidly with the number n, to balance the above-mentioned exponential multipliers. We shall define the norm of a function $f(x) \in F$ in terms of its Fourier coefficients f_n. Let

$$f(z) = \sum_{n=1}^{\infty} f_n \sin(n\pi z).$$

We define

$$\|f\|_F = \left[\frac{1}{2} \sum_{n=1}^{\infty} f_n^2 e^{2n^2\pi^2 T}\right]^{1/2}, \qquad (2.2.10)$$

$$\|\varphi\|_U = \left[\int_0^1 \varphi^2(z)\,dz\right]^{1/2} = \|\varphi\|_{L_2[0,1]}.$$

Now the inverse problem is correct. Indeed, if two functions $f(z)$ and $f_\delta(z)$ differ from each other by the value

$$f(z) - f_\delta(z) = \sum_{n=1}^{\infty} (f_n - f_{\delta n}) \sin(n\pi z),$$

then the corresponding solutions $\varphi(z)$ and $\varphi_\delta(z)$ to the inverse problem differ by the value

$$\varphi(z) - \varphi_\delta(z) = \sum_{n=1}^{\infty} (f_n - f_{\delta n}) e^{n^2\pi^2 T} \sin(n\pi z).$$

It is clear that

$$\|\varphi - \varphi_\delta\|_U = \left[\frac{1}{2} \sum_{n=1}^{\infty} (f_n - f_{\delta n})^2 e^{2n^2\pi^2 T}\right]^{1/2} \leq \delta$$

when $\|f - f_\delta\|_F \leq \delta$.

Thus, if the data of the inverse problem differ "a little" from each other in the norm (2.2.10), then the solutions to the inverse problem corresponding to these data are close in the norm $\|\cdot\|_{L_2[0,1]}$. In fact, the inverse problem did not change in the result of introducing the norm (2.2.10). The point is that the assumption that the data $f(z)$ of the inverse problem are close to $f_\delta(z)$ in the norm (2.2.10) contains a very strong constraint: all the derivatives of the function $f(z)$ differ little from the corresponding derivatives of the function $f_\delta(z)$. This difference is the less, the greater the order of the derivative.

It is easy to check that if the definition of the norm in the space F involves only derivatives of finite order,

$$\|f\|_F = \sum_{n=0}^{k} C_n \left\|\frac{d^n f}{dz^n}\right\|_{L_2}, \quad C_n > 0, \qquad (2.2.11)$$

it is impossible to have the above problem with inversion of time direction as a correct one. This follows from the fact that $\exp(n^2\pi^2 T)$ grows faster than n^p as $n \to \infty$ for every $p \in \mathbb{N}$. The norms (2.2.11) for small values $k = 0, 1, 2$ are called "usual" or "natural" ones. The norm (2.2.10) will be called the "*strong*" one. Spaces with strong norms of the type (2.2.11) with $k = \infty$ are used mostly when studying analytic and, in particular, entire functions.

If we assume that we study a problem with inverse flow of time for the heat equation such that the function $f(z)$ should be an element of a space of very smooth (analytic) functions with a strong norm, then this problem turns out to be rather good. Conceptually, this inverse problem is connected, e. g., with reconstruction of the temperature of a body in the past if the temperature distribution at the present moment of time is given. This distribution is known not exactly, but with some error. If we merely solve the problem with "time inversion", we obtain a solution of a non-physical character: it may contain large negative and positive temperatures which do not exist in nature.

Assume that one can assert, based on the general qualitative considerations, that the sought temperature $u(z, 0) = \varphi(z)$ is not too large and, moreover, is a "simple" function. The function $f(z)$ is related to the function $\varphi(z)$ through the solution $u(z, t)$ to the direct problem (2.2.6)–(2.2.8), i. e., $f(z) = u(z, T)$. As is known, the solution $u(z, t)$ to the direct problem (2.2.6)–(2.2.8) is a very smooth function, moreover, the "degree of smoothness" grows with the growth of time t.

Thus, the proposal to make the inverse problem (2.2.6)–(2.2.9) a correct one in the classical sense by choosing a very narrow space of functions as the space F is not so artificial as it seems at the first glance. But the trouble is that, as was mentioned earlier, the additional information (2.2.9) is known from experiment not exactly, but approximately with some error δ, i.e., instead of the function $f(z)$ a function $f_\delta(z)$ is given such that $\|f_\delta - f\|_F \leq \delta$. Moreover, the error is such that it takes the function $f(z)$ out of the space F, i.e., $f_\delta(z) \notin F$. Unfortunately, one cannot extract the function $f(z)$ out of the function $f_\delta(z)$ rather objectively and uniquely. It was proposed by Isaacson, D. and Isaacson, E. L. (1989) in this situation to seek for the so-called "quasi-solution" instead of the exact solution. The idea is as follows.

Let the operator A in equation (2.1.1) be a continuous operator acting from a normed space U into a normed space F, and let $M \subset U$ be a given compact set. In the compact set M we shall seek for a *quasi-solution* u_q such that the distance between the elements Au_q and f is minimal. In other words, the determination of a quasi-solution is reduced to the problem of conditional extremum

$$\|Au_q - f\|_F = \min_{u \in M} \|Au - f\|_F. \qquad (2.2.12)$$

Let us establish the fact of existence of a quasi-solution. For this, consider the function $\Phi(u) = \|Au - f\|_F$. It is known from calculus that a function that is bounded from below and is continuous on a compact set achieves its minimal value on this set. The function $\Phi(u)$ is bounded from below since $\Phi \geq 0$ by the property of the norm $\|\cdot\|_F$. Let us prove that $\Phi(u)$ is a continuous function.

In virtue of the continuity of the operator A we have

$$\lim_{v \to 0} A(u+v) = Au,$$

which implies that

$$\lim_{v \to 0} \|A(u+v) - Au\|_F = 0.$$

Here we have used the property of continuity of the norm $\|\cdot\|_F$. Thus, the function $\Phi(u)$ is continuous, therefore, the quasi-solution exists.

For the quasi-solution to be meaningful in view of applications, it should possess an important property: the continuous dependence on the right-hand side. Such property makes it possible to construct stable approximate methods of searching for a quasi-solution. This property is derived under additional requirements on the compact set M. These additional requirements were formulated in Theorem 2.2.2: there should exist a function $\omega(\tau)$ continuous at zero and such that

1) $\omega(0) = 0$;

2) for any two elements $u_1, u_2 \in M$ we have

$$\|u_1 - u_2\|_U \leq \omega(\|Au_1 - Au_2\|_F). \qquad (2.2.13)$$

In this case, the compact set M will be a set of correctness of the problem (2.1.1) on the pair of normed spaces.

Let us show that the quasi-solution depends continuously on the right-hand side f. It is assumed that there exists an exact solution $\bar{u} \in M$. The exact right-hand side $\bar{f} \in R = AM$ corresponds to this solution. The right-hand side is known approximately, that is, the function f_δ is given, which satisfies the inequality $\|\bar{f} - f_\delta\|_F \leq \delta$, where δ is a small positive number. However, in a strong norm of the type (2.2.10) the error on the right-hand side is infinite. Assume that a quasi-solution u_q is found. We assert that in this case

$$\|u_q - \bar{u}\|_U \to 0 \quad \text{as} \quad \delta \to 0.$$

Indeed, let $f_q = Au_q \in R$. We have

$$\|f_q - \bar{f}\|_F \leq \|f_q - f_\delta\|_F + \|\bar{f} - f_\delta\|_F.$$

The second term on the right-hand side is estimated in terms of δ, $\|\bar{f} - f_\delta\|_F \leq \delta$. Let us evaluate the first term

$$\|f_q - f_\delta\|_F = \|Au_q - f_\delta\|_F = \min_{u \in M} \|Au - f_\delta\|_F$$
$$\leq \|A\bar{u} - f_\delta\|_F = \|\bar{f} - f_\delta\|_F \leq \delta.$$

Thus, $\|f_q - \bar{f}\|_F \leq 2\delta$, and in virtue of (2.2.13) we have $\|u_q - \bar{u}\|_U \leq \omega(2\delta)$ since $f_q, \bar{f} \in R$. By the properties of the function ω, this implies that $\|u_q - \bar{u}\|_U \to 0$ as $\delta \to 0$.

Since the concept of the set of correctness is very important, let us explain the technique of its construction on a particular example of an ill-posed problem, namely, the problem with inverse time direction for the heat equation (2.2.6)–(2.2.9). We remind that the sought function $\varphi(z)$ is the initial data in the direct problem (2.2.6)–(2.2.8) whose solution $u(z,t)$ coincides at $t = T$ with the additional information (2.2.9), $u(z,T) = f(z)$. Let us define the set M by the condition $\|\varphi'\|_{L_2[0,1]} \leq 1$. (Besides, it follows from the boundary conditions (2.2.7) that $\varphi(0) = \varphi(1) = 0$.) Applying the Fourier method, we represent the solution to the direct problem (2.2.6)–(2.2.8) as

$$u(z,t) = \sum_{n=1}^{\infty} \varphi_n e^{-tn^2\pi^2} \sin(n\pi z), \qquad (2.2.14)$$

where φ_n are the Fourier coefficients of the function $\varphi(z)$. Similarly, the function $f(z)$ can be represented as

$$f(z) = \sum_{n=1}^{\infty} f_n \sin(n\pi z),$$

where f_n are the Fourier coefficients of the function $f(z)$. Since the functions $f(z)$ and $u(z,t)$ are connected by the relation $u(z,T) = f(z)$, therefore,

$$f(z) = \sum_{n=1}^{\infty} f_n \sin(n\pi z) = \sum_{n=1}^{\infty} \varphi_n \lambda_n^{-1} \sin(n\pi z), \qquad (2.2.15)$$

where $\lambda_n = \exp(n^2\pi^2 T)$. Since the system of functions $\{\sin(n\pi z)\}_{n\in\mathbb{N}}$ is complete and orthogonal in the space $L_2[0,1]$, therefore, it follows from equality (2.2.15) that $f_n = \varphi_n \lambda_n^{-1}$. Thus, the solution to the inverse problem (2.2.6)–(2.2.9) has the form

$$\varphi(z) = \sum_{n=1}^{\infty} f_n \lambda_n \sin(n\pi z). \qquad (2.2.16)$$

Let us consider two functions $f_1(z)$ and $f_2(z)$ corresponding to functions $\varphi_1(z)$ and $\varphi_2(z)$ of the type (2.2.16). Let δ denote the distance between the functions $f_1(z)$ and $f_2(z)$ in the norm of the space $L_2[0,1]$. Applying the Parseval equality, we obtain

$$\delta^2 = \|f_1 - f_2\|_{L_2[0,1]}^2 = \frac{1}{2}\sum_{n=1}^{\infty}(f_{1n} - f_{2n})^2. \qquad (2.2.17)$$

Similarly,

$$\|\varphi_1 - \varphi_2\|_{L_2[0,1]}^2 = \frac{1}{2}\sum_{n=1}^{\infty}(\varphi_{1n} - \varphi_{2n})^2 = \frac{1}{2}\sum_{n=1}^{\infty}\lambda_n^2(f_{1n} - f_{2n})^2. \qquad (2.2.18)$$

Equality (2.2.16) implies

$$\varphi_k'(z) = \pi \sum_{n=1}^{\infty} f_{kn}\lambda_n \cdot n \cos(n\pi z), \qquad k = 1, 2.$$

Then the functions $f_1(z)$ and $f_2(z)$ will belong to the set $R = AM$, where A is the operator of the inverse problem, $A\varphi = f$, if the following inequality holds:

$$\frac{\pi^2}{2} \sum_{n=1}^{\infty} \lambda_n^2 n^2 f_{kn}^2 \leq 1, \qquad k = 1, 2, \tag{2.2.19}$$

since only in this case the functions $\varphi_1(z)$ and $\varphi_2(z)$ belong to the set M. Let us evaluate expression (2.2.18). Choose some positive integer p and use the estimate

$$\lambda_p^2 f_{kp}^2 \leq \frac{2}{\pi^2 p^2}, \qquad k = 1, 2,$$

which is a corollary of inequalities (2.2.19). We have

$$\sum_{n=1}^{\infty} \lambda_n^2 (f_{1n} - f_{2n})^2 \leq \sum_{n=1}^{p} \lambda_n^2 (f_{1n} - f_{2n})^2 + 2 \sum_{n=p+1}^{\infty} \lambda_n^2 (f_{1n}^2 + f_{2n}^2). \tag{2.2.20}$$

The right-hand side of inequality (2.2.20) is estimated as follows:

$$2 \sum_{n=p+1}^{\infty} \lambda_n^2 (f_{1n}^2 + f_{2n}^2) \leq \frac{8}{\pi^2} \sum_{n=p+1}^{\infty} \frac{1}{n^2} \leq \frac{8}{\pi^2} \cdot \frac{1}{p},$$

$$\sum_{n=1}^{p} \lambda_n^2 (f_{1n} - f_{2n})^2 \leq \lambda_p^2 \sum_{n=1}^{\infty} (f_{1n} - f_{2n})^2 \leq 2\lambda_p^2 \delta^2.$$

We obtain

$$\|\varphi_1 - \varphi_2\|_{L_2[0,1]}^2 \leq e^{2p^2 \pi^2 T} \cdot \delta^2 + \frac{4}{\pi^2} \cdot \frac{1}{p}. \tag{2.2.21}$$

Let us choose the number p such that the summands in the right-hand side of inequality (2.2.21) are equal, i. e.,

$$\delta^2 e^{2p^2 \pi^2 T} = \frac{4}{p\pi^2}.$$

Taking the logarithm of both sides of the last equality, we obtain the following equation for p:

$$g(p) \equiv \alpha p^2 + \ln\left(\frac{\delta^2 \pi^2}{4}\right) + \ln p = 0, \qquad \alpha = 2\pi^2 T. \tag{2.2.22}$$

To simplify the calculations below, consider the case $T = 10^{-2}$ and $\delta \approx 10^{-2}$. In this case $\alpha \approx 0.2$ and $\ln(\delta^2 \pi^2/4) \approx -7.5$. Apply the Newton method to approximate the root of the equation $g(p) = 0$ (the first iteration):

$$p_2 = p_1 - \frac{g(p_1)}{g'(p_1)}, \quad \text{where} \quad p_1 = \left[-\alpha^{-1} \ln\left(\frac{\delta^2 \pi^2}{4}\right)\right]^{1/2} \approx 6.$$

Obviously,
$$g(p_1) = \ln p_1, \qquad g'(p_1) = 2\alpha p_1 + 1/p_1.$$

Since we consider the values $\delta \ll 1$, and $p_1 \approx 6$, therefore, we can simplify the formulas by setting
$$p_2 = p_1 - \frac{\ln p_1}{2\alpha p_1}.$$

Let us estimate the first term in inequality (2.2.21), using the approximation p_2. We have
$$\exp(2\pi^2 T p_2^2) = \exp(\alpha p_2^2) = \exp\left\{\alpha\left[p_1^2 - \frac{1}{\alpha}\ln p_1 + \beta\right]\right\},$$

where $\beta = \left(\frac{\ln p_1}{2\alpha p_1}\right)^2 \approx 0.105$ is a small value. Thus, we have

$$\delta^2 \exp(\alpha p_2^2) \approx \delta^2 \exp(\alpha p_1^2) \exp(-\ln p_1) = \frac{\delta^2}{p_1}\exp(\alpha p_1^2)$$
$$= \frac{\delta^2}{p_1} \cdot \exp\left(-\frac{\alpha}{\alpha}\ln\frac{\delta^2 \pi^2}{4}\right) = \frac{\delta^2 \cdot 4}{\pi^2 \delta^2 p_1} = \frac{4}{\pi^2 p_1}.$$

Finally,
$$\|\varphi_1 - \varphi_2\|_{L_2[0,1]} \approx \sqrt{\frac{8}{\pi^2 p_1}} = \frac{2\sqrt{2}}{\pi\sqrt[4]{-\frac{1}{\alpha}\ln\frac{\pi^2}{4} + \frac{2}{\alpha}\ln(\delta^{-1})}} \approx c\left[\frac{1}{\ln(\delta^{-1})}\right]^{1/4}$$

for small $\delta \ll 1$. In our case this means that we have constructed (to within some constant multiplier independent of the error δ) the function $\omega(\tau) = [\ln(\tau^{-1})]^{-1/4}$. This function is continuous for $\tau > 0$, and $\omega(0) = 0$. Since $\delta = \|f_1 - f_2\|_{L_2[0,1]}$, therefore, the conditional stability estimate is valid
$$\|\varphi_1 - \varphi_2\|_{L_2[0,1]} \approx \omega(\|f_1 - f_2\|_{L_2[0,1]}).$$

Thus, we have showed that the set M of functions $\varphi(z)$ such that $\|\varphi'\|_{L_2[0,1]} \leq 1$ defines the set of correctness in the inverse problem (2.2.6)–(2.2.9).

2.3. THE METHOD OF QUASI-INVERSION

Let us consider one of approximate methods of solution of ill-posed problems: the method of quasi-inversion. This method was proposed by Lions and Lattès (1969). We shall explain the basic idea on an example of the problem with inverse time direction for the heat equation (2.2.6)–(2.2.9). As was accentuated in Section 2.2, we can make this problem a correct one in the classical sense by choosing strong norms of the type (2.2.10) or (2.2.11) with $k = \infty$. But this approach is not suitable for finding an approximate solution since experimental data of the inverse problem are known in usual spaces such as L_2 or C. Another approach consists in singling out, in usual spaces, a compact set M on which the problem becomes correct. The essence of the method of quasi-inversion is that instead of the differential operator of the problem we find an operator "close" to it and such that the problem obtained becomes correct in usual spaces such as L_2 or C.

In application to the inverse problem (2.2.6)–(2.2.9), the method of quasi-inversion consists in the following. We switch from the direct problem (2.2.6)–(2.2.8) to a problem for a differential equation of a higher order containing a small parameter. Consider the initial boundary value problem

$$\frac{\partial u_\alpha}{\partial t} = \frac{\partial^2 u_\alpha}{\partial z^2} + \alpha \frac{\partial^4 u_\alpha}{\partial z^4}, \quad z \in (0,1), \quad t \in (0,T], \quad (2.3.1)$$

$$u_\alpha(0,t) = u_\alpha(1,t) = 0, \quad t \in [0,T], \quad (2.3.2)$$

$$\frac{\partial^2 u_\alpha}{\partial z^2}(0,t) = \frac{\partial^2 u_\alpha}{\partial z^2}(1,t) = 0, \quad t \in [0,T], \quad (2.3.3)$$

$$u_\alpha(z,0) = \varphi_\alpha(z), \quad z \in [0,1], \quad (2.3.4)$$

where α is a small positive parameter.

Inverse Problem 2.3.1 (IP 2.3.1). *The inverse problem will be understood as the problem of finding the function $\varphi_\alpha(z)$ if we know the following additional information on the solution to the direct problem* (2.3.1)–(2.3.4):

$$u_\alpha(z,T) = f(z), \quad z \in [0,1]. \quad (2.3.5)$$

Let us show that under certain assumptions the function $\varphi_\alpha(z)$ can be treated as an approximate solution to the problem with inverse time direction (2.2.6)–(2.2.9). This problem consists in solving the equation

$$A\varphi = f, \quad (2.3.6)$$

where A is the operator of the problem (2.2.6)–(2.2.9). We shall assume that the operator A acts from the space $L_2[0,1]$ into the space $L_2[0,1]$. Let us consider the problem of solving equation (2.3.6) in the case where the right-hand side $f(z)$ is given approximately. Let us assume that for the function $\bar{f}(z) \in L_2[0,1]$ there exists the exact solution $\bar{\varphi}(z) \in L_2[0,1]$ to equation (2.3.6). But the function $\bar{f}(z)$ is unknown, and instead of it the function $f_\delta(z) \in L_2[0,1]$ and the value of the error δ are given such that

$$\|\bar{f} - f_\delta\|_{L_2[0,1]} \leq \delta. \tag{2.3.7}$$

In this case one cannot take the function $\varphi_\delta(z) = A^{-1} f_\delta(z)$ as an approximate solution to the problem (i.e., one cannot apply formula (2.2.16)) since the inverse operator is defined not for all $f_\delta(z) \in L_2[0,1]$ and is not continuous (note that formula (2.2.16) contains rapidly growing, as $n \to \infty$, terms with the cofactors $\exp(n^2\pi^2 T)$).

Let us apply the Fourier method to the problem (2.3.1)–(2.3.4). We have

$$u_\alpha(z,t) = \sum_{n=1}^{\infty} \varphi_{\alpha n} \exp\{\pi^2 n^2 t (\alpha \pi^2 n^2 - 1)\} \sin(n\pi z), \tag{2.3.8}$$

where $\varphi_{\alpha n}$ are the Fourier coefficients of the function $\varphi_\alpha(z)$. Setting $t = T$, we obtain

$$u_\alpha(z,T) = f(z) = \sum_{n=1}^{\infty} \varphi_{\alpha n} \exp\{\pi^2 n^2 T(\alpha \pi^2 n^2 - 1)\} \sin(n\pi z)$$

$$= \sum_{n=1}^{\infty} f_n \sin(n\pi z), \tag{2.3.9}$$

where f_n are the Fourier coefficients of the function $f(z)$. Since the system of functions $\{\sin(n\pi z)\}_{n \in \mathbb{N}}$ is complete and orthogonal in the space $L_2[0,1]$, therefore, it follows from equality (2.3.9) that

$$\varphi_{\alpha n} = f_n \exp\{\pi^2 n^2 T(1 - \alpha \pi^2 n^2)\},$$

i.e.,

$$\varphi_\alpha(z) = \sum_{n=1}^{\infty} f_n \exp\{\pi^2 n^2 T(1 - \alpha \pi^2 n^2)\} \sin(n\pi z). \tag{2.3.10}$$

This formula gives the solution to the inverse problem (2.3.1)–(2.3.5). Let us consider the linear operators R_α defined by formula (2.3.10)

$$R_\alpha f = \sum_{n=1}^{\infty} f_n \exp\{\pi^2 n^2 T(1 - \alpha \pi^2 n^2)\} \sin(n\pi z). \tag{2.3.11}$$

For every $\alpha > 0$ the operator R_α is defined on the *whole* of the space $L_2[0,1]$ and is continuous if it is considered as acting from the space $L_2[0,1]$ into the space $L_2[0,1]$. Let us show that the operators R_α are continuous. We need to check whether the following condition of continuity is satisfied:

An operator A acting from a normed space U into a normed space F is called continuous on an element $u_0 \in U$ if for every $\varepsilon > 0$ there exists $\delta > 0$ such that for all $u \in U$ satisfying the inequality $\|u - u_0\|_U \leq \delta$ the inequality $\|Au - Au_0\|_F \leq \varepsilon$ holds true.

We check this property for the operator R_α. Let $\|R_\alpha f - R_\alpha f_0\|_{L_2[0,1]} \leq \varepsilon$. In virtue of the linearity of R_α and the Parseval equality we have

$$\|R_\alpha f - R_\alpha f_0\|^2_{L_2[0,1]} = \frac{1}{2}\sum_{n=1}^{\infty}(f_n - f_{0n})^2 \exp\{2n^2\pi^2 T(1 - \alpha\pi^2 n^2)\},$$

where f_n and f_{0n} are the Fourier coefficients of the functions $f(z), f_0(z) \in L_2[0,1]$, respectively. The following inequalities are true:

$$\frac{1}{2}\sum_{n=1}^{\infty}(f_n - f_{0n})^2 \exp\{2n^2\pi^2 T(1 - \alpha\pi^2 n^2)\}$$

$$\leq \frac{1}{2}\sum_{n=1}^{\infty}(f_n - f_{0n})^2 \cdot \sum_{n=1}^{\infty}\exp\{2n^2\pi^2 T(1 - \alpha\pi^2 n^2)\}$$

$$\leq \frac{\delta^2}{2}\sum_{n=1}^{\infty}\exp\{2n^2\pi^2 T(1 - \alpha\pi^2 n^2)\} \leq \varepsilon^2.$$

Obviously the series

$$\sum_{n=1}^{\infty}\exp\{2n^2\pi^2 T(1 - \alpha\pi^2 n^2)\}$$

converges. Let its sum be equal to some positive number γ. Then for a given ε we can choose $\delta = \varepsilon/\gamma$; in this case the conditions of the definition of the continuity of the operator on the element $f_0(z) \in L_2[0,1]$ are satisfied. Since $f_0(z)$ was chosen arbitrarily, this implies the continuity of the operator R_α on the whole of the space $L_2[0,1]$.

Now let us show that the function $\varphi_\alpha(z) = R_\alpha f_\delta$ can be treated as an approximate solution to the inverse problem (2.2.6)–(2.2.9).

Theorem 2.3.1. *Let the function $\alpha(\delta)$ be such that $\alpha(\delta) > 0$ for $\delta > 0$ and, in addition, $\alpha(\delta) \to 0$ and $\delta \exp\{T/(4\alpha(\delta))\} \to 0$ as $\delta \to 0$. Then $\|\varphi_{\alpha(\delta)} - \bar{\varphi}\|_{L_2[0,1]} \to 0$ as $\delta \to 0$, where $\varphi_{\alpha(\delta)} = R_{\alpha(\delta)} f_\delta$.*

Proof. Consider the element $R_\alpha f_\delta$, where $\alpha > 0$, and evaluate the value of $\|R_\alpha f_\delta - \bar{\varphi}\|_{L_2[0,1]}$. From the triangle inequality we get

$$\|R_\alpha f_\delta - \bar{\varphi}\|_{L_2[0,1]} \leq \|R_\alpha f_\delta - R_\alpha \bar{f}\|_{L_2[0,1]} + \|R_\alpha \bar{f} - \bar{\varphi}\|_{L_2[0,1]}. \qquad (2.3.12)$$

Let us estimate the first term on the right-hand side of this inequality. Taking into account the fact that the operator R_α is linear and using inequality (2.3.7), we obtain

$$\|R_\alpha f_\delta - R_\alpha \bar{f}\|_{L_2[0,1]} \leq \|R_\alpha\| \cdot \|f_\delta - \bar{f}\|_{L_2[0,1]} \leq \|R_\alpha\| \cdot \delta, \qquad (2.3.13)$$

where $\|R_\alpha\|$ is the norm of the operator $R_\alpha : L_2[0,1] \to L_2[0,1]$. Let us estimate $\|R_\alpha\|$. For this, we use the Parseval equality and formula (2.3.11). We have

$$\|R_\alpha f\|^2_{L_2[0,1]} = \frac{1}{2} \sum_{n=1}^{\infty} f_n^2 \exp\{2n^2\pi^2 T(1 - \alpha\pi^2 n^2)\}.$$

Since for every positive integer n we have

$$n^2\pi^2 T(1 - \alpha\pi^2 n^2) \leq \frac{T}{4\alpha},$$

therefore,

$$\|R_\alpha f\|^2_{L_2[0,1]} \leq \frac{1}{2} \exp\left\{\frac{T}{2\alpha}\right\} \cdot \sum_{n=1}^{\infty} f_n^2 = \exp\left\{\frac{T}{2\alpha}\right\} \cdot \|f\|^2_{L_2[0,1]},$$

and, consequently,

$$\|R_\alpha\|_{L_2[0,1]} \leq \exp\left\{\frac{T}{4\alpha}\right\}.$$

Using the above estimate in (2.3.13), we obtain

$$\|R_\alpha f_\delta - R_\alpha \bar{f}\|_{L_2[0,1]} \leq \delta \exp\left\{\frac{T}{4\alpha}\right\}. \qquad (2.3.14)$$

Now let us estimate the second term on the right-hand side of inequality (2.3.12). Since, in virtue of our assumption on the exact right-hand side $\bar{f}(z)$, there exists the solution $\bar{\varphi}(z)$ to equation (2.3.6), therefore, the Fourier

coefficients $\bar{f}_n$ and $\bar{\varphi}_n$ of the functions $\bar{f}(z)$ and $\bar{\varphi}(z)$ are connected by the formula
$$\bar{f}_n = \bar{\varphi}_n \exp\{-n^2\pi^2 T\}, \quad n \in \mathbb{N}.$$
Using this formula, we get
$$R_\alpha \bar{f} = \sum_{n=1}^{\infty} \bar{\varphi}_n \exp\{-\alpha \pi^4 n^4 T\} \sin(n\pi z),$$
$$R_\alpha \bar{f} - \bar{\varphi} = \sum_{n=1}^{\infty} \bar{\varphi}_n \left[\exp(-\alpha \pi^4 n^4 T) - 1\right] \sin(n\pi z).$$
By the Parseval equality, we have
$$\|R_\alpha \bar{f} - \bar{\varphi}\|_{L_2[0,1]}^2 = \frac{1}{2} \sum_{n=1}^{\infty} \bar{\varphi}_n^2 \left[1 - \exp(-\alpha \pi^4 n^4 T)\right]^2.$$
Consider the function $p(\alpha) = \|R_\alpha \bar{f} - \bar{\varphi}\|_{L_2[0,1]}^2$. Let us show that
$$p(\alpha) = \|R_\alpha \bar{f} - \bar{\varphi}\|_{L_2[0,1]}^2 \to 0 \quad \text{as} \quad \alpha \to 0.$$
Indeed, since the series
$$\sum_{n=1}^{\infty} \bar{\varphi}_n^2 = 2 \|\bar{\varphi}\|_{L_2[0,1]}^2$$
converges, therefore, for every $\varepsilon > 0$ there exists $N > 0$ such that
$$\sum_{n=N+1}^{\infty} \bar{\varphi}_n^2 \left[1 - \exp(-\alpha \pi^4 n^4 T)\right]^2 \leq \frac{\varepsilon}{2}$$
for all $\alpha > 0$. On the other hand, since $\exp(-\alpha \pi^4 n^4 T) \to 1$ as $\alpha \to 0$, therefore, there exists a number $\alpha(\varepsilon)$ such that for $0 < \alpha < \alpha(\varepsilon)$ we have
$$\sum_{n=1}^{N} \bar{\varphi}_n^2 \left[1 - \exp(-\alpha \pi^4 n^4 T)\right]^2 \leq \frac{\varepsilon}{2}.$$
This implies that $p(\alpha) \to 0$ as $\alpha \to 0$. Inequalities (2.3.12) and (2.3.14) imply that
$$\|R_\alpha f_\delta - \bar{\varphi}\|_{L_2[0,1]} \leq \delta \cdot \exp\left\{\frac{T}{4\alpha}\right\} + \sqrt{p(\alpha)}.$$
Thus, if the function $\alpha = \alpha(\delta)$ satisfies the assumptions of the theorem, then $\delta \exp\{T/(4\alpha(\delta))\} \to 0$ and $\sqrt{p(\alpha(\delta))} \to 0$ as $\delta \to 0$. This means that
$$\|R_\alpha f_\delta - \bar{\varphi}\|_{L_2[0,1]} \to 0 \quad \text{as} \quad \delta \to 0.$$

□

2.4. REGULARIZATION METHODS

The concept of to solution of inverse problems presented in Section 2.2 was based on applying the *a priori* information that the solution belongs to a compact set M. But in many inverse problems we often encounter the situation where such *a priori* information is absent. Moreover, the situation occurs where the class M of possible solutions is not a compact set, and, besides, changes in the right-hand side f of the equation

$$Au = f \qquad (2.4.1)$$

which are connected with approximate solution of the problem can take the solution beyond the limits of the set $R = AM$. A new method to study such problems has been proposed and developed by Tikhonov and Arsenin (1977). This approach allows us to construct approximate solutions to equation (2.4.1), the solutions obtained by this method being stable to small variations of the initial data, i.e., the right-hand side of equation (2.4.1). This approach is based on the fundamental concept of a regularizing operator. To simplify the discussion in this section, we shall assume that only the right-hand side f can be an approximate entity in equation (2.4.1), while the operator A is known exactly.

Let us consider the problem of solving equation (2.4.1), where the operator A maps a normed space U into a normed space F. Let the operator A be such that the inverse operator A^{-1} is not continuous on the set $R = AM$, and the set of possible solutions M is not compact. Let there exist a unique solution $\bar{u}$ to the equation for the exact right-hand side $\bar{f}$, i.e., $A\bar{u} = \bar{f}$. But the element $\bar{f}$ is unknown, and instead of it the element f_δ and the value of the error δ are given such that $\|f_\delta - \bar{f}\|_F \leq \delta$. It is required to construct an approximate solution to the equation, i.e., an element u_δ which would tend to the exact solution $\bar{u}$ as the value of the error δ, with which the initial information is given, tends to zero. Obviously, one cannot take the exact solution to this equation with the approximate right-hand side $f = f_\delta$, i.e., the element

$$u_\delta = A^{-1} f_\delta,$$

as the approximate solution u_δ to equation (2.4.1) since a solution does not exist in case of an arbitrary element $f \in F$ and does not possess the property of stability to small changes of the right-hand side f.

The numerical parameter δ characterizes the error of the right-hand side of equation (2.4.1). Therefore, it is natural to determine the element u_δ with the aid of an operator depending on some parameter. In this case the value of this parameter should be balanced with the error δ of the initial data f_δ. This "agreement" should be such that as $\delta \to 0$, i.e., as the right-hand side f_δ of equation (2.4.1) tends to the exact element $\bar{f}$, the approximate solution u_δ would tend to the sought exact solution $\bar{u}$.

Definition 2.4.1. An operator $R(f, \delta)$ acting from a normed space F into a normed space U is called a *regularizing operator* of the equation $Au = f$ (with respect to the element $\bar{f}$) if it has the following properties:

1) there exists a number $\delta_1 > 0$ such that the operator $R(f, \delta)$ is defined for all $\delta \in [0, \delta_1]$ and for any element $f_\delta \in F$ such that

$$\|f_\delta - \bar{f}\|_F \leq \delta;$$

2) for every $\varepsilon > 0$ there exists a number $\delta_0(\varepsilon, f_\delta) \leq \delta_1$ such that the inequality

$$\|f_\delta - \bar{f}\|_F \leq \delta \leq \delta_0$$

implies the inequality $\|u_\delta - \bar{u}\|_U \leq \varepsilon$, where $u_\delta = R(f_\delta, \delta)$.

In this definition, the multivalence of the operator A is allowed. The symbol u_δ denotes an arbitrary element from the set of values of the operator $R(f, \delta)$.

In the theory of approximate methods, it is typical when a concrete method of approximate solution depends on some parameter. This parameter can be the grid step, the iteration number, etc. Therefore, the following scheme of construction of a regularizing operator is often used. One specifies some family of operators $R(f, \alpha)$ which act from the space F into the space U and depend on some parameter α. Then the parameter α is chosen according to δ and f_δ, $\alpha = \alpha(f_\delta, \delta)$, in such way that

$$R(f_\delta, \alpha(f_\delta, \delta)) \to \bar{u} \quad \text{as} \quad \delta \to 0.$$

The problem of approximate solution of equation (2.4.1) can be considered in a more general case where not only the right-hand side f but also the operator A are given with an error. In this case the general principles of construction of an approximate solution to equation (2.4.1) for this statement of the problem remain, as a rule, the same as in the case of the operator A given exactly.

Below we present an example showing how regularizing operators are used in concrete problems.

Example 2.4.1. Consider the classical problem of approximate calculation of the derivative $u(z) = f'(z)$ from the approximate values of the function $f(z)$; here the distance between the initial data and the exact function is understood in the norm of the space C. This problem is solved with the aid of the regularizing operator

$$R(f, \alpha) = \frac{f(z+\alpha) - f(z)}{\alpha}.$$

Indeed, assume that instead of the exact values of the function $f(z)$ we have approximate values $f_\delta(z) = f(z) + v(z)$, where $|v(z)| \leq \delta$ for all z. Then

$$R(f_\delta, \alpha) = \frac{f(z+\alpha) - f(z)}{\alpha} + \frac{v(z+\alpha) - v(z)}{\alpha}.$$

The first term on the right-hand side of this equality tends to the derivative $f'(z)$ as α tends to zero. Let us estimate the second term:

$$\left| \frac{v(z+\alpha) - v(z)}{\alpha} \right| \leq \frac{2\delta}{\alpha}.$$

If we take $\alpha = \delta/\eta(\delta)$, where $\eta(\delta) \to 0$ as $\delta \to 0$, then $2\delta/\alpha = 2\eta(\delta) \to 0$ as $\delta \to 0$ and, consequently, if $\alpha = \alpha_1(\delta) = \delta/\eta(\delta)$, then $R(f_\delta, \alpha_1(\delta))$ tends to $f'(x)$ as δ tends to zero.

In solving ill-posed problems, the most widely used method is the method of construction of regularizing operators which is based on variational principles and is called *the method of the Tikhonov regularization*. An important part in this method is the concept of a *stabilizing functional (stabilizer)* which is a criterion of selection of solutions to the problem (2.4.1).

Definition 2.4.2. A *stabilizer* is a nonnegative functional $\Omega(u)$ defined on a normed space U such that the condition of its boundedness, $\Omega(u) \leq C$ ($C \geq 0$), implies that all functions u satisfying this condition form a set M which is compact in U, and the exact solution $\bar{u}$ to the problem (2.4.1) belongs to the set M, i.e., $\bar{u} \in M$.

Using the stabilizer, one can obtain the conditional extremum problem

$$\inf_{u \in U} \|Au - f_\delta\|_F^2, \qquad \Omega(u) = C, \qquad (2.4.2)$$

where the parameter C should be balanced with the error δ, with which the right-hand side f_δ is given, $\|\bar{f} - f_\delta\|_F \leq \delta$. In other words, we look for an element u_r in the normed space U such that the distance between the elements Au_r and f_δ (under the condition $\Omega(u_r) = C$) is minimal.

It is convenient to pass from the problem (2.4.2) to the unconditional extremum problem for the so-called *smoothing* functional

$$\Phi_\alpha(u) = \|Au - f_\delta\|_F^2 + \alpha\Omega(u), \qquad \alpha > 0. \qquad (2.4.3)$$

Solving the problem

$$\inf_{u \in U} \Phi_\alpha(u) \qquad (2.4.4)$$

we find the element u_α which can be taken as an approximate solution if the parameter α is chosen according to the error level δ, i.e., $\alpha = \alpha(\delta)$.

In contrast to Section 2.2, where the stability (convergence) of approximate solutions was achieved by narrowing the class of possible solutions to a compact set M, in the Tikhonov regularization method this property of approximate solutions is recovered by adding to the main functional of the unstable variational problem $\inf_{u \in U} \|Au - f_\delta\|_F^2$ a stabilizing functional $\Omega(u)$ with a small parameter α matched to the error level δ, i.e., $\alpha = \alpha(\delta)$. The element $u_{\alpha(\delta)}$ which yields the minimum in the problem (2.4.4) can be treated as the result of applying some operator R depending on the parameter δ to the right-hand side $f = f_\delta$ of equation (2.4.1), i.e.,

$$u_{\alpha(\delta)} = R(f_\delta, \delta).$$

Assume that such an element exists for any $\delta > 0$ and $f_\delta \in F$. Then the element

$$u_{\alpha(\delta)} = R(f_\delta, \delta)$$

can be treated as an approximate solution to the equation $Au = f_\delta$ since under these conditions the following theorem is valid:

Theorem 2.4.1. *The operator $R(f, \delta)$ is a regularizing operator for equation (2.4.1).*

One can show that if A is a linear bounded one-to-one operator and $\Omega(u) = \|u\|_U^2$, then the problem (2.4.4) is solvable and has the unique solution $u_{\alpha(\delta)}$ for any $\alpha > 0$ and $f_\delta \in F$. In addition, the following theorem is true:

Theorem 2.4.2. *If $\alpha(\delta) > 0$ for $\delta > 0$ and, in addition, $\alpha(\delta) \to 0$ and $\delta^2/\alpha(\delta) \to 0$ as $\delta \to 0$, then $\|u_{\alpha(\delta)} - \bar{u}\|_U \to 0$ as $\delta \to 0$.*

Chapter 2. Approximate methods

The Tikhonov regularization method described above can be applied both to linear and nonlinear problems. Besides, one can consider the case where not only the right-hand side f but also the operator A of the problem (2.4.1) are given approximately.

As an example of application of the regularization method consider the inverse problem of exploration geophysics: the problem of reconstructing the interface of two media. In the simplest two-dimensional case the interface is described by the equation $z = u(x)$.

For $0 < z < u(x)$ we have the medium characterized by the parameter $q_1 = \text{const}$, and for $z > u(x)$ we have the medium with $q_2 = \text{const}$. It is required to determine the function $u(x)$ if we know q_1, q_2, and the characteristic of the field $f_\delta(\lambda)$, $\lambda \in [\lambda_1, \lambda_2]$, measured experimentally (with error δ) along some profile. Let the operator of the direct problem $A_\lambda u = f(\lambda)$ be known. This operator allows us to calculate the field characteristic $f(\lambda)$ on the Earth's surface if we know the function $z = u(x)$ and the characteristics q_1 and q_2 (in the gravitational exploration it suffices to know only the value $\Delta q = q_2 - q_1$). Solution of the inverse problem by the regularization method will consist in determining the function $u(x)$ which realizes the minimum of the functional

$$\Phi_{\alpha\beta}(u) = \int_{\lambda_1}^{\lambda_2} |A_\lambda u - f_\delta(\lambda)|^2 \, d\lambda + \alpha \int_{-\infty}^{\infty} |u(x) - u_0(x)|^2 \, dx$$
$$+ \beta \int_{-\infty}^{\infty} |u'(x) - u_0'(x)|^2 \, dx, \quad (2.4.5)$$

where $z = u_0(x)$ is the assumed interface and α and β are the regularization parameters that are to be defined. In this example, the sought function $u(x)$ is smooth, and when solving the problem one seeks for the most slowly varying function $u(x)$ (the interface between two media) such that the field characteristic $f(\lambda)$ calculated for this function coincides with the experimental data to within the error δ, i.e., $\|f - f_\delta\|_F \leq \delta$. In this case the criterion of selection of a solution is the slowness of (spatial) changes of the interface.

The case is possible where a piecewise smooth function will be a solution to the inverse problem. When constructing an approximate solution in the class of piecewise smooth functions one cannot use formula (2.4.5). It is necessary to impose constraints, basing on some *a priori* information, on the number of discontinuities and seek for a solution in the class of piecewise smooth functions with the specified number of discontinuities. Then the smoothing functional will have the form

$$\Phi_{\alpha\beta}(u) = \int_{\lambda_1}^{\lambda_2} |A_\lambda(u) - f_\delta(\lambda)|^2 \, d\lambda + \sum_{n=0}^{N} \Big\{ \alpha \int_{x_n}^{x_{n+1}} |u_n(x) - u_{0n}(x)|^2 \, dx$$
$$+ \beta \int_{x_n}^{x_{n+1}} |u'_n(x) - u'_{0n}(x)|^2 \, dx \Big\}, \quad (2.4.6)$$

where $x_0 = -\infty$, $x_{N+1} = +\infty$, and N is the number of discontinuities of the interface. Here the regularization parameters are not only α and β but N as well. The sought entities are the locations of discontinuities, i.e., the numbers x_n, $n = 1, 2, \ldots, N$, and the equations of the interface between the discontinuities: $u_n(x)$, $n = 0, 1, 2, \ldots, N$.

Chapter 3.

Integral geometry problems

3.1. STATEMENT OF INTEGRAL GEOMETRY PROBLEMS

Let $x \in \mathbb{R}^n$, $x = (x_1, \ldots, x_n)$; $y \in \mathbb{R}^m$, $y = (y_1, \ldots, y_m)$; and let $S(y)$ be a family of manifolds in $\mathbb{R}^n$ depending on the parameter y of dimension m, $\dim S = p$. Further, let $u(x)$ be a function defined in some domain $D \subset \mathbb{R}^n$, $\rho(x, y)$ be a function of the variables x and y, and $\omega(y)$ be a measure on the manifold $S(y)$.

Consider the function

$$\int_{S(y)} \rho(x, y)\, u(x)\, d\omega = f(y). \tag{3.1.1}$$

Integral geometry is the field of mathematics that deals with various relations between the elements entering in (3.1.1).

We shall assume that the functions $S(y)$, $\rho(x, y)$, and $f(x)$ in (3.1.1) are given and consider equation (3.1.1) as a linear operator equation with respect to the function $u(x)$.

The problem of solution of equation (3.1.1) can be considered as the inverse problem. A lot of inverse problems are reduced to equation (3.1.1).

3.2. THE RADON PROBLEM

We consider the Radon problem for compactly supported functions of two arguments.

Let $u(x,y)$ be a continuous function vanishing outside the unit disk

$$D = \{(x,y) \mid x^2 + y^2 \leq 1\}.$$

It is required to determine the function u if we know the integrals of this function along all straight lines that intersect D.

Let us point out two representations of the Radon problem in the form of problems of solving the linear operator equations that correspond to the classical formulas of parametrization of families of straight lines on the plane:

$$\int_{-\infty}^{\infty} u(x + s\cos\alpha, y + s\sin\alpha)\,\mathrm{d}s = f(x,y,\alpha), \qquad (3.2.1)$$

$$\iint_D u(\xi,\eta)\,\delta(x_0\xi + y_0\eta - p)\,\mathrm{d}\xi\,\mathrm{d}\eta = \varphi(x_0, y_0, p). \qquad (3.2.2)$$

Since the integrals of the function u along the straight lines that do not intersect D are equal to zero, therefore, we can think that the functions f and φ are given for all values of the variables (x,y,α) and (x_0, y_0, p).

The right-hand sides of equations (3.2.1) and (3.2.2) are functions of three arguments; however, they are defined in terms of functions of two arguments.

The function f satisfies the differential equation

$$f'_x \cdot \cos\alpha + f'_y \cdot \sin\alpha = 0,$$

which implies

$$f(x,y,\alpha) = f_0(\sin\alpha \cdot x - \cos\alpha \cdot y, \alpha).$$

By the properties of the δ-function, the function φ satisfies the relation

$$\varphi(x_0, y_0, p) = \frac{1}{r}\varphi\left(\frac{x_0}{r}, \frac{y_0}{r}, \frac{p}{r}\right) = \frac{1}{r}\psi(\beta, \tau),$$

where $\tau = p/r$, $x_0 = r\cos\beta$, and $y_0 = r\sin\beta$.

Representation (3.2.1) is the most natural from the viewpoint of applications, while representation (3.2.2) yields the simplest inversion formula. Clearly, with the help of the change of variables it is easy to go from representation (3.2.1) to (3.2.2) and vice versa.

In virtue of the indicated property of the function φ equation (3.2.2) is equivalent to the equation

$$\iint u(\xi,\eta)\,\delta(x_0\xi + y_0\eta - \tau)\,\mathrm{d}\xi\,\mathrm{d}\eta = \psi(\beta, \tau). \qquad (3.2.3)$$

Chapter 3. Integral geometry problems

The fact that the function u vanishes outside the disk D implies

$$\psi(\beta, \tau) = 0, \quad |\tau| > 1.$$

Thus, the operator in (3.2.3) maps the function u, which is given in the disk D, to the function ψ, which is given in the rectangle

$$|\beta| \leq \pi, \quad |\tau| \leq 1.$$

The function ψ is the integral of the function u along the straight line defined by the equation

$$\xi \cos \beta + \eta \sin \beta - \tau = 0$$

with the differential equal to the length element on this straight line.

Let us establish several properties of the solution to equation (3.2.3).

1. The problem of solving equation (3.2.3) is ill-posed if u and ψ are considered as elements from the spaces L_2 or C. Let

$$u(x, y) = \begin{cases} 1, & x^2 + y^2 \leq \delta^2, \\ 0, & x^2 + y^2 > \delta^2 \end{cases} \quad (0 < \delta < 1).$$

Then

$$\psi(\beta, \tau) = \begin{cases} 2\sqrt{\delta^2 - \tau^2}, & |\tau| \leq \delta, \\ 0, & |\tau| > \delta, \end{cases}$$

$$\|u\|_{L_2} = \pi \delta^2, \quad \|\psi\|_{L_2} = 8\pi \delta^{3/2}/\sqrt{3}, \quad \|u\|_C = 1, \quad \|\psi\|_C = 2\delta.$$

Thus, for a sufficiently small δ the ratio of the norms of the functions u and ψ will be arbitrarily large; and, therefore, arbitrarily small changes of ψ may result in finite changes of u.

2. The problem of solving equation (3.2.3) is weakly ill-posed (see the Introduction).

Applying the Fourier transform with respect to the variable p to equation (3.2.2), we obtain

$$\int e^{ip} \varphi(x_0, y_0, p) \, dp = \iint e^{i(x_0 \xi + y_0 \eta)} u(\xi, \eta) \, d\xi \, d\eta$$

$$= \frac{1}{r} \int e^{ip} \psi\left(\beta, \frac{p}{r}\right) dp = \int e^{i\tau r} \psi(\beta, \tau) \, d\tau = v(\beta, r), \quad (3.2.4)$$

where v is the Fourier transform of the function u in the polar coordinates.

The Parseval equality for the functions u and v yields

$$\iint u^2(\xi,\eta)\,d\xi\,d\eta = \iint |v(\beta,r)|^2\,r\,dr\,d\beta.$$

We shall denote by $H_{0,1/2}$ the space of the functions ψ with the norm

$$\|\psi\|_H = \left(\iint |v(\beta,r)|^2\,r\,dr\,d\beta\right)^{1/2},$$

where v is the Fourier transform of the function ψ with respect to the variable τ defined by equality (3.2.4). Then

$$\|u\|_{L_2} = \|\psi\|_H.$$

Now we give another method of investigating the uniqueness and stability of the Radon problem. This method gives weaker stability estimates but admits generalization to a certain class of integral geometry problems of general form.

We consider the problem of solving equation (3.2.1). Let us denote by $F(x,y)$ the function

$$F(x,y) = \int_{-\pi}^{\pi} f(x,y,\alpha)\,d\alpha.$$

It is easy to see that the functions F and u are connected by the formula

$$F(x,y) = \iint_D \frac{1}{r} u(\xi,\eta)\,d\xi\,d\eta, \qquad r = \sqrt{(x-\xi)^2 + (y-\eta)^2}. \tag{3.2.5}$$

Equality (3.2.5) can be treated as an integral equation of first kind with a weak singularity with respect to the sought function u.

Let us derive an inversion formula for equation (3.2.5). This formula uses, in fact, a fractional power of the Laplace operator. Consider the function

$$W(x,y,z) = \iint_D \frac{1}{R} u(\xi,\eta)\,d\xi\,d\eta,$$

$$R = \sqrt{(x-\xi)^2 + (y-\eta)^2 + z^2}.$$

The function W is the potential of a simple layer with density of distribution $u(\xi,\eta)$ on the disk D.

From the known formulas of the potential theory we obtain

$$W(x,y,z) = \iint \frac{z}{R^3} F(\xi,\eta)\,d\xi\,d\eta, \qquad u(x,y) = \frac{1}{2\pi}\frac{\partial}{\partial z} W(x,y,z)\Big|_{z=0}.$$

Chapter 3. Integral geometry problems   57

The Radon problem with incomplete data. From the viewpoint of applications, of much interest is the problem of inverting the Radon transform in the case where the integrals of the sought function are known not for all straight lines intersecting the domain.

So, let us consider the problem of solving the operator equation (3.2.1) in the case where the right-hand side f is given for all x and y and for the values of the variable α on the interval

$$|\alpha| \leq \alpha_0 < \pi.$$

One can show that in this case the problem of solving equation (3.2.1) is essentially ill-posed. The character of instability in this problem is the same as in the Cauchy problem for the Laplace equation.

If we reduce solution of the Radon problem with incomplete data to equation (3.2.5), then we can think that the left-hand side $F(x,y)$ is given not for all values of x and y but only for the values belonging to a certain domain D_1,

$$D_1 \cap D = \varnothing.$$

In the space (x, y, z), the function W outside the disk D will be a solution of the Laplace equation. Obviously,

$$W'_z(x, y, 0) = 0, \quad (x, y) \in D_1,$$

and thus the Radon problem with incomplete data is reduced to the problem of determining the function $W(x, y, z)$ from the following data:

$$W(x, y, 0) = F(x, y), \quad W'_z(x, y, 0) = 0, \quad (x, y) \in D_1,$$

i.e., to the Cauchy problem for the Laplace equation.

3.3. THE PROBLEM OF GENERAL FORM ON THE PLANE

Consider a bounded open simply connected domain D with a boundary Γ on the plane (x, y). We assume that the boundary is specified by an equation of the form $x = g(s)$, $y = p(s)$, where s is the length counted from a fixed point on Γ in the positive direction coordinated with the choice of orientation on Γ, $g(s)$ and $p(s)$ are functions of the class $C^1[0, l]$, $g(0) = g(l)$, $p(0) = p(l)$, and l is the length of Γ. Moreover, we assume that in the domain D, a two-parameter family of curves $L(t_1, t_2)$ with the properties listed below is given.

1. Each pair of points of the domain $\bar{D}$ is connected by one and only one curve of the family $L(t_1, t_2)$; each curve $L(t_1, t_2)$ intersects Γ at the points (x_1, y_1) and (x_2, y_2), i.e.,

$$x_1 = g(s_1), \qquad x_2 = g(s_2),$$
$$y_1 = p(s_1), \qquad y_2 = p(s_2),$$

other points of the curve $L(t_1, t_2)$ belong to D, and the lengths of the curves $L(t_1, t_2)$ are bounded in union.

2. The equation of the curve passing through the point (x_0, y_0) in the direction $\nu^0 = (\cos\theta_0, \sin\theta_0)$ is given by the equalities

$$x = f_1(s, \theta_0, x_0, y_0) = x_0 + s\cos\theta_0 + s^2 \tilde{f}_1(s, \cos\theta_0, \sin\theta_0, x_0, y_0),$$
$$y = f_2(s, \theta_0, x_0, y_0) = y_0 + s\sin\theta_0 + s^2 \tilde{f}_2(s, \cos\theta_0, \sin\theta_0, x_0, y_0),$$
(3.3.1)

where $\tilde{f}_j(s, \cos\theta_0, \sin\theta_0, x_0, y_0)$ are continuously differentiable and bounded, along with their derivatives, functions of the curve length s counted from the point (x_0, y_0) and the parameters $\theta_0 \in [0, 2\pi]$ and $(x_0, y_0) \in \bar{D}$; $f_j(s, 0, x_0, y_0) = f_j(s, 2\pi, x_0, y_0)$; and, moreover,

$$\frac{1}{s}\frac{\partial(f_1, f_2)}{\partial(s, \theta_0)} \geq c_0 > 0 \tag{3.3.2}$$

in the whole domain of variation of the parameters (s, θ_0, x_0, y_0).

We note that for s close to zero, the fulfilment of inequality (3.3.2) follows evidently from (3.3.1), therefore inequality (3.3.2) is essential for finite s. In this case it is equivalent to the positiveness of the Jacobian $\partial(f_1, f_2)/\partial(s, \theta_0)$.

For the family of the curves L, the following lemma holds.

Lemma 3.3.1. *Suppose that the family of the curves satisfies the conditions indicated above and, moreover, the functions $f_j(s, \theta_0, x_0, y_0)$ have continuous and bounded derivatives up to the order $m \geq 1$. Then equality (3.3.1) defines s and ν^0 as one-valued and m times continuously differentiable functions of the points (x_0, y_0) and (x, y) for all $(x_0, y_0), (x, y) \in \bar{D}$, $(x_0, y_0) \neq (x, y)$; and the following estimates hold for these functions in a neighborhood of the set $(x_0, y_0) = (x, y)$, $(x_0, y_0) \in \bar{D}$:*

$$|D^\alpha s(x_0, y_0, x, y)| \leq \frac{c}{[(x-x_0)^2 + (y-y_0)^2]^{(|\alpha|-1)/2}},$$
$$|D^\alpha \nu^0(x_0, y_0, x, y)| \leq \frac{c}{[(x-x_0)^2 + (y-y_0)^2]^{|\alpha|/2}}, \quad |\alpha| \leq m.$$
(3.3.3)

Chapter 3. Integral geometry problems

Proof. To prove the lemma we note that (3.3.1) implies the equalities

$$s = \frac{[(x-x_0)^2 + (y-y_0)^2]^{1/2}}{[(\cos\theta_0 + s\tilde{f}_1)^2 + (\sin\theta_0 + s\tilde{f}_2)^2]^{1/2}},$$

$$\cos\theta_0 = \frac{x-x_0}{[(x-x_0)^2+(y-y_0)^2]^{1/2}}\left[(\cos\theta_0+s\tilde{f}_1)^2+(\sin\theta_0+s\tilde{f}_2)^2\right]^{1/2}$$

$$+ [(x-x_0)^2+(y-y_0)^2]^{1/2}\frac{\tilde{f}_1}{[(\cos\theta_0+s\tilde{f}_1)^2+(\sin\theta_0+s\tilde{f}_2)^2]^{1/2}},$$

$$\sin\theta_0 = \frac{y-y_0}{[(x-x_0)^2+(y-y_0)^2]^{1/2}}\left[(\cos\theta_0+s\tilde{f}_1)^2+(\sin\theta_0+s\tilde{f}_2)^2\right]^{1/2}$$

$$+ [(x-x_0)^2+(y-y_0)^2]^{1/2}\frac{\tilde{f}_2}{[(\cos\theta_0+s\tilde{f}_1)^2+(\sin\theta_0+s\tilde{f}_2)^2]^{1/2}}.$$

Hence, using the implicit function theorem, we infer that in a sufficiently small domain $(x-x_0)^2 + (y-y_0)^2 < \delta^2$ the functions s and ν^0 have the following structure:

$$s = [(x-x_0)^2+(y-y_0)^2]^{1/2}\{1+[(x-x_0)^2+(y-y_0)^2]^{1/2}\varphi\},$$

$$\cos\theta_0 = \frac{x-x_0}{[(x-x_0)^2+(y-y_0)^2]^{1/2}}\{1+[(x-x_0)^2+(y-y_0)^2]^{1/2}\varphi\}$$
$$+ [(x-x_0)^2+(y-y_0)^2]^{1/2}\psi_1,$$

$$\sin\theta_0 = \frac{y-y_0}{[(x-x_0)^2+(y-y_0)^2]^{1/2}}\{1+[(x-x_0)^2+(y-y_0)^2]^{1/2}\varphi\}$$
$$+ [(x-x_0)^2+(y-y_0)^2]^{1/2}\psi_2,$$

where

$$\varphi = \varphi\Big(x_0, y_0, [(x-x_0)^2+(y-y_0)^2]^{1/2},$$
$$\frac{x-x_0}{[(x-x_0)^2+(y-y_0)^2]^{1/2}}, \frac{y-y_0}{[(x-x_0)^2+(y-y_0)^2]^{1/2}}\Big),$$

$$\psi_j = \psi_j\Big(x_0, y_0, [(x-x_0)^2+(y-y_0)^2]^{1/2},$$
$$\frac{x-x_0}{[(x-x_0)^2+(y-y_0)^2]^{1/2}}, \frac{y-y_0}{[(x-x_0)^2+(y-y_0)^2]^{1/2}}\Big)$$

are m times continuously differentiable functions of their arguments. In particular, this implies the estimates (3.3.3). For $(x-x_0)^2 + (y-y_0)^2 > \delta^2$, the inequality $s > \delta$ is evidently true and, consequently, the inequality $\partial(f_1, f_2)/\partial(s, \theta_0) \geq c_0\delta > 0$ is also holds. But in this case the smoothness of the functions s and ν^0 asserted in the lemma follows from the smoothness of the functions $f_j(s, \theta_0, x_0, y_0)$; and their existence and single-valuedness follow from condition 1 for the family of curves. □

Let us denote by $L(x_0, y_0, x, y)$ the segment of the curve of the family passing through the points (x_0, y_0) and (x, y) and enclosed between them and by $\nu = (\cos\theta, \sin\theta)$ the unit vector tangent to the curve $L(x_0, y_0, x, y)$ at the point (x, y). The vector ν can also be considered as a function of the points (x_0, y_0) and (x, y) with the same smoothness properties as the vector ν^0. Indeed, if $\nu^0 = (h_1(x, y, x_0, y_0), h_2(x, y, x_0, y_0))$, then, in virtue of equal rights of the points (x_0, y_0) and (x, y), it is evident that $\nu = (-h_1(x_0, y_0, x, y), -h_2(x_0, y_0, x, y))$.

Lemma 3.3.2. *The following inequality is true:*

$$\frac{\partial}{\partial s_1} \theta(g(s_1), p(s_1), x, y) \geq 0, \quad (x, y) \in D, \quad s_1 \in [0, l]. \tag{3.3.4}$$

Proof. To prove the lemma we consider the function $\theta(x_0, y_0, x, y)$ for fixed $(x, y) \in D$ as a function of (x_0, y_0). The level lines of this function are the segments of the curves of the family L passing through the point (x, y) and enclosed between (x, y) and the boundary Γ. Consequently, $\operatorname{grad} \theta(x_0, y_0, x, y)$ is directed along the normal to the curve $L(x_0, y_0, x, y)$ at the point (x_0, y_0) in the direction of increasing θ. Since the curves $L(x_0, y_0, x, y)$ for fixed x and y intersect with each other only at the point (x, y), therefore, on every closed curve enclosing the point $(x, y) \in D$ and, in particular, on Γ, the positive direction corresponds to the increase of θ. Setting $x_0 = g(s_1)$ and $y_0 = p(s_1)$, we conclude that the greater s_1 yields the greater values of $\theta(g(s_1), p(s_1), x, y)$. This yields inequality (3.3.4). □

Now we consider for $u(x, y) \in C^1(\overline{D})$ and $\rho(x_0, y_0, x, y) \in C^1(\overline{D} \times \overline{D})$ the function $w(x_0, y_0, x, y)$ defined by the formula

$$w(x_0, y_0, x, y) = \int_{L(x_0, y_0, x, y)} \rho(x_0, y_0, x_1, y_1) u(x_1, y_1) \, ds \tag{3.3.5}$$

and set the following problem: find $u(x, y)$ for given functions $\rho(x_0, y_0, x, y)$ and $v(t_1, t_2)$

$$v(t_1, t_2) = w(g(t_1), p(t_1), g(t_2), p(t_2)), \quad t_1, t_2 \in [0, l]. \tag{3.3.6}$$

Chapter 3. Integral geometry problems

Theorem 3.3.1. *If the family of curves has properties 1 and 2 and the weight function* $\rho(x_0, y_0, x, y) \in C^1(\overline{D} \times \overline{D})$ *satisfies the conditions*

$$\rho(x_0, y_0, x, y) \geq \rho_0 > 0, \quad (x_0, y_0) \in \Gamma, \quad (x, y) \in \overline{D}, \quad (3.3.7)$$

$$\left|\frac{\partial}{\partial t_1} \ln \rho_0(g(t_1), p(t_1), x, y)\right| \leq q \frac{\partial}{\partial t_1} \theta(g(t_1), p(t_1), x, y), \quad 0 \leq q < 1, \quad (3.3.8)$$

then the solution of the above problem is unique for $u \in C^1(\overline{D})$ *and the following stability estimate holds for this solution:*

$$\|u\|_{L_2(D)} \leq \frac{1}{\rho_0\sqrt{1-q^2}} \frac{1}{2\sqrt{\pi}} \|\mathrm{grad}_{t_1,t_2} v(t_1, t_2)\|_{L_2(Q)}, \quad (3.3.9)$$

$$Q = [0, l] \times [0, l].$$

Proof. At first we prove the estimate (3.3.9) for the case where the smoothness of the functions u and ρ and the functions $\tilde{f}_j$, which enter (3.3.1), is higher by one than in the assumption of the theorem. In this case, the following lemma holds true for the function $w(x_0, y_0, x, y)$:

Lemma 3.3.3. *Let the functions* $\tilde{f}_j$, ρ, *and* u *be twice continuously differentiable with respect to their arguments and bounded together with their partial derivatives. Then the function* $w(x_0, y_0, x, y)$ *is twice continuously differentiable with respect to* (x_0, y_0, x, y) *everywhere except for the points* $(x_0, y_0) = (x, y)$; *and in a neighborhood of these points the following estimate holds:*

$$|D^\alpha w(x_0, y_0, x, y)| \leq c\left[(x - x_0)^2 + (y - y_0)^2\right]^{(1-|\alpha|)/2}, \quad |\alpha| \leq 2. \quad (3.3.10)$$

Proof. The validity of the lemma follows from the representation

$$w(x_0, y_0, x, y) = \int_0^{s(x_0, y_0, x, y)} \rho(x_0, y_0, f_1, f_2) u(f_1, f_2) \, ds,$$

where $f_j = f_j(s, \theta_0(x_0, y_0, x, y), x, y)$, $j = 1, 2$, and from Lemma 3.3.1. □

Now let us derive a differential equation for the function $w(x_0, y_0, x, y)$. For this, we note that

$$w(x_0, y_0, f_1(s, \theta_0(x_0, y_0, x, y), x, y), f_2(\cdot)) = \int_0^s \rho(x_0, y_0, f_1, f_2) u(f_1, f_2) \, ds.$$

Differentiating this equality with respect to s and using the fact that

$$(f_1(s, \theta_0, x_0, y_0), f_2(\cdot)) = (x, y), \quad (f_{1s}(\cdot), f_{2s}(\cdot)) = \nu(x_0, y_0, x, y),$$

we find that

$$(\text{grad}_{x,y} w(x_0, y_0, x, y), \nu(x_0, y_0, x, y)) = \rho(x_0, y_0, x, y) u(x, y). \quad (3.3.11)$$

Dividing both sides of this equality by $\rho(x_0, y_0, x, y)$, setting $x_0 = g(t_1)$ and $y_0 = p(t_1)$, and differentiating the equality obtained with respect to t_1, we obtain the equation for the function $w(g(t_1), p(t_1), x, y) = \tilde{w}(t_1, x, y)$

$$\frac{\partial}{\partial t_1}\left\{\frac{1}{\tilde{\rho}(t_1, x, y)}(\text{grad}_{x,y}\tilde{w}, \tilde{\nu})\right\} = 0, \quad (3.3.12)$$

where

$$\tilde{\rho}(t_1, x, y) = \rho(g(t_1), p(t_1), x, y), \quad \tilde{\nu}(t_1, x, y) = \nu(g(t_1), p(t_1), x, y).$$

Equation (3.3.12) belongs to equations of mixed type, namely, hyperbolo-parabolic type. This equation is satisfied in the cylindrical domain $[0, l] \times \overline{D}$. On the boundary of this domain we have the condition of periodicity

$$\tilde{w}(0, x, y) = \tilde{w}(l, x, y), \quad (x, y) \in \overline{D}, \quad (3.3.13)$$

and the condition resulting from formula (3.3.6)

$$\tilde{w}(t_1, g(t_2), p(t_2)) = v(t_1, t_2). \quad (3.3.14)$$

Thus, the function $\tilde{w}$ is a solution of the nonclassical problem (3.3.12)–(3.3.14). It should be noted that formulas (3.3.5) and (3.3.6) imply that $v(t_1, t_1) = 0$ ($t_1 \in [0, l]$). If this condition is fulfilled, the problem (3.3.12)–(3.3.14) and the integral geometry problem which was formulated above are equivalent. Therefore, it suffices to find $\tilde{w}(t_1, x, y)$; and then we can find $u(x, y)$ from formula (3.3.11). To investigate the uniqueness and stability of the problem (3.3.12)–(3.3.14) we use the method of energy estimates.

For $x \neq g(t_1)$, $y \neq p(t_1)$, and $\beta = (-\sin\tilde{\theta}, \cos\tilde{\theta})$, $\tilde{\theta} = \tilde{\theta}(t_1, x, y) = \theta(g(t_1), p(t_1), x, y)$, we have two evident identities

$$\tilde{\rho}(\nabla_{x,y}\tilde{w}, \beta)\frac{\partial}{\partial t_1}\left[\frac{1}{\tilde{\rho}}(\nabla_{x,y}\tilde{w}, \tilde{\nu})\right] = (\nabla_{x,y}\tilde{w}, \beta)(\nabla_{x,y}\tilde{w}, \tilde{\nu}_{t_1})$$
$$+ (\nabla_{x,y}\tilde{w}, \beta)(\nabla_{x,y}\tilde{w}_{t_1}, \tilde{\nu}) - (\nabla_{x,y}\tilde{w}, \beta)(\nabla_{x,y}\tilde{w}, \tilde{\nu})\frac{\partial}{\partial t_1}\ln\tilde{\rho},$$

$$\tilde{\rho}(\nabla_{x,y}\tilde{w}, \beta)\frac{\partial}{\partial t_1}\left[\frac{1}{\tilde{\rho}}(\nabla_{x,y}\tilde{w}, \tilde{\nu})\right] = \frac{\partial}{\partial t_1}\left[(\nabla_{x,y}\tilde{w}, \beta)(\nabla_{x,y}\tilde{w}, \tilde{\nu})\right]$$
$$- (\nabla_{x,y}\tilde{w}, \tilde{\nu})(\nabla_{x,y}\tilde{w}, \beta_{t_1}) - (\nabla_{x,y}\tilde{w}, \tilde{\nu})(\nabla_{x,y}\tilde{w}_{t_1}, \beta)$$
$$- (\nabla_{x,y}\tilde{w}, \beta)(\nabla_{x,y}\tilde{w}, \tilde{\nu})\frac{\partial}{\partial t_1}\ln\tilde{\rho}.$$

Chapter 3. Integral geometry problems

Summing up side by side the above identities and taking into account the relations

$$\tilde{\nu}_{t_1} = \beta \frac{\partial \tilde{\theta}}{\partial t_1}, \qquad \beta_{t_1} = -\tilde{\nu} \frac{\partial \tilde{\theta}}{\partial t_1},$$

$$(\nabla_{x,y}\tilde{w}, \beta)(\nabla_{x,y}\tilde{w}_{t_1}, \tilde{\nu}) - (\nabla_{x,y}\tilde{w}, \tilde{\nu})(\nabla_{x,y}\tilde{w}_{t_1}, \beta)$$
$$= \frac{\partial}{\partial x}(\tilde{w}_{t_1}\tilde{w}_y) - \frac{\partial}{\partial y}(\tilde{w}_{t_1}\tilde{w}_x),$$

we find that

$$2\tilde{\rho}\,(\nabla_{x,y}\tilde{w}, \beta) \frac{\partial}{\partial t_1}\left(\frac{1}{\tilde{\rho}}\nabla_{x,y}\tilde{w}, \tilde{\nu}\right)$$
$$= \left\{[(\nabla_{x,y}\tilde{w}, \tilde{\nu})^2 + (\nabla_{x,y}\tilde{w}, \beta)^2]\frac{\partial \tilde{\theta}}{\partial t_1} - 2(\nabla_{x,y}\tilde{w}, \tilde{\nu})(\nabla_{x,y}\tilde{w}, \beta)\frac{\partial}{\partial t_1}\ln\tilde{\rho}\right\}$$
$$+ \frac{\partial}{\partial t_1}[(\nabla_{x,y}\tilde{w}, \tilde{\nu})(\nabla_{x,y}\tilde{w}, \beta)] + \frac{\partial}{\partial x}(\tilde{w}_{t_1}\tilde{w}_y) - \frac{\partial}{\partial y}(\tilde{w}_{t_1}\tilde{w}_x). \quad (3.3.15)$$

The left-hand side of the above identity vanishes on the solutions of equation (3.3.12). Let us assume that $\tilde{w}(t_1, x, y)$ is the solution of equation (3.3.12) and integrate identity (3.3.15) over the domain G_ε that is obtained from the domain $G = [0, l] \times \overline{D}$ by removing the set

$$\{(t_1, x, y) \mid t_1 \in [0, l], (x, y) \in \overline{D}, (x - g(t_1))^2 + (y - p(t_1))^2 \leq \varepsilon^2\}$$

for a sufficiently small $\varepsilon > 0$. Then, using the Gauss–Ostrogradskii formula, we obtain

$$\int_{G_\varepsilon} \left\{[(\nabla_{x,y}\tilde{w}, \tilde{\nu})^2 + (\nabla_{x,y}\tilde{w}, \beta)^2]\frac{\partial \tilde{\theta}}{\partial t_1}\right.$$
$$\left. - 2(\nabla_{x,y}\tilde{w}, \tilde{\nu})(\nabla_{x,y}\tilde{w}, \beta)\frac{\partial}{\partial t_1}\ln\tilde{\rho}\right\}dx\,dy\,dt_1$$
$$+ \int_{S_\varepsilon}\left\{(\nabla_{x,y}\tilde{w}, \tilde{\nu})(\nabla_{x,y}\tilde{w}, \beta)\cos(\widehat{n, t_1})\right.$$
$$\left. + \tilde{w}_{t_1}[\tilde{w}_y\cos(\widehat{n, x}) - \tilde{w}_x\cos(\widehat{n, y})]\right\}dS = 0.$$

Here S_ε is the boundary of the set G_ε, n is the external normal vector to S_ε, and dS is the area element.

We pass to the limit as $\varepsilon \to 0$ in the last equality. In this case, by the estimates from Lemma 3.3.3, the integrals over G_ε and S_ε converge, as singular integrals, to the integrals over G and S (here S is the boundary of G). The integral over the surface S can be split in two parts: the integral over the upper and lower bases of the cylinder G and the integral over the lateral surface $[0, l] \times \Gamma$. But $\cos(\widehat{n, x}) = \cos(\widehat{n, y}) = 0$ on the upper and lower bases of the cylinder and $\cos(\widehat{n, t_1})$ has opposite signs at the respective points. By the periodicity condition (3.3.13), the sum of the integrals over the upper and lower bases of the cylinder vanishes. On the lateral surface we have $\cos(\widehat{n, t_1}) = 0$ and for $x = g(t_2)$ and $y = p(t_2)$ we have

$$\tilde{w}_{t_1} = v_{t_1}(t_1, t_2),$$
$$\tilde{w}_y \cos(\widehat{n, x}) - \tilde{w}_x \cos(\widehat{n, y}) = \frac{\partial}{\partial t_2} \tilde{w}(t_1, g(t_2), p(t_2)) = v_{t_2}(t_1, t_2)$$

and $dS = dt_1\, dt_2$. Thus, we finally obtain

$$\int_G \left\{ [(\nabla_{x,y}\tilde{w}, \tilde{\nu})^2 + (\nabla_{x,y}\tilde{w}, \beta)^2] \frac{\partial \tilde{\theta}}{\partial t_1} - 2(\nabla_{x,y}\tilde{w}, \tilde{\nu})(\nabla_{x,y}\tilde{w}, \beta) \frac{\partial}{\partial t_1} \ln \tilde{\rho} \right\} dx\, dy\, dt_1 = -\int_0^l \int_0^l v_{t_1} v_{t_2}\, dt_1\, dt_2. \quad (3.3.16)$$

By Lemma 3.3.2, $\partial \tilde{\theta}/\partial t_1 \geq 0$. At all points where $\partial \tilde{\theta}/\partial t_1 = 0$ the expression in curly brackets vanishes in virtue of condition (3.3.8). For the points where $\partial \tilde{\theta}/\partial t_1 > 0$ we have

$$[(\nabla_{x,y}\tilde{w}, \tilde{\nu})^2 + (\nabla_{x,y}\tilde{w}, \beta)^2] \frac{\partial \tilde{\theta}}{\partial t_1} - 2(\nabla_{x,y}\tilde{w}, \tilde{\nu})(\nabla_{x,y}\tilde{w}, \beta) \frac{\partial}{\partial t_1} \ln \tilde{\rho}$$
$$= (\nabla_{x,y}\tilde{w}, \tilde{\nu})^2 \left[\frac{\partial \tilde{\theta}}{\partial t_1} - \left(\frac{\partial}{\partial t_1} \ln \tilde{\rho} \right)^2 \left(\frac{\partial \tilde{\theta}}{\partial t_1} \right)^{-1} \right]$$
$$+ \left[(\nabla_{x,y}\tilde{w}, \tilde{\nu}) \left(\frac{\partial}{\partial t_1} \ln \tilde{\rho} \right) \left(\frac{\partial \tilde{\theta}}{\partial t_1} \right)^{-1/2} - (\nabla_{x,y}\tilde{w}, \beta) \left(\frac{\partial \tilde{\theta}}{\partial t_1} \right)^{1/2} \right]^2$$
$$\geq (\nabla_{x,y}\tilde{w}, \tilde{\nu})^2 (1 - q^2) \frac{\partial \tilde{\theta}}{\partial t_1} \geq \rho_0^2 (1 - q^2) u^2(x, y) \frac{\partial \tilde{\theta}}{\partial t_1}.$$

Therefore, sharpening inequality (3.3.16), we obtain

$$\rho_0^2 (1 - q^2) \int_D u^2(x, y)\, dx\, dy \cdot \int_0^{2\pi} \frac{\partial \tilde{\theta}}{\partial t_1}\, dt_1 \leq \frac{1}{2} \int_0^l \int_0^l |\nabla_{t_1, t_2} v|^2\, dt_1\, dt_2,$$

which implies the estimate (3.3.9).

To complete the proof of the theorem we note that the expression on the right-hand side of equality (3.3.9) has sense for $u \in C^1(\overline{D})$, $\rho(x_0, y_0, x, y) \in C^1(\Gamma \times \overline{D})$, and the family of curves L satisfying conditions 1 and 2. Therefore, approximating the functions u, ρ, and $\tilde{f}$ that satisfy the assumptions of the theorem by the functions u_n, ρ_n, and $\tilde{f}_n$ for which the estimate (3.3.9) is already established and passing to the limit as $n \to \infty$ in that estimate, we obtain the estimate (3.3.9) under the conditions of the theorem. In particular, this estimate implies the uniqueness of solution of the integral geometry problem in the space $C^1(\overline{D})$. $\square$

3.4. PROBLEMS OF GENERAL FORM IN THE SPACE

Let us consider the following problem in the space $\mathbb{R}^n$, $x = (x_1, \ldots, x_n)$. Let $D \subset \mathbb{R}^n$ be a bounded domain and let Ω be an open domain belonging to D. In the case where the dimension of the space is even we shall assume that the distance between the boundaries of the domains D and Ω is no less than some positive quantity h and, thus, Ω lies strictly inside of D. In the case where the dimension of the space is odd, D and Ω may coincide. In addition, we assume that for every point $x \in D$ and every unit vector $\nu = (\nu_1, \ldots, \nu_n)$ there exists a unique smooth hypersurface $S(x, \nu)$ that passes through the point x and has the normal ν at this point.

Let us denote by $\mathcal{U}$ the class of functions $u(x)$ compactly supported in Ω and such that $u(x) \in L_2(\Omega)$. We consider the problem of finding a function $u(x) \in \mathcal{U}$ from the equation

$$v(x, \nu) = \int_{S(x,\nu)} \rho(\xi, x, \nu) u(\xi) \, dS, \quad x \in D, \quad |\nu| = 1, \qquad (3.4.1)$$

where $\rho(\xi, x, \nu)$ is a given smooth function and dS is the surface area element.

This problem is, generally speaking, overdetermined since the function $u(x)$ depends on n variables while the function v depends on $2n - 1$ variables. However, let us mention one important case where the number of essential variables of the functions u and v coincides. Consider the surface $S(x, \nu)$ and an arbitrary point $x^0 \in S(x, \nu)$. Let ν^0 be the normal vector to $S(x, \nu)$ at the point x^0. Then $S(x^0, \nu^0) = S(x, \nu)$. But if $\rho(\xi, x^0, \nu^0) = \rho(\xi, x, \nu)$ for $\xi \in S(x, \nu)$, then $v(x^0, \nu^0) = v(x, \nu)$. Since every point of the hypersurface $S(x, \nu)$ is characterized by $n-1$ parameters, therefore, the above equality shows that there are only n essential parameters on which $v(x, \nu)$ depends in this case. So, if the weight function ρ depends only on the point $\xi \in D$ and the surface $S(x,\nu)$, then the functions u and v have the same number of essential variables. This reason is in favor of the fact that, apparently, there exists a "better" statement of the integral geometry problem with the same family of surfaces, this statement being connected with another parametrization of this family for which the problem is not overdetermined.

The problem of solving equation (3.4.1) will be considered in this section under the assumption that the diameter of the domain D is small enough. In what follows we shall assume that the equation of the surface $S(x,\nu)$ can be characterized with the help of a smooth function $\varphi(\xi, x, \nu)$:

$$S(x,\nu) = \{\xi \mid \varphi(\xi, x, \nu) = 0\}; \qquad (3.4.2)$$

moreover, we assume that at the points of $S(x,\nu)$ we have $|\nabla_\xi \varphi| \neq 0$.

The assumptions about the method of parametrization lead to the following properties for the function φ:

$$\varphi(x, x, \nu) = 0, \qquad \nabla_\xi \varphi|_{\xi=x} = \nu |\nabla_\xi \varphi|_{\xi=x}. \qquad (3.4.3)$$

Evidently, one can always suppose that

$$|\nabla_\xi \varphi|_{\xi=x} = 1, \qquad (3.4.4)$$

dividing, if necessary, φ by $|\nabla_\xi \varphi|_{\xi=x}$. Therefore, in the sequel we shall assume that equality (3.4.4) holds true. Then the following representation is true for the function φ:

$$\varphi(\xi, x, \nu) = (\nu, \xi - x) + \sum_{i,j=1}^{n} a_{ij}(\xi, x, \nu)(\xi_i - x_i)(\xi_j - x_j), \qquad (3.4.5)$$

where

$$a_{ij}(\xi, x, \nu) = \int_0^1 \varphi_{\xi_i \xi_j}(x + t(\xi - x), x, \nu)(1 - t)\, dt.$$

We shall use this representation below.

At first we formulate and prove the result for the case of an odd-dimensional space and only then we prove it for the case of an even-dimensional space.

Theorem 3.4.1. *Let n be odd, $s = (n-1)/2 \geq 1$, the function $\varphi(\xi, x, \nu)$ be continuous in the domain $G = \{(\xi, x, \nu) \mid \xi \in \overline{D},\ x \in \overline{D},\ |\nu| = 1\}$ together with its partial derivatives of order up to $n + 2$, and the weight function*

Chapter 3. Integral geometry problems

$\rho(\xi, x, \nu)$ be continuous in G together with its partial derivatives of order up to n and satisfy in G the condition

$$\rho(x, x, \nu) \geq \rho_0 > 0. \tag{3.4.6}$$

Then there exist a number $d^* > 0$, $d^* = d^*(\varphi, \rho)$, such that if $\operatorname{diam} D < d^*$, then the solution of equation (3.4.1) in the class of functions $\mathcal{U}$ is unique and the following stability estimate is true:

$$\|u\|_{L_2(\Omega)} \leq C\|\Delta^s v_1\|_{L_2(\Omega)}, \quad v_1(x) = \int_{|\nu|=1} \frac{v(x,\nu)}{\rho(x,x,\nu)} \, d\omega_\nu. \tag{3.4.7}$$

Here $d\omega_\nu$ is the area element of the unit sphere and Δ^s is the s-th power of the Laplace operator.

Proof. Using the δ-function and the fact that the function $u(x)$ is compactly supported, we write equation (3.4.1) in the form

$$\int_\Omega \rho(\xi, x, \nu) \, u(\xi) \, |\nabla_\xi \varphi(\xi, x, \nu)| \, \delta(\varphi(\xi, x, \nu)) \, d\xi = v(x, \nu), \tag{3.4.8}$$

divide equality (3.4.8) by $\rho(x, x, \nu)$, and average it for fixed $x \in D$ over all ν. With account of the notation (3.4.7) equality (3.4.8) takes the form

$$\int_\Omega K(x, \xi) \, u(\xi) \, d\xi = v_1(x). \tag{3.4.9}$$

Here

$$K(x, \xi) = \int_{|\nu|=1} \tilde{\rho}(\xi, x, \nu) \, \delta(\varphi(\xi, x, \nu)) \, d\omega_\nu, \tag{3.4.10}$$

$$\tilde{\rho}(\xi, x, \nu) = \frac{\rho(\xi, x, \nu)}{\rho(x, x, \nu)} |\nabla_\xi \varphi(\xi, x, \nu)|.$$

The kernel (3.4.10) can be investigated rather easily for points ξ and x close to each other. But this is sufficient since in proving the theorem we can consider the diameter of the domain D to be small. Let us use the invariance of the measure $d\omega_\nu$ under orthogonal transformations of the space $\nu_1, \ldots, \nu_n$ and make a transformation with an orthogonal matrix Q. Then the formula for the kernel becomes

$$K(x, \xi) = \int_{|\nu|=1} \tilde{\rho}(\xi, x, Q\nu) \, \delta(\varphi(\xi, x, Q\nu)) \, d\omega_\nu. \tag{3.4.11}$$

Introduce the unit vector $\nu^0 = (\xi - x)/|\xi - x|$, $\nu^0 = (\nu_1^0, \ldots, \nu_n^0)$. In calculation of the integral (3.4.11) we shall use two different orthogonal transforms according to the position of the vector ν^0 on the unit sphere. Namely, let $e^1 = (1, 0, \ldots, 0)$. Then for $(\nu^0, e^1) \geq 0$ we take the transform defined by the equality

$$Q\nu = \nu + 2(\nu, e^1)\nu^0 - \frac{(\nu, \nu^0) + (\nu, e^1)}{1 + (\nu^0, e^1)}(\nu^0 + e^1). \tag{3.4.12}$$

It is easy to check that for every unit vector ν the equality $(Q\nu, Q\nu) = 1$ is true. Moreover,

$$Qe^1 = \nu^0. \tag{3.4.13}$$

For $(\nu^0, e^1) < 0$, in calculation of the integral we use the transform (3.4.12) in which e^1 is replaced by $-e^1$ and, consequently, $Qe^1 = -\nu^0$.

Evidently, it suffices to study the case of such position of the points x and ξ for which $(\nu^0, e^1) \geq 0$. In what follows we restrict ourselves to this very case. The equation $\varphi(\xi, x, Q\nu) = 0$ for fixed x and ξ defines a surface of dimension $n - 2$ on the surface of the sphere. Indeed, in virtue of (3.4.5), we can write this equality divided by $|x - \xi|$ in the form

$$(\nu^0, Q\nu) + |x - \xi| \sum_{i,j=1}^n a_{ij}(x + |x - \xi|\nu^0, x, Q\nu)\nu_i^0 \nu_j^0 = 0. \tag{3.4.14}$$

We represent the vector ν as

$$\nu = qe^1 + \sqrt{1 - q^2}\,\bar{\nu}, \tag{3.4.15}$$

where $\bar{\nu}$ is the unit vector orthogonal to the vector e^1. Then, by (3.4.13),

$$(\nu^0, Q\nu) = (Q^*\nu^0, \nu) = (e^1, \nu) = q,$$

and equality (3.4.14) becomes

$$q + |x - \xi| \sum_{i,j=1}^n a_{ij}(x + |x - \xi|\nu^0, x, q\nu^0 + \sqrt{1 - q^2}\,Q\bar{\nu})\nu_i^0 \nu_j^0 = 0. \tag{3.4.16}$$

The contraction mapping principle implies the following lemma.

Lemma 3.4.1. *If the function φ satisfies the assumptions of the theorem, then for every $q_0 \in (0, 1)$ there exists $d_0 = d_0(\varphi, q_0)$ such that for domains Ω with $\operatorname{diam}\Omega < d_0$ the equation $\varphi(\xi, x, Q\nu) = 0$ for $x, \xi \in \Omega$*

defines q as a one-valued, continuous, and bounded function of the arguments x, $|x - \xi|$, ν^0, and $\bar{\nu}$. Moreover, this function q has continuous and bounded derivatives up to the order n with respect to these arguments and satisfies the inequality

$$|q(x, |x-\xi|, \nu^0, \bar{\nu})| \leq q_0 < 1. \qquad (3.4.17)$$

Proof. Indeed, since φ satisfies the assumptions of the theorem, functions $a_{ij}(\xi, x, \nu)$ are uniformly bounded for $x, \xi \in \Omega$ and $|\nu| = 1$ by some constant c_0. Hence, writing equality (3.4.16) in the form $q = A(q)$, we ascertain that under the condition $c_0 d_0 \leq q_0$ the operator A maps the set of continuous functions which satisfy inequality (3.4.17) into itself. The condition of uniform boundedness of the derivatives of a_{ij} with respect to $\nu_1, \ldots, \nu_n$ leads to the conclusion that if $|x - \xi|$ is small, the operator A is a contraction operator on this set. Thus, equation (3.4.16) defines q as a continuous function of the arguments x, $|x - \xi|$, ν^0, and $\bar{\nu}$ which satisfies inequality (3.4.17). The existence and boundedness of the derivatives of order up to n of the function q with respect to its arguments follows easily from equality (3.4.16) since the functions a_{ij} and matrix $Q = Q(\nu^0)$ defined by formula (3.4.12) have the corresponding derivatives. $\square$

It also follows from equality (3.4.16) that $q \to 0$ for $|x - \xi| \to 0$.

Now we choose a fixed $q_0 \in (0, 1)$; in the sequel we shall assume that $\operatorname{diam} \Omega < d_0(\varphi, q_0)$. Formula (3.4.15) in which q is the solution of equation (3.4.16) for an arbitrary unit vector $\bar{\nu}$, $(\bar{\nu}, e^1) = 0$, defines (for fixed x and ξ) ν as a smooth one-valued function of $\bar{\nu}$ and thus defines a surface of dimension $n-2$ on the sphere $|\nu| = 1$. Let us denote it by $\Sigma(x, \xi)$. If $x \to \xi$ in such a way that $(\xi - x)/|\xi - x| \to \nu^0$, then the surface $\Sigma(x, \xi)$ coincides in the limit with the cross-section of the sphere $|\nu| = 1$ by the plane passing through the center of the sphere and orthogonal to the vector e^1.

Let n be the unit vector of the normal to $\Sigma(x, \xi)$ lying in the plane tangent to the sphere $|\nu| = 1$ at the point ν. Taking the above into account, we can write formula (3.4.11) for the kernel $K(x, \xi)$ of equation (3.4.9) in the form

$$K(x, \xi) = \int_{\Sigma(x,\xi)} \tilde{\rho}(\xi, x, Q\nu) \left|\frac{\partial \varphi}{\partial n}\right|^{-1} d\sigma, \qquad (3.4.18)$$

where $d\sigma$ is the area element of the surface $\Sigma(x, \xi)$. Since on $\Sigma(x, \xi)$ the vector n is parallel to the vector

$$\nabla_\nu \varphi(\xi, x, Q\nu) - \nu(\nu, \nabla_\nu \varphi(\xi, x, Q\nu)),$$

therefore,
$$\left|\frac{\partial \varphi}{\partial n}\right| = [|\nabla_\nu \varphi(\xi, x, Q\nu)|^2 - (\nu, \nabla_\nu \varphi(\xi, x, Q\nu))^2]^{1/2}. \tag{3.4.19}$$

From formula (3.4.5) we find
$$\frac{1}{|x-\xi|}\nabla_\nu \varphi(\xi, x, Q\nu) = e^1 + |x-\xi|\sum_{i,j=1}^n \nabla_\nu a_{ij}(\xi, x, Q\nu)\nu_i^0 \nu_j^0.$$

Using formula (3.4.19) and the fact that on the surface $\Sigma(x,\xi)$ we have $\nu = \nu(x, |x-\xi|, \nu^0, \bar{\nu})$, for $|\partial \varphi/\partial n|$ on $\Sigma(x,\xi)$ we obtain the representation
$$\left|\frac{\partial \varphi}{\partial n}\right|^{-1} = \frac{1}{|x-\xi|} + b(x, |x-\xi|, \nu^0, \bar{\nu}), \tag{3.4.20}$$

where b is a continuous and bounded function of its arguments together with its partial derivatives of order up to $n-1$. Similarly, the following representation is true for the function $\tilde{\rho}$ on $\Sigma(x,\xi)$:
$$\tilde{\rho}(\xi, x, Q\nu) = 1 + |x-\xi|\, c(x, |x-\xi|, \nu^0, \bar{\nu}), \tag{3.4.21}$$

where the function c has the same properties as the function b.

The vector $\bar{\nu}$ is orthogonal to the vector e^1 and has unit length; therefore, it is completely characterized by the angular spherical coordinates $\psi_1, \ldots, \psi_{n-2}$ in the coordinate plane orthogonal to the vector e^1. The surface element $d\sigma$ is calculated by the formula
$$d\sigma = \left[\Gamma\left(\frac{\partial \nu}{\partial \psi_1}, \ldots, \frac{\partial \nu}{\partial \psi_{n-2}}\right)\right]^{1/2} d\psi, \tag{3.4.22}$$

where $d\psi = d\psi_1 \cdots d\psi_{n-2}$ and Γ stands for the Gram determinant constructed for the vectors $\partial \nu/\partial \psi_1, \ldots, \partial \nu/\partial \psi_{n-2}$:

$$\Gamma\left(\frac{\partial \nu}{\partial \psi_1}, \ldots, \frac{\partial \nu}{\partial \psi_{n-2}}\right) = \begin{vmatrix} \left(\frac{\partial \nu}{\partial \psi_1}, \frac{\partial \nu}{\partial \psi_1}\right) & \cdots & \left(\frac{\partial \nu}{\partial \psi_1}, \frac{\partial \nu}{\partial \psi_{n-2}}\right) \\ \vdots & \vdots & \vdots \\ \left(\frac{\partial \nu}{\partial \psi_{n-2}}, \frac{\partial \nu}{\partial \psi_1}\right) & \cdots & \left(\frac{\partial \nu}{\partial \psi_{n-2}}, \frac{\partial \nu}{\partial \psi_{n-2}}\right) \end{vmatrix}.$$

It follows from formulas (3.4.18) and (3.4.20)–(3.4.22) that the kernel $K(x,\xi)$ can be represented in the form
$$K(x,\xi) = \frac{\omega_{n-1}}{|x-\xi|} + K_0(x, |x-\xi|, \nu^0), \qquad \omega_{n-1} = \frac{2\pi^s}{\Gamma(s)}, \tag{3.4.23}$$

where K_0 is a continuous and bounded function of its arguments together with its derivatives of order up to $n-1$ and $\Gamma(s)$ is the gamma-function. But then, applying to equation (3.4.9) the operator Δ^s, $s = (n-1)/2$, with respect to the variable x and using

$$\Delta^s \left(\frac{1}{|x-\xi|} \right) = (-1)^s (4\pi)^s (s-1)! \delta(x-\xi),$$

we obtain the following equation for $u(x)$:

$$u(x) + \int_\Omega \tilde{K}(x,\xi) u(\xi) \, d\xi = \frac{(-1)^s}{2^n \pi^{2s}} \Delta^s v_1(x). \qquad (3.4.24)$$

Here

$$\tilde{K}(x,\xi) = \frac{(-1)^s}{2^n \pi^{2s}} \Delta_x^s K_0(x, |x-\xi|, \nu^0).$$

The kernel $\tilde{K}(x,\xi)$ of equation (3.4.24) has an integrable singularity. Indeed, calculation of the operator Δ^s leads to differentiation of K_0 up to the order $2s$ with respect to the variables x, $|x-\xi|$, and ν^0 and differentiation up to the order $2s$ of the arguments $|x-\xi|$ and ν^0 with respect to x. But $2s = n-1$ and, consequently, the derivatives with respect to the arguments of the function K_0 are continuous and bounded. At the same time it is clear that the following estimates are true for the derivatives of $|x-\xi|$ and ν^0 with respect to x:

$$\left| \frac{\partial^k}{\partial x_i^k} |x-\xi| \right| \leq \frac{c_1}{|x-\xi|^{k-1}}, \qquad \left| \frac{\partial^k}{\partial x_i^k} \nu^0 \right| \leq \frac{c_1}{|x-\xi|^k}, \qquad 1 \leq k \leq 2s.$$

These considerations imply that the kernel $\tilde{K}(x,\xi)$ is continuous everywhere except for the set of points x, ξ such that $x = \xi$. In a neighborhood of this set, the following estimate is valid for $\tilde{K}(x,\xi)$:

$$|\tilde{K}(x,\xi)| \leq c_2 / |x-\xi|^{n-1}.$$

It is known that equation (3.4.24) with a kernel of such type has a unique solution from $L_2(\Omega)$ if the diameter of the domain Ω is small enough. In this case, the estimate (3.4.17) for the solution of equation (3.4.24) is evident. □

Now we consider the case of an even-dimensional space. As it was said earlier, in this case we assume that Ω lies strictly inside of D so that the distance between Ω and the boundary of the domain D is greater than a certain $h > 0$. In addition, we assume that $h > \operatorname{diam} \Omega$. Let Ω_h be the open set obtained as the union of all open balls of radius h centered at the points $x \in \Omega$.

Theorem 3.4.2. Let $n = 2s$, $s \geq 1$, and let the functions φ and ρ satisfy all the assumptions of Theorem 3.4.1 with n replaced by $n+1$. Then there exists a number $d^* > 0$, $d^* = d^*(\varphi, \rho, h)$, such that if $\operatorname{diam} \Omega < d^*$, then the solution of equation (3.4.1) in the class of functions $\mathcal{U}$ is unique and the following stability estimate holds:

$$\|u\|_{L_2(\Omega)} \leq 0 \|\Delta^s v_2\|_{L_2(\Omega)}, \tag{3.4.25}$$

where

$$v_2(x) = \int_{|x-y| \leq h} \frac{v_1(y)}{|x-y|^{n-1}} \, dy \tag{3.4.26}$$

and $v_1(x)$ is calculated in terms of $v(x, \nu)$ by formula (3.4.7).

Proof. The scheme of the proof is as follows. At first, just as in the case of odd n, we derive equation (3.4.9). For the kernel $K(x, \xi)$ of this equation, representation (3.4.23) is true; moreover, by the assumptions of the theorem, the smoothness of the function K_0 is higher by one, namely, K_0 has continuous bounded derivatives of order up to n with respect to the arguments x, $|x - \xi|$, and $\nu^0 = (\xi - x)|x - \xi|^{-1}$. Now we apply the operator of averaging over the ball of radius h defined by formula (3.4.26) to both sides of the equality. This yields the equation

$$\int_\Omega T(x, \xi) \, u(\xi) \, d\xi = v_2(x), \quad x \in \Omega, \tag{3.4.27}$$

where

$$T(x, \xi) = \int_{|x-y| \leq h} \frac{K(y, \xi)}{|x-y|^{n-1}} \, dy. \tag{3.4.28}$$

Next we show that $T(x, \xi)$, $x, \xi \in \Omega$, admits the representation

$$T(x, \xi) = -\omega_{n-1} \omega_n \ln |x - \xi| + T_0(\xi, |\xi - x|, \nu^0), \tag{3.4.29}$$

where $T_0(\xi, \rho, \nu^0)$ (here ρ is merely the notaion of the argument) is a function continuous and bounded together with its derivatives of order up to n everywhere except for the set $\rho = 0$ and in a neighborhood of this set the function $T_0(\xi, \rho, \nu^0)$ satisfies the inequalities ($|\alpha| + k \leq n$)

$$\left| \frac{\partial^k}{\partial \rho^k} D^\alpha_{\xi, \nu^0} T_0(\xi, \rho, \nu^0) \right| \leq c_0 \begin{cases} \ln \rho, & k = 1, \\ \rho^{1-k}, & k \geq 2. \end{cases} \tag{3.4.30}$$

Applying the operator Δ^s to equation (3.4.27) and using the relation

$$\Delta^s_x \ln |x - \xi| = (-1)^{s-1} 2^{n-1} \pi^s (s-1)! \, \delta(x - \xi),$$

Chapter 3. Integral geometry problems

we obtain an equation of Fredholm type which is similar to equation (3.4.24) and has a singularity of the form $|x-\xi|^{n-1} \ln|x-\xi|$, which implies the validity of the theorem.

Thus, it remains to ascertain that the kernel T has the properties indicated above. For this, we calculate the integral (3.4.28) separately for each of the summands of the function $K(x,\xi)$, see (3.4.23). Calculation of the integral of the first summand is equivalent to calculation of the integral

$$f_1(\rho) = \int_{|x-y|\leq h} \frac{1}{|x-y|^{n-1}} \frac{1}{|y-\xi|} dy$$

$$= \int_{|\nu|=1} d\omega_\nu \int_0^h \frac{dr}{[r^2 + \rho^2 + 2r\rho(\nu^0,\nu)]^{1/2}}$$

$$= \int_{|\nu|=1} \left[\ln\left|h + \rho(\nu^0,\nu) + [h^2 + \rho^2 + 2h\rho(\nu^0,\nu)]^{1/2}\right| - \ln|1 + (\nu,\nu^0)|\right] d\omega_\nu$$

$$- \omega_n \ln\rho.$$

Here we introduced the spherical system of coordinates $y = x + r\nu$; and $(\rho, -\nu^0)$ denotes the spherical coordinates of the point ξ in this system, i.e., $\rho = |\xi - x|$, $\nu^0 = (x-\xi)/|\xi - x|$. One can see from the formula obtained that calculation of the integral (3.4.28) of the first summand of the function $K(x,\xi)$ leads to appearance of the principal part of the kernel $T(x,\xi)$, namely, $\ln\rho$, and some smooth function (even analytic for $\rho < h$) that depends only on ρ. Here we used essentially the fact that $\rho \leq \text{diam }\Omega < h$ for $x \in \Omega$ and $\xi \in \Omega$. Now let us show that calculation of the integral (3.4.28) of the second summand of the function $K(x,\xi)$ leads to a function with the properties of the function T_0. For calculation of the integral in this case it is convenient to introduce the spherical system centered at the point ξ ($y = \xi + r\nu$, $\nu = (\nu_1, \ldots, \nu_n)$, $x = \xi + \rho\nu^0$):

$$f_2(\xi, \rho, \nu^0) = \int_{|x-y|\leq h} \frac{K_0\left(y, |y-\xi|, \frac{\xi-y}{|\xi-y|}\right)}{|x-y|^{n-1}} dy$$

$$= \int_{|\nu|=1} d\omega_\nu \int_0^{r(\rho,(\nu^0,\nu))} \frac{r^{n-1} K_0(\xi + r\nu, r, -\nu)}{[r^2 + \rho^2 - 2r\rho(\nu^0,\nu)]^{(n-1)/2}} dr,$$

$$r(\rho, (\nu^0, \nu)) = \rho(\nu^0,\nu) + [h^2 - \rho^2[1 - (\nu,\nu^0)^2]]^{1/2}.$$

Let us introduce (similarly to that as it was done earlier) an orthogonal transformation of the system of coordinates with a matrix Q such that we have $Q^*\nu^0 = e^1 = (1,0,\ldots,0)$ (under the condition $(\nu^0, e^1) \geq 0$). Then, denoting $K_0(\xi + rQ\nu, r, -Q\nu) = \tilde{K}(\xi, r, \nu, \nu^0)$ and $(\nu^0, \nu) = \nu_1$, we find

$$f_2(\xi, \rho, \nu^0) = \int_{|\nu|=1} d\omega_\nu \int_0^{r(\rho,\nu_1)} \frac{r^{n-1} \tilde{K}(\xi, r, \nu, \nu^0)}{(r^2 + \rho^2 - 2r\rho\nu_1)^{(n-1)/2}} dr.$$

One can easily see from the above formula that the derivatives $D^\alpha_{\xi,\nu^0} f_2$, $|\alpha| \leq n$, are continuous and bounded for $\xi, x \in \Omega$. Indeed,

$$D^\alpha_{\xi,\nu^0} f_2 = \int_{|\nu|=1} d\omega_\nu \int_0^{r(\rho,\nu_1)} \frac{r^{n-1} D^\alpha_{\xi,\nu^0} \tilde{K}(\xi, r, \nu, \nu^0)}{(r^2 + \rho^2 - 2r\rho\nu_1)^{(n-1)/2}} dr. \quad (3.4.31)$$

Now it remains to examine the derivatives that contain the differentiation with respect to the variable ρ. The change of variable $r = \rho r_1$ in the internal integral shows at once that for $\rho > 0$ the integral f_2 depends continuously on ξ, ρ, and ν^0 together with the derivatives of order up to n. However, when $\rho \to 0$, the derivatives containing the differentiation with respect to the variable ρ are unbounded. Indeed, for sufficiently small ρ we can always choose $\delta > 0$ such that

$$\rho(1 + \delta) < r(\rho, \nu_1), \quad \nu_1 \in [-1, 1].$$

Then we split the internal integral into two integrals: the first is taken over the interval $[0, \rho(1 + \delta)]$ and the second is taken over the interval $[\rho(1 + \delta), r(\rho, \nu_1)]$. Making the change of variables $r = \rho r_1$ in the first integral, we obtain

$$f_2(\xi, \rho, \nu^0) = \rho \int_{|\nu|=1} d\omega_\nu \int_0^{1+\delta} \frac{r_1^{n-1} \tilde{K}(\xi, \rho r_1, \nu, \nu^0)}{(r_1^2 + 1 - 2r_1\nu_1)^{(n-1)/2}} dr_1$$

$$+ \int_{|\nu|=1} d\omega_\nu \int_{\rho(1+\delta)}^{r(\rho,\nu_1)} \frac{r^{n-1} \tilde{K}(\xi, r, \nu, \nu^0)}{(r^2 + \rho^2 - 2r\rho\nu_1)^{(n-1)/2}} dr. \quad (3.4.32)$$

The first integral in this formula is a function continuous and bounded together with the derivatives of order up to n. All the singularities in the derivatives are generated by the second integral. Now we note that the multiplier standing at $\tilde{K}$ in the second integral can be represented in the form

$$\frac{r^{n-1}}{(r^2 + \rho^2 - 2r\rho\nu_1)^{(n-1)/2}} = \Phi\left(\frac{\rho}{r}, \nu_1\right),$$

where the function $\Phi(z, \nu_1)$ as a function of the arguments z and ν_1 is bounded together with any finite number of derivatives in the domain

$|z| \leq (1+\delta)^{-1}$, $\nu_1 \in [-1,1]$. Therefore, the following estimates are true for its derivatives with respect to ρ:

$$\left|\frac{\partial^k}{\partial \rho^k} \Phi\left(\frac{\rho}{r}, \nu_1\right)\right| \leq \frac{c_0}{r^k}, \qquad r \geq \rho(1+\delta). \tag{3.4.33}$$

When calculating the derivatives $(\partial^k/\partial\rho^k)D^\alpha_{\xi,\nu^0}$ of the second term in formula (3.4.32) we can bring the symbol D^α_{ξ,ν^0} under the symbol of the internal integral. Calculation of the derivative $\partial^k/\partial\rho^k$ of the internal integral yields, first, several terms due to calculation of the derivatives with respect to the upper and lower limits of integration and, second, a certain integral due to differentiation of the integrand. The first terms are evidently bounded since the function $\Phi(\rho/r, \nu_1)$ is bounded and independent of ρ on the lower limit and coincides with the analytic function $h^{1-n}[r(\rho,\nu_1)]^{n-1}$ on the upper limit. The integral appearing in the result of differentiation of the integrand has the form

$$\int_{\rho(1+\delta)}^{r(\rho,\nu_1)} \frac{\partial^k}{\partial \rho^k} \Phi\left(\frac{\rho}{r}, \nu_1\right) D^\alpha_{\xi,\nu^0} \tilde{K}(\xi, r, \nu, \nu^0) \, dr$$

and, by inequality (3.4.33), is estimated by the integral

$$c_1 \int_{\rho(1+\delta)}^{r(\rho,\nu_1)} \frac{dr}{r^k} = c_1 \begin{cases} \ln \dfrac{\nu_1 + [(h/\rho)^2 - (1-\nu_1^2)]^{1/2}}{1+\delta}, & k = 1, \\ \dfrac{1}{1-k}\{[r(\rho,\nu_1)]^{1-k} - [\rho(1+\delta)]^{1-k}\}, & k > 1. \end{cases}$$

This implies the estimate (3.4.30). Thus, the proof of Theorem 3.4.2 is complete. □

Remark 3.4.1. Let $D(x,\nu)$ be the set of those points of the domain D for which $\varphi(\xi, x, \nu) \geq 0$. A more general problem than the problem of solving equation (3.4.1) is the problem of determining $u(x)$ from the equation

$$\int_{S(x,\nu)} \rho(\xi, x, \nu) \, u(\xi) \, dS + \int_{D(x,\nu)} \rho_1(\xi, x, \nu) \, u(\xi) \, d\xi = v(x, \nu),$$

$$x \in D, \quad |\nu| = 1.$$

If the smoothness of the function $\rho_1(\xi, x, \nu)$ is less not more than by one than the smoothness of the function $\rho(\xi, x, \nu)$, then Theorems 3.4.1 and 3.4.2 hold true for this problem.

Remark 3.4.2. If in formula (3.4.1) we replace the weight function $\rho(\xi, x, \nu)$ by a matrix weight function $R(\xi, x, \nu)$ of arbitrary finite dimension $m \times m$ and the function $u(\xi)$ by a vector function $\mathbf{u}(\xi)$ with the components $u_1, \ldots, u_m$, then we obtain a vector problem of integral geometry. For this problem, Theorems 3.4.1 and 3.4.2 hold true with natural replacement of condition (3.4.6) by the condition

$$|\det R(x, x, \nu)| \geq \rho_0 > 0$$

and the function ρ^{-1} in the estimate (3.4.7) by the matrix function $R^{-1}(x, x, \nu)$.

Remark 3.4.3. From Theorems 3.4.1 and 3.4.2 it follows that estimate

$$\|u\|_{L_2(\Omega)} \leq c_0 \sup_{|\nu|=1} \|v(x, \nu)\|_{W_2^l(\overline{D})}, \quad l = 2\left[\frac{n}{2}\right] \tag{3.4.34}$$

is fulfilled for every $n \geq 2$.

In this section we considered the statement of the integral geometry problem in the case where for every point x and every direction ν there exists a smooth hypersurface $S(x, \nu)$ that passes through the point x and has the normal ν at this point. A more general statement of this problem is possible in which the surfaces $S(x, \nu)$ passing through the point x exist only for ν belonging to a certain set $\omega(x)$ of the points of the unit sphere. Here the set $\omega(x)$ does not coincide, in general, with the unit sphere. It turns out that the question about the character of stability of the integral geometry problem is closely connected with the structure of the set $\omega(x)$ (Bukhgeim, 1975; Lavrentiev, Romanov, and Shishatskii, 1986).

For simplicity, we consider the case where $\omega(x)$ is independent of x: $\omega(x) = \omega_0$. Let $\omega_{\delta\alpha} = \{\nu \mid |\nu| = 1, |(\nu, \alpha)| \leq \delta\}$, i.e., $\omega_{\delta\alpha}$ is a spherical belt. Then if there are a unit vector α and $\delta > 0$ such that $\omega_{\delta\alpha} \cap \omega_0 \neq \varnothing$, then estimates of the type (3.4.34) for the integral geometry problem (with the set $|\nu| = 1$ under the symbol sup replaced by the set ω_0) cannot exist for any l. In particular, this implies that the estimates of the form

$$\|u\|_{W_2^k(\Omega)} \leq c_0 \sup_{\nu \in \omega_0} \|v(x, \nu)\|_{W_2^l(\overline{D})} \tag{3.4.35}$$

with any finite k and l and a constant $c_0 > 0$ cannot exist as well.

We show this for a special case $n = 2$. We set $\varphi(\xi, x, \nu) = \nu(\xi - x)$ and $\rho = 1$. The idea of the proof is completely clear from this example. Let

$\omega_{\delta\alpha} \cap \omega_0 = \varnothing$ for $\delta > 0$ and $\alpha = (0,1)$. Take a function $u_\lambda \in C_0^\infty(\Omega)$ of the following form:
$$u_\lambda = \sin(\lambda x_1)\,\psi(x),$$
where $\psi(x)$ is an infinitely differentiable function compactly supported in Ω and is not an identical zero and λ is a sufficiently large numerical parameter. Then
$$\|u_\lambda\|^2_{L_2(\Omega)} = \frac{1}{2}\|\psi\|^2_{L_2(\Omega)} - \frac{1}{2}\int_\Omega \cos(2\lambda x_1)\,\psi^2(x)\,\mathrm{d}x.$$

Integrating by parts sufficiently many times in this formula, we obtain
$$\left|\int_\Omega \cos(2\lambda x_1)\,\psi^2(x)\,\mathrm{d}x\right| \le c_k/|\lambda|^k.$$

On the other hand, for $x \in D$ and $\nu \in \omega_0$ we have
$$v_\lambda(x,\nu) = \frac{1}{|\nu_2|}\int_{-\infty}^\infty u_\lambda\left(\xi_1, x_2 - \frac{\nu_1}{\nu_2}(\xi_1 - x_1)\right)\mathrm{d}\xi_1$$
$$= \frac{1}{|\nu_2|}\int_{-\infty}^\infty \sin(\lambda\xi_1)\,\psi\left(\xi_1, x_2 - \frac{\nu_1}{\nu_2}(\xi_1 - x_1)\right)\mathrm{d}\xi_1.$$

Since for $\nu \in \omega_0$ we have $|(\nu,\alpha)| = |\nu_2| \ge \delta$, it follows that $|\nu_2|^{-1} \le \delta^{-1}$ and, consequently,
$$|D^\alpha_{x,\nu} v_\lambda(x,\nu)| \le \frac{c_{\alpha k}}{|\lambda|^k}, \quad x \in D,\quad \nu \in \omega_0.$$

Passing λ to infinity, we obtain $\|v_\lambda\|_{W_2^l(D)} \to 0$ uniformly over all $\nu \in \omega_0$ and for every finite l and $\|u_\lambda\|_{L_2(\Omega)} \to \|\psi\|_{L_2(\Omega)}/\sqrt{2} \ne 0$. This implies the assertion that was hypothesized earlier.

3.5. PROBLEMS OF VOLTERRA TYPE WITH MANIFOLDS INVARIANT UNDER THE MOTION GROUP

In this section, we consider in detail the intergal geometry problems on the plane. It should be noted that the results of this section are carried over to the intergal geometry problems of this type in the space of arbitrary dimension.

Let $u(x,y)$ be a function of two variables. We shall consider the function $u(x,y)$ in the half-plane $y \ge 0$ and shall assume that $u(x,y)$ is continuous and compactly supported in the variable x:
$$u(x,y) = 0,\quad |x| \ge l > 0.$$

Further, let $\varphi(y,\eta)$ and $g(y,\eta)$, $y \geq \eta$, be given functions such that
1) $\varphi(y,\eta) = \sqrt{y-\eta}\,\varphi_0(y,\eta)$,
2) $g(y,\eta) = g_0(y,\eta)/\sqrt{y-\eta}$,

where $\varphi_0(y,\eta)$ and $g_0(y,\eta)$ are continuously differentiable functions and such that
$$\varphi_0(y,\eta) \geq \varphi^0 > 0, \qquad g_0(y,\eta) \geq g^0 > 0.$$

Consider the following equation in the function $u(x,y)$:

$$\int_0^y g(y,\eta)\left[u(x-\varphi,\eta) + u(x+\varphi,\eta)\right] d\eta = f(x,y). \qquad (3.5.1)$$

The problem of solving equation (3.5.1) is an intergal geometry problem, i.e., a problem of recovering a function if we know the integrals of this function along a given family of curves. In our case, both the family of curves defined by the function $\varphi(y,\eta)$ and the weight function $g(y,\eta)$ are invariant under the group of motions parallel to the axis x.

For investigation of equation (3.5.1) we can use two methods that are, in essence, equivalent.

The first method. We apply to (3.5.1) the Fourier transform with respect to the variable x. After the Fourier transform equation (3.5.1) becomes

$$\int_0^y g(y,\eta)\cos(\lambda\varphi)\,v(\lambda,\eta)\,d\eta = \tilde{f}(\lambda,y), \qquad (3.5.2)$$

$$v(\lambda,\eta) = \int_{-\infty}^\infty e^{i\lambda x} u(x,\eta)\,dx, \qquad \tilde{f}(\lambda,y) = \int_{-\infty}^\infty e^{i\lambda x} f(x,y)\,dx.$$

Thus, equation (3.5.1) went over into the family of Volterra integral equations of first kind. The properties of the functions $\varphi(y,\eta)$ and $g(y,\eta)$ imply that equation (3.5.2) satisfies the conditions for Volterra integral equations of first kind which have been considered by Lavrentiev, Romanov, and Shishatskii (1986). Therefore, the solution of equation (3.5.2), and, consequently, equation (3.5.1), is unique.

Applying the operator of fractional differentiation to (3.5.2), we obtain the following equation of second kind:

$$v(\lambda,\eta) + \int_0^y G(\lambda,y,\eta)\,v(\lambda,\eta)\,d\eta = \tilde{f}_1(\lambda,y), \qquad (3.5.3)$$

$$G(\cdot) = \frac{1}{\pi g_0(y,y)}\frac{\partial}{\partial y}\int_\eta^y \left[g_{0y}\cos(\lambda\varphi) - \lambda g_0 \varphi_y \sin(\lambda\varphi)\right]\sqrt{\frac{y-\eta}{\xi-\eta}}\,d\xi,$$

$$\tilde{f}_1(\lambda,y) = \frac{1}{\pi g_0(y,y)}\frac{\partial}{\partial y}\int_0^y \frac{1}{\sqrt{y-\eta}}\tilde{f}(\lambda,\eta)\,d\eta.$$

Chapter 3. Integral geometry problems

The solution of equation (3.5.3) can be represented in the form

$$v(\lambda, y) = \tilde{f}_1(\lambda, y) + \int_0^y R(\lambda, y, \eta)\tilde{f}_1(\lambda, \eta)\, d\eta. \qquad (3.5.4)$$

Generally speaking, the norm of the integral operator in equation (3.5.3) increases infinitely with the increase of the parameter λ so that the problem of solving the family of equations (3.5.3), and, consequently, equation (3.5.1), is ill-posed.

Below we give a complete investigation of the character of instability of the solution of equations (3.5.1) and (3.5.3) in one simple case.

Let in (3.5.1)

$$\varphi_0(y, \eta) = g_0(y, \eta) = 1.$$

In this case equation (3.5.1) becomes

$$\int_0^y [u(x - \sqrt{y-\eta}, \eta) + u(x + \sqrt{y-\eta}, \eta)] \frac{d\eta}{\sqrt{y-\eta}} = f(x, y). \qquad (3.5.5)$$

After the Fourier transform equation (3.5.5) takes the form

$$\int_0^y \cos(\lambda\sqrt{y-\eta})\, v(\lambda, \eta) \frac{d\eta}{\sqrt{y-\eta}} = \tilde{f}(\lambda, y). \qquad (3.5.6)$$

Let us show that the solution of equation (3.5.6) is given by the formula

$$v(\lambda, y) = \frac{1}{\pi} \frac{\partial}{\partial y} \int_0^y \cosh(\lambda\sqrt{y-\eta})\tilde{f}(\lambda, \eta) \frac{d\eta}{\sqrt{y-\eta}}. \qquad (3.5.7)$$

Indeed, applying to (3.5.6) the Volterra operator with the kernel $\cosh(\lambda\sqrt{z-y})/\sqrt{z-y}$, we obtain

$$\int_0^z \cosh(\lambda\sqrt{z-y})\tilde{f}(\lambda, y) \frac{dy}{\sqrt{z-y}}$$

$$= \int_0^z \int_\eta^z \cos(\lambda\sqrt{y-\eta})\cosh(\lambda\sqrt{z-y}) \frac{dy}{\sqrt{(y-\eta)(z-y)}} v(\lambda, \eta)\, d\eta.$$

Using the well-known formula

$$\int_0^1 \cos(p\sqrt{t})\cosh(p\sqrt{1-t}) \frac{dt}{\sqrt{t(1-t)}} = \pi$$

we obtain (3.5.7).

Formula (3.5.7) implies that the character of instability of the solution of equation (3.5.5) is the same as in the Cauchy problem for the Laplace equation. One can obtain similar conditional stability estimates for the general equation (3.5.1) and the corresponding equation (3.5.2).

Namely, one can show that the function in (3.5.4) satisfies the inequality

$$|R(\lambda, y, \eta)| \leq a e^{b\sqrt{y-\eta}\,\lambda} \frac{1}{\sqrt{y-\eta}},$$

where a and b are certain constants.

The second method. In equation (3.5.1), we shall consider the function $u(x, y)$ for fixed y as an element of the Hilbert space W:

$$u(x, y) = w(y) \in W.$$

Then equation (3.5.1) can be considered as a Volterra operator equation

$$\int_0^y A(y, \eta) w(\eta) \, d\eta = \psi(y). \tag{3.5.8}$$

The family of operators $A(y, \eta)$ in (3.5.8) is defined as follows:

$$A(y, \eta) w(\eta) = g(y, \eta) \big[u(x - \varphi, \eta) + u(x + \varphi, \eta) \big].$$

It is easy to show that the Volterra operator equation (3.5.8) defined in this way is reduced to an equation satisfying the conditions of the following theorem (Bukhgeim, 1999; Lavrentiev, Romanov, and Shishatskii, 1986):

Theorem 3.5.1. *Let $u(t)$ be a continuous function of the scalar argument t with values in a Hilbert space U; let $A(t, \tau)$ be a family of continuous operators with the domain of definition U and range of values in U; and let B be a continuous operator with $N(B) = 0$ (here N is the kernel). We assume that the family of operators $A(t, \tau)$ depends continuously on the variables t and τ. Let the operator B be self-adjoint and permutable with all operators $A(t, \tau)$. Then the solution of the Volterra integral equation of second kind*

$$Bu(t) + \int_0^t A(t, \tau) u(\tau) \, d\tau = f(t)$$

is unique.

In the capacity of the operator B appearing in that theorem we can take, for example, the integral operator

$$Bw(y) = \int_{-l}^{l} p(x,\xi)\, u(\xi,y)\, \mathrm{d}\xi,$$

$$p(x,\xi) = \begin{cases} (l+\xi)(l-x)/2l, & \xi \leq x, \\ (l-\xi)(l+x)/2l, & x \leq \xi. \end{cases}$$

The same results can be obtained for the integral geometry problems in a space of higher dimension.

3.6. INTEGRAL GEOMETRY PROBLEMS WITH PERTURBATION ON THE PLANE

Consider the operator equation

$$\int_0^y [u(x+h,\eta) + u(x-h,\eta)] \frac{\mathrm{d}\eta}{\sqrt{y-\eta}}$$
$$+ \int_0^y \int_{x-h}^{x+h} K(x,y,\xi,\eta)\, u(\xi,\eta)\, \mathrm{d}\xi\, \mathrm{d}\eta = f(x,y), \quad (3.6.1)$$
$$h = \sqrt{y-\eta},$$

for a function $u(\xi,\eta)$. We assume the function $u(\cdot)$ to be twice continuously differentiable and compactly-supported in the rectangle $-l_0 \leq \xi \leq l_0$, $0 \leq \eta \leq b$. The function $K(\cdot)$ has continuous derivatives up to the second order and satisfies the condition $K(x,y,\xi,\eta) = 0$ for $|\xi + x| \leq h$. The function $f(x,y)$ is assumed to be given in the strip $0 \leq y \leq b$.

Equation (3.6.1) corresponds to the integral geometry problem with perturbation.

The first summand on the left-hand side of (3.6.1), i.e.,

$$\int_0^y [u(x+h,\eta) + u(x-h,\eta)] \frac{\mathrm{d}\eta}{\sqrt{y-\eta}} = f_0(x,y),$$

is the set of the integrals of the sought function $u(\cdot)$ over the family of parabolas each with vertex (x,y). The second summand

$$f_1(x,y) = f(x,y) - f_0(x,y)$$

represents the integrals with weight $K(\cdot)$ over the parts of the half-plane bounded by the parabolas.

Theorem 3.6.1. *A solution to equation (3.6.1) is unique.*

Proof. Let the right-hand side of (3.6.1) be identical zero:
$$f(x,y) \equiv 0.$$
Following Lavrentiev (1989), we consider the function
$$v(x,y,t) = \frac{\partial}{\partial t} \int_0^y [u(x+gh,\eta) + u(x-gh,\eta)] \frac{d\eta}{\sqrt{y-\eta}},$$
$$g(t) = \sqrt{1-t}, \quad t \in [0,1].$$
This function satisfies the integro-differential equation
$$\frac{\partial}{\partial y} v(x,y,t) = -\frac{1}{4} \int_0^t \frac{\partial^2}{\partial x^2} v(x,y,\tau)\, d\tau + \varphi(x,y), \tag{3.6.2}$$
$$\varphi(x,y) = -\frac{1}{4} \frac{\partial^2}{\partial x^2} f_0(x,y) = \frac{1}{4} \frac{\partial^2}{\partial x^2} f_1(x,y).$$

Put $v_\sigma(x,y,t) = e^{\sigma(b-y)^2} v(x,y,t)$. From (3.6.2) we infer that $v_\sigma(\cdot)$ satisfies the equation
$$\frac{\partial}{\partial y} v_\sigma(x,y,t) = -\frac{1}{4} \int_0^t \frac{\partial^2}{\partial x^2} v_\sigma(x,y,\tau)\, d\tau - 2\sigma(b-y) v_\sigma(x,y,t) + \varphi_\sigma(x,y), \tag{3.6.3}$$
$$\varphi_\sigma(x,y) = e^{\sigma(b-y)^2} \varphi(x,y).$$

Lemma 3.6.1. *There exists a constant M such that*
$$\int_0^b \int_{-l}^l \varphi_\sigma^2(x,y)\, dx\, dy \leq M \int_0^b \int_{-l}^l \int_0^1 v_\sigma^2(x,y,t)\, dx\, dy\, dt, \tag{3.6.4}$$
where $l = l_0 + \sqrt{b}$.

It follows from the definition of $v(\cdot)$ and $f(\cdot)$ and the properties of $K(\cdot)$ that
$$2 \int_0^y u(x,\eta) \frac{d\eta}{\sqrt{y-\eta}} + f_1(x,y) = \int_0^1 v(x,y,t)\, dt,$$
$$u(x,y) + \int_0^y \int_{x-h}^{x+h} K_1(x,y,\xi,\eta)\, u(\xi,\eta)\, d\xi\, d\eta$$
$$= \frac{1}{2\pi} \int_0^y \int_0^1 v'_\eta(x,\eta,t)\, dt\, \frac{d\eta}{\sqrt{y-\eta}}, \tag{3.6.5}$$

Chapter 3. Integral geometry problems

$$K_1(x,y,\xi,\eta) = \frac{1}{2\pi} \int_{\eta+(x-\xi)^2}^{y} K'_{y_1}(x,y_1,\xi,\eta) \frac{dy_1}{\sqrt{y-y_1}}.$$

The left-hand side of (3.6.5) results from the application of a Volterra operator of second kind to the function $u(\cdot)$. Inverting the indicated operator, we obtain

$$u(x,y) = \frac{1}{2\pi} \int_0^y \int_0^1 v'_\eta(x,\eta,t)\,dt\,\frac{d\eta}{\sqrt{y-\eta}}$$
$$+ \int_0^y \int_{x-h}^{x+h} K_2(x,y,\xi,\eta) v'_\eta(\xi,\eta,t)\,dt\,d\xi\,d\eta, \quad (3.6.6)$$

where $K_2(\cdot)$ is a continuous function.

From (3.6.6), carrying over the operator of differentiation with respect to the variable y from the function $v(\cdot)$ (see (3.6.2)) to the functions $K(\cdot)$ and $K_2(\cdot)$, we obtain the equality

$$\varphi(x,y) = \int_0^y \int_{x-h}^{x+h} \int_0^1 K_3(x,y,\xi,\eta) v(\xi,\eta,t)\,dt\,d\xi\,d\eta, \quad (3.6.7)$$

where $K_3(\cdot)$ is a continuous function depending on $K(\cdot)$ and its derivatives.

It follows from (3.6.7) that

$$\varphi_\sigma(x,y) = \int_0^y \int_{x-h}^{x+h} \int_0^1 K_{3\sigma}(x,y,\xi,\eta) v(\xi,\eta,t)\,dt\,d\xi\,d\eta, \quad (3.6.8)$$

$$K_{3\sigma}(x,y,\xi,\eta) = e^{\sigma(b-y)^2} K_3(x,y,\xi,\eta).$$

Obviously, inequality (3.6.4) follows from (3.6.8).

Now, we apply the Fourier transform with respect to x to equations (3.6.2) and (3.6.3). It is easy to see that the functions

$$w(\lambda,y,t) = \frac{1}{\sqrt{2\pi}} \int_{-\infty}^{\infty} e^{i\lambda x} v(x,y,t)\,dx$$

and

$$w_\sigma(\lambda,y,t) = e^{\sigma(b-y)^2} w(\lambda,y,t)$$

solve equations

$$\frac{\partial}{\partial y} w(\lambda,y,t) = \frac{\lambda^2}{4} \int_0^t w(\lambda,y,\tau)\,d\tau + \psi(\lambda,y), \quad (3.6.9)$$

$$\frac{\partial}{\partial y} w_\sigma(\lambda,y,t) = \frac{\lambda^2}{4} \int_0^t w_\sigma(\lambda,y,\tau)\,d\tau - 2\sigma(b-y)w_\sigma(\lambda,y,t) + \psi_\sigma(\lambda,y). \quad (3.6.10)$$

with
$$\psi(\lambda, y) = \frac{1}{\sqrt{2\pi}} \int_{-\infty}^{\infty} e^{i\lambda x} \varphi(x, y) \, dx.$$

Note that the function w in (3.6.9) is determined only for $t \leq 1$. In order to extend the domain of determination for $t \geq 0$ let us introduce the function
$$w_1(\lambda, t) = \begin{cases} w(\lambda, b, t), & t \leq 1, \\ 0, & t > 1 \end{cases}$$
and consider for equation (3.6.9) the Cauchy problem
$$w(\lambda, b, t) = w_1(\lambda, t) \qquad (3.6.11)$$
in the half-strip $0 \leq y \leq b$, $0 \leq t < \infty$ and extend the function $w(\lambda, y, t)$ to $t > 1$ as a solution to this Cauchy problem.

It is obvious that a solution to the Cauchy problem (3.6.11) for equation (3.6.9) coincides with the above-defined function $w(\cdot)$ for $t \leq 1$.

It is easy to verify that a solution to the problem (3.6.11) for equation (3.6.9) is given by the formula
$$w(\lambda, y, t) = \int_0^t \frac{\cos \lambda \sqrt{(t-\tau)(b-y)}}{\sqrt{t-\tau}} w_2(\lambda, \tau) \, d\tau$$
$$+ \frac{1}{\lambda \sqrt{t}} \int_y^b \sin \lambda \sqrt{t(\eta - y)} \, \psi_1(\lambda, \eta) \, d\eta, \qquad (3.6.12)$$
$$w_2(\lambda, t) = \frac{1}{\pi} \frac{\partial}{\partial t} \int_0^t w_1(\lambda, \tau) \frac{d\tau}{\sqrt{t-\tau}},$$
$$\psi_1(\lambda, y) = \frac{1}{2\pi} \frac{\partial}{\partial y} \int_y^b \psi(\lambda, \eta) \frac{d\eta}{\sqrt{\eta - y}}.$$

Indeed, it is well known that a solution to the problem (3.6.11) for equation (3.6.9) is unique due to the properties of a solution to evolution equations; and the fact that the function $w(\cdot)$ given by (3.6.12) is a solution to this problem is easily verified. Below, we need one auxiliary proposition.

Lemma 3.6.2. *Let $g(x, y)$ be a twice continuously differentiable function in the rectangle $0 \leq x \leq a$, $0 \leq y \leq b$. Moreover, it is possible that $a = \infty$. Furthermore, let A be an integral operator of the form*
$$Ag = \int_0^a R(x, \xi) \, g(\xi, y) \, d\xi,$$

where $R(x,\xi)$ is a continuous function. In the case $a = \infty$ we suppose that the integrals

$$\int_0^\infty |R(x,\xi)|\,\mathrm{d}\xi, \qquad \int_0^\infty |R(x,\xi)|\,\mathrm{d}x$$

converge and the function $g(x,y)$ is bounded. Then the following identity holds:

$$\left[-\left(\frac{\partial}{\partial y}+A^*+\sigma y\right)\left(\frac{\partial}{\partial y}-A-\sigma y\right)+\left(\frac{\partial}{\partial y}-A-\sigma y\right)\left(\frac{\partial}{\partial y}+A^*+\sigma y\right)\right]g$$
$$= [2\sigma + (A^*A - AA^*)]\,g. \quad (3.6.13)$$

Here A^* is the integral operator with the kernel $R^*(x,\xi) = R(\xi,x)$.

To prove (3.6.13), it suffices to take the product of the operators on the left-hand side of (3.6.13). Afterwards, the left-hand side of (3.6.13) represents the sum of 18 summands all disappearing except those on the right-hand side of (3.6.13).

An identity similar to (3.6.13) is presented, for instance, in the monograph by Maurin (1967) in the exposition of Tréves' method for studying the Cauchy problem.

Denote by J the Volterra integral operator

$$Jw = \int_0^t w(\lambda,y,\tau)\,\mathrm{d}\tau$$

and consider the functional

$$F_\delta(w_\sigma) = \int_0^b \int_0^\infty e^{-\delta t}\left[\left(\frac{\partial}{\partial y}-\frac{\lambda^2}{4}J+2\sigma(b-y)\right)w_\sigma(\cdot)\right]^2 \mathrm{d}t\,\mathrm{d}y$$
$$= -\int_0^b \int_0^\infty e^{-\delta t}\left[\left(\frac{\partial}{\partial y}+\frac{\lambda^2}{4}J^*-2\sigma(b-y)\right)\right.$$
$$\left.\times\left(\frac{\partial}{\partial y}-\frac{\lambda^2}{4}J+2\sigma(b-y)\right)w_\sigma(\cdot)\,w_\sigma(\cdot)\right]\mathrm{d}t\,\mathrm{d}y$$
$$+\int_0^\infty e^{-\delta t}\left[\frac{\partial}{\partial y}w_\sigma^2(\lambda,b,t)-\frac{\lambda^2}{4}Jw_\sigma(\lambda,b,t)\,w_\sigma(\lambda,b,t)+2\sigma b w_\sigma^2(\lambda,b,t)\right]\mathrm{d}t,$$

where $\delta > 0$ is a parameter and J^* is the adjoint operator of J in the Hilbert space of functions defined on the half-axis $0 \le t < \infty$ with the inner product

$$(g,p) = \int_0^\infty e^{-\delta t}g(t)\,p(t)\,\mathrm{d}t.$$

It follows from (3.6.10) and (3.6.13) that

$$F_\delta(w_\sigma) = -\int_0^b \int_0^\infty e^{-\delta t}\left[\left(\frac{\partial}{\partial y} + \frac{\lambda^2}{4} J^* - 2\sigma(b-y)\right)\right.$$
$$\left. \times \left(\frac{\partial}{\partial y} - \frac{\lambda^2}{4} J + 2\sigma(b-y)\right) w_\sigma(\cdot) w_\sigma(\cdot)\right] dt\, dy$$
$$+ \int_0^\infty e^{-\delta t}\left[\frac{\partial}{\partial y} w_\sigma^2(\lambda, b, t) - \frac{\lambda^2}{4} J w_\sigma(\cdot) w_\sigma(\cdot) + 2b\sigma w_\sigma^2(\cdot)\right] dt$$
$$+ \int_0^b \int_0^\infty e^{-\delta t}\left\{\left[4\sigma(b-y) + \frac{\lambda^4}{16}(J^*J - JJ^*)\right] w_\sigma(\cdot)\right\} w_\sigma(\cdot)\, dt\, dy$$
$$= \int_0^b \int_0^\infty e^{-\delta t}\left[\left(\frac{\partial}{\partial y} - \frac{\lambda^2}{4} J^* + 2\sigma(b-y)\right) w_\sigma(\cdot)\right]^2 dt\, dy$$
$$- \int_0^\infty e^{-\delta t}\left[\frac{\lambda^2}{4}(J+J^*)w_\sigma(\cdot)w_\sigma(\cdot)\right] dt$$
$$+ \int_0^b \int_0^\infty e^{-\delta t}\left\{\left[\frac{\partial}{\partial y} + \frac{\lambda^4}{16}(J^*J - JJ^*)\right] w_\sigma(\cdot)\right\} w_\sigma(\cdot)\, dt\, dy$$
$$= \frac{1}{\sigma}\int_0^b \psi_\sigma^2(\lambda, y)\, dy. \qquad (3.6.14)$$

Lemma 3.6.3. *Let $g(t)$ be a bounded continuous function on the half-line $0 \le t < \infty$, $|g(t)| \le g_0$. Then the following equality is valid:*

$$(J^*J - JJ^*)g(\cdot) = \frac{1}{\delta}\int_0^\infty e^{-\delta(t+\tau)}g(\tau)\, d\tau, \qquad (3.6.15)$$

where J^* is defined by the equality

$$J^*g = e^{\delta t}\int_t^\infty e^{-\delta\tau}g(\tau)\, d\tau.$$

Hence, we infer that

$$J^*Jg = \int_0^\infty G(t,\tau)g(\tau)\, d\tau, \qquad G(t,\tau) = \begin{cases} e^{-\delta t}/\delta, & \tau \le t, \\ e^{-\delta\tau}/\delta, & \tau \ge t, \end{cases}$$

$$JJ^*g = \int_0^\infty R(t,\tau)g(\tau)\, d\tau, \qquad R(t,\tau) = \begin{cases} (1-e^{-\delta\tau})e^{-\delta t}/\delta, & \tau \le t, \\ (1-e^{-\delta t})e^{-\delta\tau}/\delta, & \tau \ge t, \end{cases}$$

$$G(t,\tau) - R(t,\tau) = \frac{1}{\delta}e^{-\delta(t+\tau)}, \qquad (J^*J - JJ^*)g = \frac{1}{\delta}\int_0^\infty e^{-\delta(t+\tau)}g(\tau)\, d\tau.$$

Using (3.6.14) and (3.6.15) we validate the inequality

$$2\sigma \int_0^b \int_0^\infty w_\sigma^2(\lambda, y, t)\, dt\, dy$$
$$\leq \int_0^\infty e^{-\delta t}\left[\frac{\lambda^2}{4}(J+J^*)w_\sigma(\cdot) - 4\sigma b w_\sigma(\cdot)\right] w_\sigma(\cdot)\, dt + \frac{1}{\sigma}\int_0^b \psi_\sigma^2(\lambda, y)\, dy. \quad (3.6.16)$$

Passing from the Fourier images of the functions $w_\sigma(\cdot)$ and $\psi_\sigma(\cdot)$ to their preimages in (3.6.16), we obtain

$$2\sigma \int_0^b \int_0^\infty \int_{-\infty}^\infty e^{-\delta t} v_\sigma^2(x, y, t)\, dx\, dy\, dt$$
$$\leq \int_0^\infty \int_{-\infty}^\infty e^{-\delta t}\left[\frac{1}{4}(J+J^*)\frac{\partial^2}{\partial x^2} v_\sigma(x, b, t) - 4\sigma b v_\sigma(x, b, t)\right] v_\sigma(x, b, t)\, dx\, dt$$
$$+ \frac{1}{\sigma}\int_0^\infty \int_{-\infty}^\infty \varphi_\sigma(x, y)\, dx\, dy. \quad (3.6.17)$$

Consider the first term on the right-hand side of (3.6.17). It is the integral of the quadratic form involving the functions $v_\sigma(\cdot)$ and $(J+J^*)v''_{\sigma xx}(\cdot)$ over the plane $y = b$ in the space (x, y, t).

The conditions imposed on the function $u(\cdot)$ and the definition of $v_\sigma(\cdot)$ imply that there are constants M_0 and M_1 such that

$$\int_0^\infty \int_{-\infty}^\infty e^{-\delta t}\left[\frac{1}{4}(J+J^*)\frac{\partial^2}{\partial x^2} v_\sigma(x, b, t) - 4\sigma b v_\sigma(x, b, t)\right] v_\sigma(x, b, t)\, dx\, dt$$
$$\leq (M_0 + M_1 \sigma)\, e^{-2\sigma b^2}. \quad (3.6.18)$$

Inequalities (3.6.4), (3.6.17), and (3.6.18) imply

$$(2\sigma - M_2)\int_0^b \int_0^\infty \int_{-\infty}^\infty v_\sigma^2(x, y, t)\, dx\, dy\, dt \leq (M_0 + M_1 \sigma)\, e^{-2\sigma b^2}, \quad (3.6.19)$$

where M_2 is some constant. It is easy to see that if the function $u(\cdot)$ and, hence, $v_\sigma(\cdot)$ are not identically zero, then there is a number q, $0 < q < 1$, such that

$$\int_0^b \int_0^\infty \int_{-\infty}^\infty v_\sigma^2(x, y, t)\, dx\, dy\, dt \geq M_3 e^{-2\sigma b^2 q}. \quad (3.6.20)$$

From (3.6.19) and (3.6.20) we obtain the inequality

$$M_3(2\sigma - M_2)\, e^{-2\sigma b^2 q} \leq (M_0 + M_1 \sigma)\, e^{-2\sigma b^2}. \quad (3.6.21)$$

This inequality obviously fails for sufficiently large values of the constant σ, which proves the assertion of the theorem. □

3.7. MATHEMATICAL PROBLEMS OF TOMOGRAPHY AND HYPERBOLIC MAPPINGS

In this section, we consider a new class of mathematical problems related to interpretation of tomography data (Lavrentiev, 2001). The main assumption is that the sought distribution of absorption is an identically one function in the domain to be determined. These problems are connected with three known directions of mathematical physics: the Dirichlet problems for hyperbolic equations, the problems of small oscillations of a rotating fluid, and the problems of supersonic flows of an ideal gas.

The first relates to studying the Dirichlet problem for the d'Alembert equation. As far as the authors know, the most complete results in this direction are presented by John (1941) wherein it was noted that the research relates to no physical problem.

The second direction is connected with boundary value problems for the system of equations describing small oscillations of a rotating fluid. Statements of problems and the first results are due to S. L. Sobolev (see also Aleksandryan, 1950; Zelenyak 1970).

Some questions in this direction lead to the Dirichlet problem for the d'Alembert equation.

The third direction is connected with gas dynamics. Here the statements of the problems are due to M. A. Lavrentiev and the first publications, to B. V. Shabat and M. M. Lavrentiev (see Lavrentiev, M. M., 1956; Shabat, 1956). This direction was later developed by Lavrentiev, M. A. and Shabat (1977), Shabat (1970).

As is well known, the two-dimensional stationary subsonic flow of an ideal gas is described by a system of elliptic equations and the supersonic flow, by a system of hyperbolic equations.

The simplest system of elliptic equations is the Cauchy–Riemann system:

$$\frac{\partial u}{\partial x} = \frac{\partial v}{\partial y}, \qquad \frac{\partial u}{\partial y} = -\frac{\partial v}{\partial x}.$$

Solutions to the Cauchy–Riemann system carry out conformal mappings of plane domains.

The simplest system of hyperbolic equations is the d'Alembert system

$$\frac{\partial u}{\partial y} = 0, \qquad \frac{\partial v}{\partial x} = 0.$$

The mappings of plane domains performed by solutions to this system are called *h-conformal mappings* by Lavrentiev, M. A. and Shabat (1977). We

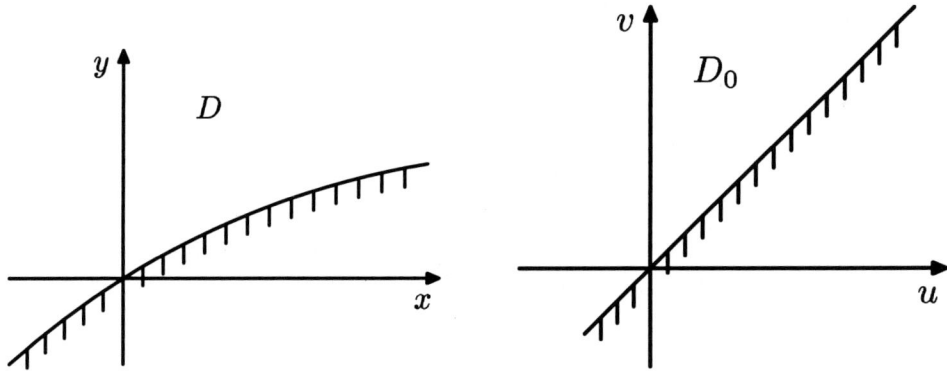

Figure 3.7.1

call these mappings simply *h-mappings*. *Hyperbolic mappings* are those performed by solutions to systems of hyperbolic equations.

One of the central questions in regard to mapping is that of the possibility of carrying domains of a certain class onto some canonical domain. The classical result for conformal mappings of simply connected domains is Riemann's theorem. We formulate two results of Lavrentiev, M. A. and Shabat (1977) from the theory of h-mappings onto canonical domains (mappings of domains like a half-plane).

Let $f(x)$ be a continuously differentiable function such that

$$0 < a < f'(x) < b, \qquad f(0) = 0,$$

where a and b are some constants. Consider the domain $D = \{(x, y) \mid y > f(x)\}$.

Theorem 3.7.1. *There exists an h-mapping of D onto the half-plane $D_0 = \{(u, v) \mid v > u\}$ which satisfies the condition $u(0,0) = v(0,0) = 0$, i.e., which leaves the origin fixed (Figure 3.7.1).*

For the mapping performed by (u, v) to be an h-mapping, it is necessary and sufficient that these functions be representable as $u = \varphi(x)$ and $v = \psi(y)$, where φ and ψ are continuously differentiable monotone functions.

We consider a somewhat more general class of mappings: φ and ψ are arbitrary continuous monotone functions.

Suppose that $f_1(x)$ and $f_2(x)$ are continuously differentiable functions satisfying the conditions

$$0 < a < f'_k(x) < b, \quad k = 1, 2, \qquad f_1(0) = 0, \qquad f_2(x) > f_1(x) + c,$$

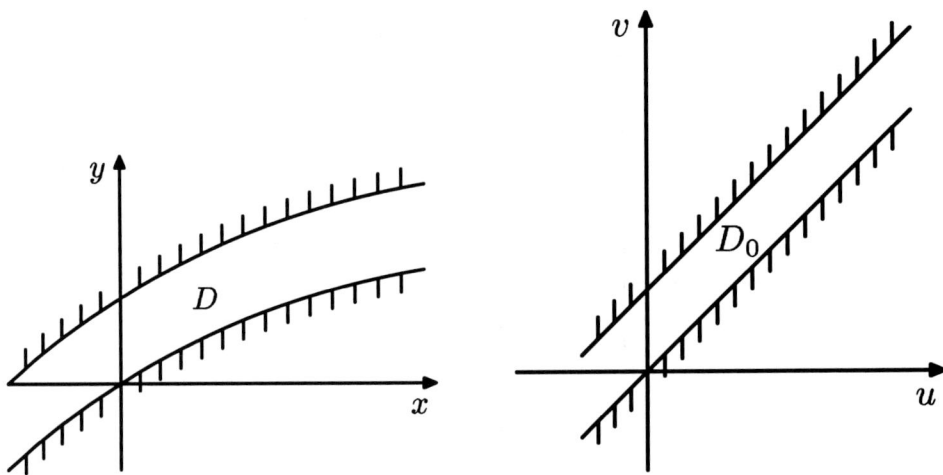

Figure 3.7.2

where a, b, and c are some constants such that $c > 0$. Consider the domain $D = \{(x, y) \mid f_1(x) < y < f_2(x)\}$.

Theorem 3.7.2. *There is an h-mapping of D onto the strip $D_0 = \{(u, v) \mid u < v < u + 1\}$ which satisfies the condition $u(0, 0) = v(0, 0) = 0$ (Figure 3.7.2).*

3.7.1. Mappings of bounded domains

In the paper by John (1941) there were considered h-mappings of domains bounded by Jordan curves convex with respect to the x- and y-axes, i.e., curves such that every straight line parallel to the x- or y-axis intersects the curve at most twice. We give some definitions and results from that article.

Suppose that C is a Jordan curve convex with respect to the x- and y-axes. A point of C such that one (or both) of the two straight lines passing through this point parallel to the x- or y-axis intersects C at none other point is called a *vertex* of C. The curve C may have two, three, or four vertices.

Let P be a point on C with coordinates (x, y): $P(x, y) \in C$. Denote by AP the point on C with coordinates (x, y_1) (i.e., the second point where the straight line passing through P parallel to the y-axis intersects C or, if the point of intersection is only one, the point P itself), by BP the point on C with coordinates (x_1, y), and by T the following transformation of C onto itself: $TP = BAP$.

The vertices of C are fixed points of the transformations A or B. The sequence of points P, AP, TP, ATP, T^2P, ..., T^kP, AT^kP, $T^{k+1}P$, ... and the segments between them constitute the λ-*polygon* determined by P. If there is $n > 0$ for which $T^nP = P$, then the least such n is called the *period* of P and P is a *periodic point* of C. The transformation T is *even* if it preserves the positive direction on C.

Relative to the transformation T, the following four cases are possible for the curve C under consideration:

1. All points of C are periodic (in this case C is called *periodic*).

2. C has periodic as well as aperiodic points (we say that C is *semiperiodic*).

3. C has no periodic points and there is no point P such that the set of the points P, TP, T^2P, ... is everywhere dense on C (in this case C is called *intransitive*).

4. C has no periodic points and for some point P the set of the points P, TP, T^2P, ... is everywhere dense on C (in this case C is called *transitive*).

In case 1 all points have the same period n.

In case 2 all periodic points have the same period n. The set F of periodic points is closed. The complementary set $C \setminus F$ comprises countably many arcs with endpoints in F each of which is invariant under the action of the operator T^n.

Consider case 3. Suppose that Q is an arbitrary point of C and σ is the set of limit points of the set Q, TQ, T^2Q, ... The set σ is a nowhere dense perfect set independent of Q.

In case 4 the transformation T is topologically equivalent to a rotation of a circle; i.e., there exist a real ξ and a continuous mapping $t = f(P)$ of the points P of C to the points $e^{2\pi i t}$ of the unit circle on the complex plane such that $f(TP) = t + \xi$. The constant ξ is irrational and uniquely determined by the curve C. This constant is called the *modulus* of C. Evidently, we may think that $\xi \in (0, 1)$. It was also proved by John (1941) that in case 4 the set of the points P, TP, T^2P, ... is everywhere dense on C for every point $P \in C$.

Theorem 3.7.3. *Suppose that C_1 and C_2 are transitive curves with the same modulus ξ and let D_1 and D_2 be the domains bounded by these curves. Then there is an h-mapping of D_1 onto D_2.*

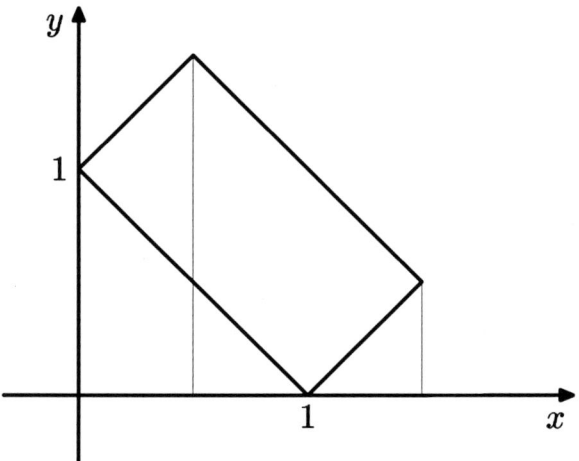

Figure 3.7.3

Corollary. Suppose that C is a transitive curve with modulus $\xi \in (0,1)$ and D is the domain bounded by C. Let $\eta = 1/(1-\xi)$. Then there is an h-mapping of D onto the rectangle with sides $y = -x+1$, $x \in [0,1]$; $y = x-1$, $x \in [1, \eta]$; $y = x+1$, $x \in [0, \eta-1]$; and $y = -x-1+2\eta$, $x \in [\eta-1, \eta]$ (Figure 3.7.3).

3.7.2. Mappings of domains with two vertices

It is easy to see that, using an h-mapping, we can transform a domain with two vertices into the domain

$$D = \{(x,y) \mid f_1(x) < y < f_2(x),\ x \in (0,1)\},$$

where the functions $f_k(x)$, $k = 1, 2$, satisfy the following conditions:

(1) $f_k(0) = 0$ and $f_k(1) = 1$;
(2) the functions $f_k(x)$ are continuous and monotone increasing;
(3) $f_1(x) < f_2(x)$, $0 < x < 1$.

We call domains of this type *slit-like domains* (Figure 3.7.4).

Theorem 3.7.4. For an arbitrary slit-like domain D, there is an h-mapping of this domain onto the domain $D_0 = \{(x,y) \mid x^2 < y < x,\ x \in (0,1)\}$ (Figure 3.7.5).

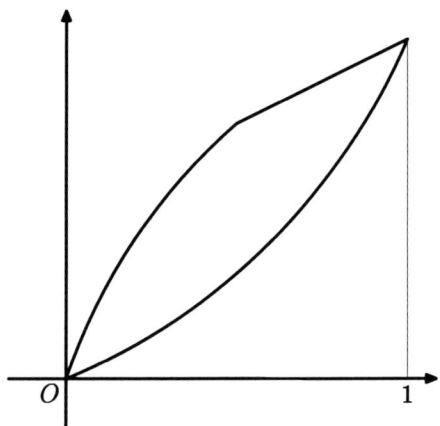

Figure 3.7.4. The domain D

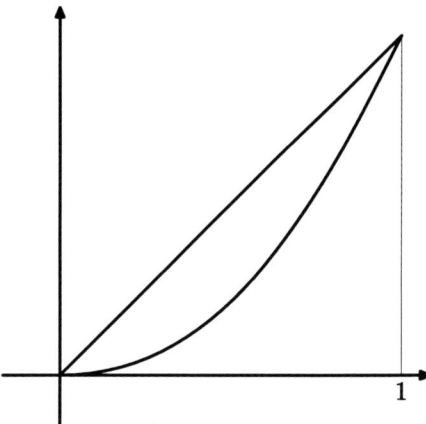

Figure 3.7.5. The domain D_0

Proof. Consider the h-mapping performed by the functions $x_1 = \varphi_1(x) = f_2(x)$ and $y_1 = y$. This mapping takes D into the domain

$$D_1 = \{(x_1, y_1) \mid f_3(x_1) < y_1 < x_1\}, \quad f_3(x_1) = f_1[\tilde{f}_2(x_1)],$$

where the tilde denotes the inverse of a function, i.e., $\tilde{f}_2[f_2(x)] = x$. Now, consider the h-mapping of D_1 performed by the functions

$$x_2 = \varphi_2(x_1), \qquad y_2 = \varphi_2(y_1),$$

where the function $\varphi_2(t)$ is continuous and monotone increasing on the interval $t \in [0, 1]$, $\varphi_2(0) = 0$, $\varphi_2(1) = 1$, and is defined by the relation

$$\varphi_2\{f_3[\tilde{\varphi}_2(x_2)]\} = x_2^2,$$

where $\tilde{\varphi}_2[\varphi_2(t)] = t$. Clearly, the h-mapping given by the composition of these two h-mappings takes the original domain D into the domain D_0 of the theorem. We can obtain a similar result for domains with three extreme points. □

3.7.3. Mappings of regular domains

Suppose that a domain D is convex with respect to the x- and y-axes and has four vertices. Denote the coordinates of these vertices by (a_1, b_1), (a_2, b_2), (a_3, b_3), and (a_4, b_4) and suppose that

$$a_1 < a_2 \leq a_3 < a_4, \qquad b_2 < b_1 \leq b_4 < b_3$$

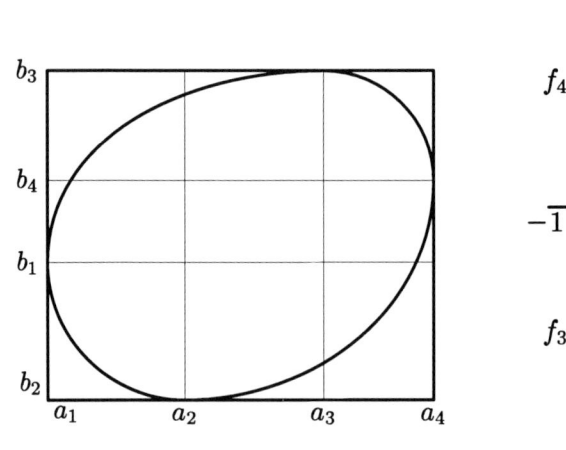
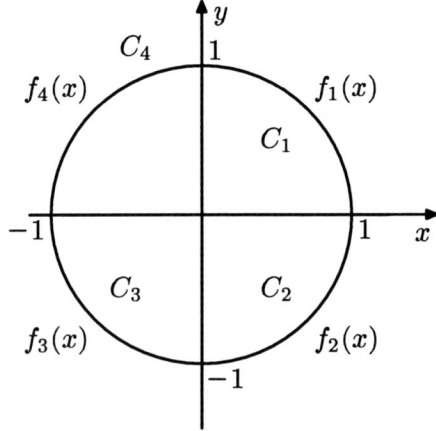

Figure 3.7.6   Figure 3.7.7

(Figure 3.7.6). We shall call this domain *regular* if $a_2 = a_3$ and $b_1 = b_4$. It is easy to see that in this case there is an h-mapping of D onto a domain with

$$a_1 = -1, \quad a_2 = a_3 = 0, \quad a_4 = 1, \quad b_2 = -1, \quad b_1 = b_4 = 0, \quad b_3 = 1$$

(Figure 3.7.7).

Thus, let D be a domain bounded by a curve C consisting of four parts: $C = C_1 \cup C_2 \cup C_3 \cup C_4$, where

$$C_1 = \{(x,y) \mid y = f_1(x), \ x \in [0,1], \ f_1(0) = 1, \ f_1(1) = 0\},$$
$$C_2 = \{(x,y) \mid y = f_2(x), \ x \in [0,1], \ f_2(0) = -1, \ f_2(1) = 0\},$$
$$C_3 = \{(x,y) \mid y = f_3(x), \ x \in [-1,0], \ f_3(-1) = 0, \ f_3(0) = -1\},$$
$$C_4 = \{(x,y) \mid y = f_4(x), \ x \in [-1,0], \ f_4(-1) = 0, \ f_4(0) = 1\}.$$

All functions $f_k(x)$ are monotone and continuous. Consider the h-mapping

$$x_1 = \varphi^1(x), \quad y_1 = \psi^1(y)$$

of D with the functions φ^1 and ψ^1 defined as follows:

(1°) $\varphi^1(x) = \varphi^1_+(x) = x$, $x \in [0,1]$;

(2°) $\psi^1(y) = \psi^1_+(y)$, $y \in [0,1]$, where the function ψ^1_+ is determined from the relation $\psi^1_+[f_1(x)] = -x + 1$, $x \in [0,1]$;

(3°) $\psi^1(y) = \psi^1_-(y)$, $y \in [-1,0]$, where the function ψ^1_- is determined from the relation $\psi^1_-[f_2(x)] = x - 1$, $x \in [0,1]$;

(4°) $\varphi^1(x) = \varphi^1_-(x)$, $x \in [-1,0]$, where the function φ^1_- is determined from the relation $-\varphi^1_-(x) - 1 = \psi^1_-[f_3(x)]$, $x \in [-1,0]$.

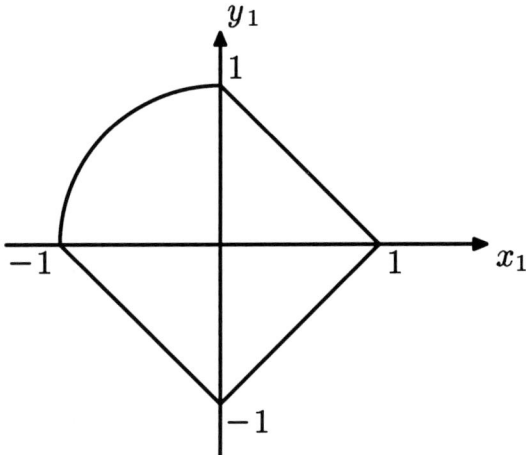

Figure 3.7.8

This mapping takes D into the domain D_0 of the (x_1, y_1)-plane which is bounded by three sides of a square and the curve $y_1 = f(x_1)$ (Figure 3.7.8), where the function f is determined from the relation $\psi_+^1[f_4(x)] = f[\varphi_-^1(x)]$.

Now, consider the h-mappings taking D_0 onto a domain D_1 of the (u,v)-plane of the same type as D_0, i.e., a domain bounded by three sides of a square and some curve:

$$v = F(u), \quad u \in [-1, 0], \quad F(-1) = 0, \quad F(0) = 1,$$
$$F'(u) > 0, \quad u \in (-1, 0).$$

Suppose that such mapping is defined by the functions $u = \varphi(x_1)$ and $v = \psi(y_1)$. Since the mapping takes D_0 onto D_1, it follows that these functions are connected by the following relations:

$$\psi(1-x) = -\varphi(x) + 1, \quad \psi(x-1) = \varphi(x) - 1, \quad x \in [0,1];$$
$$\varphi(x) = -\psi(-x-1) - 1, \quad x \in [-1, 0]. \qquad (3.7.1)$$

Thus, the values of $\varphi(x)$ on the interval $[-1,0]$ and $\psi(y)$ on the interval $[-1,1]$ are determined by the values of the function $\varphi(x)$ on the interval $[0,1]$. Now, suppose that the functions $f(x)$ and $F(u)$ describing parts of the boundaries of D_0 and D_1 satisfy the inequalities

$$f(x) > x + 1, \quad x \in (-1, 0), \qquad F(u) > u + 1, \quad u \in (-1, 0). \qquad (3.7.2)$$

The functions $f(x)$, $F(u)$, and $\varphi(x)$ are connected by the following functional relation:

$$\psi[f(x)] = F[\varphi(x)], \quad x \in [-1, 0].$$

Using relations (3.7.1), we rewrite the above formula in a more convenient form, i.e., in terms of the function φ only, on the interval $x \in [0, 1]$:

$$-\varphi[1 - f(-x)] + 1 = F[-\varphi(x)], \quad x \in [0, 1].$$

Treating the functions $f(x)$ and $F(u)$ as given, consider this relation as a functional equation in $\varphi(x)$. It is easy to demonstrate that, under conditions (3.7.2), this equation has a unique solution satisfying

$$\varphi(0) = 0, \quad \varphi(1) = 1, \quad \varphi'(x) > 0, \quad x \in (0, 1).$$

We can take the function $F(u)$ to be, for example, the function $F(u) = \sqrt{1+u}$, $u \in [-1, 0]$. We have thus proven the following theorem:

Theorem 3.7.5. *For every regular domain D of the (x, y)-plane such that the function $f(x)$ satisfies (3.7.2), there is an h-mapping of D onto the domain D_1 of the (u, v)-plane which is bounded by three sides of the square (as in Figure 3.7.8) and a part of the parabola $v = \sqrt{1+u}$.*

We have thus established theorems on existence of h-mappings onto canonical domains for two classes of unbounded domains and three classes of bounded domains convex with respect to the x- and y-axes. In the author's opinion, it is interesting to study the question of existence of h-mappings of arbitrary convex domains onto canonical domains.

3.7.4. Mappings of the spatial domains

We discuss the question of analogs of h-mappings in the 3D space. As is well known, conformal mappings in the 3D space lead to a system of five equations in three functions. The set of these mappings includes motions, similarity transformations, and inversion (Liouville's theorem). In a series of works M. A. Lavrentiev considered some mappings which he called harmonic. These mappings are connected with a system of four equations and describe a stationary motion of an ideal incompressible fluid in the 3D space. In the author's opinion, a natural analog of h-mappings in the 3D space is represented by mappings that solve the following system of three equations:

$$\frac{\partial u}{\partial x} = \frac{\partial v}{\partial y} = 0, \quad \frac{\partial w}{\partial z} = C,$$

where C is some constant.

Chapter 3. Integral geometry problems

The general solution to this system is representable as

$$u = u(y, z), \quad v = v(x, z), \quad w = w(x, y) + cz.$$

We call these mappings *H-mappings*.

Suppose that D_0 is a convex domain in the (x, y)-plane with continuously differentiable boundary C and $f(x, y)$ is a twice continuously differentiable function in D_0 satisfying the following conditions:

(1°) $f(x, y) = 0$, $(x, y) \in C$, and $f(x, y) > 0$, $(x, y) \in D_0$;

(2°) $f'''_{xx}(x, y)\xi^2 + 2f'''_{xy}(x, y)\xi\eta + f'''_{yy}(x, y)\eta^2 \leq -\rho(\xi^2 + \eta^2)$, $\rho > 0$.

Denote by D the domain $\{(x, y, z) \mid f(x, y) > z,\ z > 0\}$ in the 3D space and by D_z the domain $\{(x, y) \mid f(x, y) > z\}$ in the (x, y)-plane. If for each $z > 0$ the domain D_z admits an h-mapping onto some canonical domain, then for the domain D of the 3D space there is an H-mapping onto a canonical domain which is performed by the functions

$$u = u(y, z), \quad v = v(x, z), \quad w = cz.$$

3.7.5. The inverse problem for a slit-like domain

We turn to considering the inverse tomography problems mentioned in the beginning of this section.

Let D be the following domain: $D = \{(x, y) \mid f_1(x) < y < f_2(x),\ x \in (0, 1)\}$, where the functions $f_k(x)$ satisfy conditions (1)–(3) for Theorem 3.7.4.

It is obvious that if $\varphi_k(y)$, $k = 1, 2$, are the inverse functions of $f_k(x)$, i.e., $\varphi_k[f_k(x)] = x$, then

$$D = \{(x, y) \mid \varphi_2(y) < x < \varphi_1(y),\ y \in (0, 1)\}.$$

Let $u(x) = f_2(x) - f_1(x)$ and $v(y) = \varphi_1(y) - \varphi_2(y)$.

Inverse Problem 3.7.1 (IP 3.7.1). *Given functions $u(x)$ and $v(y)$, find the functions $f_k(x)$ or $\varphi_k(y)$.*

Theorem 3.7.6. *A solution to the IP 3.7.1 is unique.*

Proof. Let x_0 be a point in the interval $(0, 1)$. Put $f_1(x_0) = y_0$ and $x_1 = x_0 - v(y_0)$. Then $f_2(x_1) = y_0$ and $f_1(x_1) = y_0 - u(x_1)$.

Consider the sequences

$$x_{k+1} = x_k - v(y_k), \quad y_{k+1} = y_k - u(x_{k+1}), \quad k = 0, 1, \ldots. \quad (3.7.3)$$

Obviously,

$$x_{k+1} < x_k, \quad y_{k+1} < y_k, \quad \lim_{k \to \infty} x_k = \lim_{k \to \infty} y_k = 0.$$

We can consider relations (3.7.3) as a functional equation in the value $f_1(x_0) = y_0$.

It is easy to see that a solution to this functional equation is unique; whence the assertion of the theorem ensues. □

3.7.6. The inverse problem for regular domains

We confine exposition to inverse problems for domains satisfying the conditions of Theorem 3.7.5. Let $\varphi_1(x)$, $\varphi_2(x)$, $\psi_1(y)$, and $\psi_2(y)$ be continuously differentiable functions defined on the intervals $[a_1, a_3]$ and $[b_1, b_3]$ such that

(1°) $\varphi_1(x) < \varphi_2(x), \quad x \in (a_1, a_3); \quad \psi_1(y) < \psi_2(y), \quad y \in (b_1, b_3);$

(2°) $\varphi_1(a_1) = \varphi_2(a_1) = \varphi_1(a_3) = \varphi_2(a_3) = b_2;$
$\psi_1(b_1) = \psi_2(b_1) = \psi_1(b_3) = \psi_2(b_3) = a_2;$

(3°) $\varphi_1'(x) < 0, \quad x \in (a_1, a_2); \qquad \varphi_1'(x) > 0, \quad x \in (a_2, a_3);$
$\varphi_2'(x) > 0, \quad x \in (a_1, a_2); \qquad \varphi_2'(x) < 0, \quad x \in (a_2, a_3);$
$\psi_1'(y) < 0, \quad y \in (b_1, b_2); \qquad \psi_1'(y) > 0, \quad y \in (b_2, b_3);$
$\psi_2'(y) > 0, \quad y \in (b_1, b_2); \qquad \psi_2'(y) < 0, \quad y \in (b_2, b_3);$

(4°) $\varphi_1[\psi_1(y)] = y, \quad y \in (b_1, b_2); \qquad \varphi_1[\psi_2(y)] = y, \quad y \in (b_1, b_2);$
$\varphi_2[\psi_1(y)] = y, \quad y \in (b_2, b_3); \qquad \varphi_2[\psi_2(y)] = y, \quad y \in (b_2, b_3).$

Denote by D the domain $\{(x, y) \mid \varphi_1(x) < y < \varphi_2(x)\} = \{(x, y) \mid \psi_1(y) < x < \psi_2(y)\}$ (Figure 3.7.9).

Moreover, suppose that D satisfies the conditions of Theorem 3.7.5 and denote by $u(x)$ and $v(y)$ the functions

$$u(x) = \varphi_2(x) - \varphi_1(x), \quad v(y) = \psi_2(y) - \psi_1(y). \quad (3.7.4)$$

Inverse Problem 3.7.2 (IP 3.7.2). *Given functions $u(x)$ and $v(y)$, find the functions $\varphi_1(x)$, $\varphi_2(x)$, $\psi_1(y)$, and $\psi_2(y)$ and the points (a_1, b_2), (a_2, b_3), (a_3, b_2), and (a_2, b_1) — the vertices of D.*

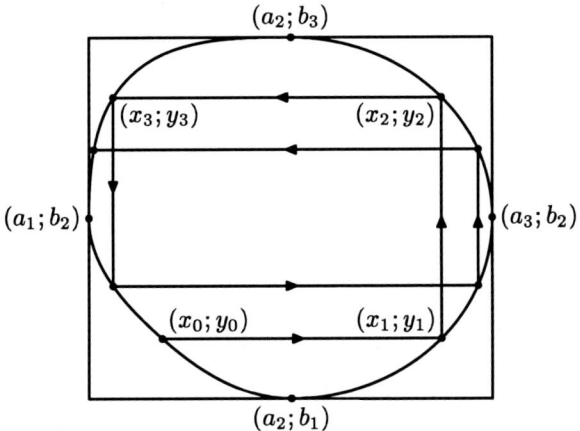

Figure 3.7.9

It follows from (3.7.4) that $b_3 = b_1 + u(a_2)$ and $a_3 = a_1 + v(b_2)$. It is easy to see that the numbers a_k and b_k, $k = 1, 2, 3$, are uniquely determined by the functions $u(x)$ and $v(y)$. Indeed,

$$u(x) = 0, \ x < a_1, \ x > a_3; \quad u(x) > 0, \ x \in (a_1, a_3); \quad u(a_2) = \max u(x);$$
$$v(y) = 0, \ y < b_1, \ y > b_3; \quad v(y) > 0, \ y \in (b_1, b_3); \quad v(b_2) = \max v(y).$$

Let (x_0, y_0) be a point on the boundary of D. For example, assume that $y_0 = \varphi_1(x_0)$ and $a_1 < x_0 < a_2$. Then the points (x_k, y_k) defined by the equalities

$$\begin{aligned} x_{4k+1} &= x_{4k} + v(y_{4k}), & y_{4k+1} &= y_{4k}, \\ x_{4k+2} &= x_{4k+1}, & y_{4k+2} &= y_{4k+1} + u(x_{4k+1}), \\ x_{4k+3} &= x_{4k+2} - v(y_{4k+2}), & y_{4k+3} &= y_{4k+2}, \\ x_{4(k+1)} &= x_{4k+3}, & y_{4(k+1)} &= y_{4k+3} - u(x_{4k+3}) \end{aligned} \quad (3.7.5)$$

(see Figure 3.7.9) lie on the boundary of D. The assumption that D satisfies the conditions of Theorem 3.7.5 yields the equalities

$$\lim_{k \to \infty} x_{4k} = \lim_{k \to \infty} x_{4k+3} = a_1, \quad \lim_{k \to \infty} x_{4k+1} = \lim_{k \to \infty} x_{4k+2} = a_3,$$
$$\lim_{k \to \infty} y_k = b_2. \quad (3.7.6)$$

Formulas (3.7.5) and (3.7.6) for other choice of the initial point on the boundary of D are exactly the same if we reindex the initial point properly, e.g., (x_1, y_1) for the initial point on the lower right arc of the boundary, etc. Note also that if we had constructed our sequence "in the clockwise direction", i.e., $x_1 = x_0$, $y_1 = y_0 + u(x_0)$, $x_2 = x_1 + v(y_1)$, $y_2 = y_1, \ldots$, then (3.7.6) would be

$$\lim_{k \to \infty} x_k = a_2, \quad \lim_{k \to \infty} y_{4k} = \lim_{k \to \infty} y_{4k+3} = b_1,$$
$$\lim_{k \to \infty} y_{4k+1} = \lim_{k \to \infty} y_{4k+2} = b_3.$$

We can consider (3.7.5) and (3.7.6) as an equation for determination of $\varphi_1(x_0)$. It is easy to see that, in view of the properties of the function $\varphi_k(x)$, a solution to this equation is unique. We have thus proved the following theorem:

Theorem 3.7.7. *A solution to the IP3.7.2 is unique.*

3.7.7. The inverse problem for domains with transitive boundary

Let D be a bounded domain convex with respect to the x- and y-axes, with continuously differentiable boundary. Denote the vertices of D by (a_1, b_1), (a_2, b_2), (a_3, b_3), and (a_4, b_4) (see Figure 3.7.6) and suppose that $a_1 < a_2 < a_3 < a_4$ and $b_2 < b_1 < b_4 < b_3$. Then there exist functions $f_1(x)$, $f_2(x)$, $g_1(y)$, and $g_2(y)$ continuously differentiable on the intervals $x \in (a_1, a_4)$ and $y \in (b_2, b_3)$ and such that

$$D = \{(x, y) \mid f_1(x) < y < f_2(x),\ x \in (a_1, a_4)\}$$
$$= \{(x, y) \mid g_1(y) < x < g_2(y),\ y \in (b_2, b_3)\};$$

(1) $f_1(a_1) = f_2(a_1) = b_1, \quad f_1(a_4) = f_2(a_4) = b_4,$
    $g_1(b_2) = g_2(b_2) = a_2, \quad g_1(b_3) = g_2(b_3) = a_3;$

(2) $f_1'(x) < 0, \quad x \in (a_1, a_2); \qquad f_1'(x) > 0, \quad x \in (a_2, a_4);$
    $f_2'(x) > 0, \quad x \in (a_1, a_3); \qquad f_2'(x) < 0, \quad x \in (a_3, a_4)$
    (analogous inequalities for g_1' and g_2' are evident);

(3) $g_1[f_1(x)] = x, \quad x \in (a_1, a_2); \qquad g_1[f_2(x)] = x, \quad x \in (a_1, a_3);$
    $g_2[f_1(x)] = x, \quad x \in (a_2, a_4); \qquad g_2[f_2(x)] = x, \quad x \in (a_3, a_4).$

Denote by $u(x)$ and $v(y)$ the functions

$$u(x) = f_2(x) - f_1(x), \qquad v(y) = g_2(y) - g_1(y). \qquad (3.7.7)$$

Inverse Problem 3.7.3 (IP 3.7.3). *Given functions $u(x)$ and $v(y)$, find the functions $f_1(x)$, $f_2(x)$, $g_1(y)$, and $g_2(y)$.*

In addition, suppose that the curve C bounding D is transitive and the vertices of C are known.

Theorem 3.7.8. *A solution to the IP3.7.3 is unique.*

Proof. Consider the left vertex (a_1, b_1) of C. By (3.7.7), the point (x_1, b_1) belongs to C, where $x_1 = a_1 + v(b_1)$. Now, $(x_1, y_1) \in C$, where $y_1 = b_1 + u(x_1)$. Consider the sequences (x_k, y_k) and (x_{k+1}, y_k), $k = 1, 2, \ldots$, of the points defined as follows:

$$x_{k+1} = x_k \pm v(y_k), \qquad y_{k+1} = y_k \pm u(x_{k+1}). \tag{3.7.8}$$

The signs in front of $v(y_k)$ and $u(x_{k+1})$ in (3.7.8) depend on the mutual position of the points (x_k, y_k) and (x_{k+1}, y_k) and the vertices of C.

Since the curve C is transitive, the points (x_k, y_k) and (x_{k+1}, y_k), $k = 1, 2, \ldots$, constitute an everywhere dense set on C, which validates the assertion of the theorem. $\square$

Chapter 4.
One dimensional inverse problems

In this chapter we discuss such Inverse Problems that unknown function(s) depend on a single variable. These problems are of special interest in the area of inverse and ill-posed problems because of two basic reasons. First, even in terms of the rather simple statements it is possible to describe a long variety of physical phenomenon both at qualitative and quantitative levels. Second, simplified models, which contain a few parameters, are convenient for theoretical analysis and numerical realization. Examples of 1D Inverse Problems are collected in this chapter. Physical statements, selection of theorems (at the level of formulation), elementary considerations with the complete proofs have been brought together in order to supply the reader with a sort of entire impression of the discussed area. Chapter could be considered to be more or less independent of the rest of the book, therefore, we start with the brief introduction.

Problems of determination of the coefficients of hyperbolic differential equations from some additional information about their solutions are very important in applications. Indeed, let us suppose that we are interested in the structure of the certain physical media and it is impossible to arrange direct measurements in the desired space area. So, we have to use a sort of remote sensing. For example, the inner structure of a certain physical body should be studied through the surface measurements of physical field(s). Basic characteristics of media under study could be: Lamé parameters and density — in case of seismic sensing (elastic waves propagation); dielectric and magnetic permeability and electric conductivity — in case of electromagnetic waves, and other.

On the basis of additional information about the solution to direct problem we can divide inverse problems for hyperbolic equations into three basic groups: kinematic, spectral and dynamical problems. Here we deal with the Dynamical Inverse Problems, in which case an additional information about solution to the corresponding direct problem, as a rule, is given on a sort of time–like surface. That is, for example, trace of solution at the given point space as the function of time. The first inverse dynamical problems for hyperbolic differential equations have been formulated and investigated by Alekseev (1962, 1967), Lavrentiev and Romanov (1966), Blagovestchensky (1966), Romanov (1969), Lavrentiev, Romanov, and Vasiliev (1970).

In this Chapter we stress on several different Inverse Problems, which are characterized by the common property that the unknown parameters depend on a *single* variable. These problems are of prime importance for applications. To explain the point we remind the usual scheme of numerical solution to Inverse Problems (*"numerical inversion"* in other words).

First of all, it is necessary to make a guess about values of unknown function(s) (structure of the media under study), that is to assume the precise form of equation coefficient(s), source term, etc. When, it is possible to solve numerically the corresponding direct problem and, therefore, to obtain the value(s) of "additional information" (*inversion data*), which has been generated by the guessed values of the problem data. Now we calculate the so-called the *Misfit*, or the *Cost, Functional*. This functional is nothing but a certain measure of the distance between the given (measured) inversion data and the simulated ones, obtained after solution to direct problem with the assumed values of the sought functions. Then it is necessary to choose the new approximation for the problem parameters, such that the value of misfit functional decrease. The minimal value of the functional is supposed to correspond to the desired solution of the inverse problem.

For example, consider the case of one of the classical inverse problems for the wave equation, namely: to find the function $c(z)$ such that solution to the problem

$$\frac{\partial u}{\partial t} = \frac{\partial}{\partial z}\left(c(z)\frac{\partial u}{\partial z}\right), \qquad (4.0.1)$$

$$u_z\big|_{z=0} = u_1(t), \qquad u\big|_{t<0} \equiv 0 \qquad (4.0.2)$$

satisfies the extra boundary condition (the additional information):

$$u\big|_{z=0} = \varphi(t). \qquad (4.0.3)$$

Chapter 4. One dimensional inverse problems

Function $\varphi(t)$ takes the position of inversion data, while the integral

$$\Phi[c(z)] = \int_0^T (u^{\text{sim}}(0,t) - \varphi(t))^2 \, dt \qquad (4.0.4)$$

often stands for the misfit functional. Here we have use notation $u^{\text{sim}}(z,t)$ for solution to problem (4.0.1)–(4.0.2) with the guessed value of the sought coefficient $c(z) = c^{\text{sim}}(z)$.

Remark 4.0.1. In view of the classical solvability theory for the second order PDE it is necessary to add an extra boundary condition in (4.0.2) in order to obtain the correct direct problem. Besides, the nonstandard requirement on initial profile that solution is trivial for *all* negative values of time could be regarded as an analogue of the boundary condition at infinity, namely as an absence of the wave, passing from infinity.

Thus, solution to the inverse problem above is nothing but a sort of optimization problems: choosing the equation coefficient, $c(z)$ for the functional in (4.0.4), which is given indirectly. Solution to such problem is based on variation of the sought function, which belongs to a certain functional space, that is to the manifold of *infinite* dimension. Solution to that type of minimization problems requires numerous solutions to direct problem (4.0.1)–(4.0.2). Indeed, for each suggested value of the sought function we calculate the value of the cost functional *after* solving the direct problem.

Therefore, one can never expect effective (that is by reasonable CPU resources) numerical inversion in case solution to direct problem requires valuable computational cost. As is well known, numerical solution to PDE in 3-D case is far nontrivial. Roughly speaking, then double the number of the grid points in each dimension we come out with the increase of the mesh size by an *order*. On the other hand, there are a number of fast numerical routines solving some classes of ODE and selected PDE.

Natural way to reduce the problem "size" is to introduce a number of simplifications, that is suggestions of the model structure. In 1-D case inverse problem (basic assumption that the problem parameters depend only on the selected single variable) it is often possible to use some symmetries and thus to reduce the direct problem to well known Helmholtz equation. In a number of special cases this equation is just an ODE. Broad variety of robust numerical codes are available for numerical solution to such object. In the example under consideration (4.0.1)–(4.0.2), the Helmholtz equation could be obtained by applying the formal Fourier transform

$$U(z,\omega) = F[u] = \int_{-\infty}^{\infty} u(z,t) e^{i\omega t}\, dt. \qquad (4.0.5)$$

Under certain assumptions, which here are out of our interest, this function $U(z,\omega)$ solves the problem

$$\frac{d}{dz}\left(c(z)\frac{dU}{dz}\right) + \omega^2 U = 0, \qquad U(0,\omega) = U_1(0,\omega). \qquad (4.0.6)$$

The point is that in view of the celebrated Parseval identity the above cost functional in (4.0.4) is nothing but

$$\Phi[c(z)] = \int_{-\infty}^{\infty} (U^{\text{sim}}(0,\omega) - F(\omega))^2\, d\omega \qquad (4.0.7)$$

and, therefore, numerical inversion could be done directly in the *frequency* domain, that is in terms of variables z and ω, rather then in time-space domain of (z,t).

The chapter is organized as follows. In Section 4.1 we discuss some inverse problems for Lamé system. Section 4.2 is dedicated to quasi-stationary approximation of Maxwell's equations. The basic result is obtained after study of relations between "classical" Maxwell's system and it's parabolic approximation. Thus, in Section 4.3 we give general reasons concerning connections among Inverse Problems with the governing equations of different types. In Sections 4.4 and 4.5 we explain in more details selected methods of investigation of Inverse Problems for hyperbolic equations. Finally, in Section 4.6 we give the recursive algorithm of solution to one-coefficient Inverse Problem for Helmholtz equation in case that the sought coefficient is piece-wise constant.

4.1. SOME INVERSE PROBLEMS FOR LAMÉ SYSTEM

Here we consider some inverse problems for Lamé system. Most attention is concentrated on the simplest model examples in order to outline the basic ideas. For one of the first systematic study of inverse *dynamical* problems we address the reader to the works by Romanov (1969, 1987), Lavrentiev, Romanov, and Shishatskii (1986), Romanov and Kabanikhin (1994). Throughout this paragraph we are following the book of Yakhno (1990).

Propagation of disturbances through elastic media is often described by the system

$$\rho \frac{\partial^2 U}{\partial t^2} = \text{Div}\, T, \qquad (4.1.1)$$

where
$$\text{Div}\, T = \Big(\sum_{j=1}^{3} \frac{\partial}{\partial x_j} T_{ij}\Big)_{i=1}^{3},$$

$U = (U_1, U_2, U_3)$ is the displacement vector depending on a point $x \in \mathbb{R}^3$ and on time variable t, T is the stress tensor, and ρ is the density of the medium. For the stress tensor T we have the following defining relation:

$$T_{ij} = \sum_{k,l=1}^{3} c_{ijkl} S_{kl}, \qquad i,j = 1,2,3,$$

where S is the strain tensor defined by the formula

$$S_{kl} = \frac{1}{2}\Big(\frac{\partial U_k}{\partial x_l} + \frac{\partial U_l}{\partial x_k}\Big), \qquad k,l = 1,2,3,$$

and the 4-rank tensor (c_{ijkl}) is the so-called *elastic modules tensor*. In view of several physical symmetries, there are only 21 independent components of this tensor (compare to the complete number of 81 components).

Moreover in case of the classical *isotropic media* (that is the media, properties of which are independent of the space direction) there are only 3 independent *elastic modules* in (4.1.1):

$$\rho \frac{\partial^2 U_i}{\partial t^2} = \sum_{j,k,l=1}^{3} \frac{\partial}{\partial x_j}\Big(c_{ijkl} \frac{\partial U_l}{\partial x_k}\Big), \qquad i = 1,2,3, \qquad (4.1.2)$$

where

$$(c_{\alpha\beta}) = \begin{pmatrix} c_{11} & c_{12} & c_{12} & 0 & 0 & 0 \\ c_{12} & c_{11} & c_{12} & 0 & 0 & 0 \\ c_{12} & c_{12} & c_{11} & 0 & 0 & 0 \\ 0 & 0 & 0 & c_{44} & 0 & 0 \\ 0 & 0 & 0 & 0 & c_{44} & 0 \\ 0 & 0 & 0 & 0 & 0 & c_{44} \end{pmatrix}. \qquad (4.1.3)$$

It often happens that $c_{44} = (c_{11} - c_{12})/2$, which case corresponds to usual isotropic elastic media. This media could be described with the help of two independent *Lamé parameters* λ and $\varkappa$: $\lambda = c_{12}$, $\varkappa = c_{44}$, and $c_{11} = \lambda + 2\varkappa$.

In a number of applications the media under study is supposed to be *stratified*, that is all the parameters depend on a selected single variable. Thus, the rough approximation of structure of the Earth crust often is reasonably described in terms of stratified media models. In case of our system (4.1.2), (4.1.3) stratification leads to assumption that Lamé parameters, as well as the density, are subject of variation only in e_3 direction: $\lambda = \lambda(x_3)$, $\varkappa = \varkappa(x_3)$, and $\rho = \rho(x_3)$. In the rest of the paragraph we consider only this situation, moreover, we assume that

$$(c_{12}, c_{11}, c_{44}, \rho) \in \Lambda;$$

$$\Lambda := \{(c_{12}, c_{11}, c_{44}, \rho) \mid c_{11}, c_{44}, \rho \in C^2(\overline{\mathbb{R}_+}),\ c_{12} \in C(\overline{\mathbb{R}_+});$$
$$c_{11} > 0,\ c_{44} > 0,\ c_{11} > c_{12},\ c_{11} + 2c_{12} > 0,\ \rho > 0\}.$$

Under these assumptions the above system (4.1.2) is *hyperbolic* in $\mathbb{R}^2 \times \mathbb{R}_+ \times \mathbb{R}$.

Consider the system (4.1.2) in case (4.1.3) with the following initial and boundary data:

$$\sum_{k,l=1}^{3} c_{3ikl} \frac{\partial U_l}{\partial x_k}\bigg|_{x_3=+0} = -\frac{1}{2}\delta(x_2)\delta(t), \qquad (4.1.4)$$

$$U_i\big|_{t<0} = 0, \qquad i = 1, 2, 3,$$

where $\delta(\cdot)$ stands for the Dirac's mass. System (4.1.2), (4.1.4) describes a seismic waves propagation through an isotropic elastic medium, caused by an instantaneous point force distributed along x_1-axis (Yakhno, 1990).

Existence of the unique solution $u(x,t)$ to system (4.1.2), (4.1.4) has been established by Yakhno (1990) under the above assumptions on the problem parameters. Generally speaking, this solution solves equations in the sense of distributions. In order to avoid introduction of rather complicated functional spaces we omit the exact formulation. Besides, in the case under study the solution $U(x,t)$ is independent of x_1 and, which is important, admits the Fourier transform with respect to variable x_2. This Fourier transform,

$$U(x_3, t, \nu) := F\big|_{x_2}[U(x,t)] \qquad (\nu \text{ being dual to } x_2),$$

could be regarded as a function, which solves the corresponding system in the classical sense.

It happens (with suitable orientation of the point force in boundary data) that equations involving the first component U_1 of the vector U are independent. Below, for short, we consider only this part of the system. Complete statements of some inverse problems along with uniqueness results are given below.

Chapter 4. One dimensional inverse problems

The function $U_1(x_3, t, \nu)$ solves equations:

$$\rho \frac{\partial^2 U_1}{\partial t^2} = \frac{\partial}{\partial x_3}\left(c_{44} \frac{\partial U_1}{\partial x_3}\right) - \nu^2 c_{44} U_1, \qquad (4.1.5)$$

$$U_1\big|_{t<0} = 0, \qquad (4.1.6)$$

$$c_{44}(0) \frac{\partial U_1}{\partial x_3}\bigg|_{x_3=+0} = -\frac{1}{2}\delta(t). \qquad (4.1.7)$$

The additional information to recover both coefficients $\rho(x_3)$ and $c_{44}(x_3)$ could be taken in the form:

$$U_1\big|_{x_3=+0,\, \nu=0} = h_1(t), \qquad (4.1.8)$$

$$U_1\big|_{x_3=+0,\, \nu=\nu_0} = h_2(t). \qquad (4.1.9)$$

Note that quantities in (4.1.8) and (4.1.9) correspond to the measurements at the Earth surface, while the density ρ and elastic module c_{44} should be defined as the functions of depth.

Inverse Problem 4.1.1 (IP 4.1.1). *Given the functions $h_i(t)$, $i = 1, 2$, $t \in [0, T]$. Find the functions*

$$c_{44}(x_3),\ \rho(x_3) \in \Lambda_1 := \{f(x_3) \in C^2(\overline{\mathbb{R}}_+) \mid f(x_3) > 0\},$$

such that solution $U_1(x_3, t, \nu)$ to the problem (4.1.5)–(4.1.7) satisfy equalities in (4.1.8), (4.1.9).

After rescalation

$$z = \tau(x_3), \qquad \tau(x_3) = \int_0^{x_3} \frac{d\xi}{w(\xi)}, \qquad w^2(x_3) = \frac{c_{44}(x_3)}{\rho(x_3)}$$

the system (4.1.5)–(4.1.9) could be rewritten as

$$\frac{\partial^2 V}{\partial t^2} = \frac{\partial^2 V}{\partial z^2} - K(z) \frac{\partial V}{\partial z} - \nu^2 d(z) V, \qquad (4.1.10)$$

$$V\big|_{t<0} = 0, \qquad (4.1.11)$$

$$\frac{\partial V}{\partial z}\bigg|_{z=0} = -\frac{1}{2} a(0)\delta(t), \qquad (4.1.12)$$

$$V\big|_{z=+0,\, \nu=0} = h_1(t), \qquad (4.1.13)$$

$$V|_{z=+0,\,\nu=\nu_0} = h_2(t), \qquad (4.1.14)$$

with $d(z) = c_{44}(\tau^{-1}(z))/\rho(\tau^{-1}(z))$, $V(z,t,\nu) = U_1(\tau^{-1}(z),t,\nu)$, function $\tau^{-1}(z)$ is reverse to $\tau(x_3)$, and

$$K(z) = \frac{d}{dz}\ln a(z), \qquad a(z) = \frac{1}{\sqrt{c_{44}(\tau^{-1}(z))\rho(\tau^{-1}(z))}}.$$

Remark 4.1.1. Dimensionless form of governing equation (like, e.g., in (4.1.10)) is very convenient in view of theoretical analysis. That is why, a long number of papers on inverse problems deal with the celebrated model equation

$$\frac{\partial^2 U}{\partial t^2} = \frac{\partial^2 U}{\partial z^2} + q(z)U. \qquad (4.1.15)$$

Note that by setting $\nu = 0$ the above system could be regarded as an inverse problem with respect to only one unknown function, namely $K(z)$. It allows us to formulate the following inverse problem.

Inverse Problem 4.1.2 (IP 4.1.2). *It is necessary to determine the function $K(z) \in C^1(0,T/2)$ from equation (4.1.10) such that solution $V(z,t,\nu=0)$ to the problem (4.1.10)–(4.1.12) satisfies relation (4.1.13) with the given function $h_1(t)$, $t \in [0,T]$.*

The following results have been established by Blagovestchensky (1971) and Romanov (1987):

Theorem 4.1.1. *Let $T > 0$ be the fixed positive number. In order the IP 4.1.2 has the unique solution from the space $K(z) \in C^1(0,T/2)$ it is necessary and sufficient that the function $h_1(t)$ satisfies the conditions:*

1) $h_1(t) \in C^2(0,T)$, $h_1(+0) = a(+0)/2$, $h'_1(+0) = a'(+0)/4$;

2) for any $z \in (0,T/2)$ and $\varphi(t) \in L_2(0,z)$, $\varphi(t) \neq 0$, the inequality $(A\varphi, \varphi) > 0$ holds, where

$$(A\varphi)(t) = \varphi(t) + \frac{2}{a(0)}\int_0^z [h'_1(t-\tau) + h'_1(t+\tau)]\varphi(\tau)\,d\tau,$$

$$h_1(-t) = -h_1(t), \qquad |t| < z;$$

symbol (ψ, φ) stands here for the scalar product in $L_2(0,z)$.

Remark 4.1.2. For any given function $h_1(t) \in C^2(0,T)$ there exists such number $T^* > 0$ that the condition 2) of the theorem is fulfilled for $T \in (0, T^*)$. Therefore, for solvability "in small" (that is for small enough time) of the IP 4.1.2 it is sufficient to claim only the compatibility conditions 1).

Theorem 4.1.2. *Let L and T be the fixed positive numbers, while $K(z), K^*(z) \in \mathcal{K}(L,T) = \{K(z) \in C^1(0,T/2) \, \|K\|_{C^1(0,T/2)} \leq L\}$ be solutions to IP 4.1.2, corresponding to additional information $h_1(t)$ and $h_1^*(t)$, respectively. Then the following estimate*

$$\|K - K^*\|_{C^1(0,T/2)} \leq C\|h_1 - h_1^*\|_{C^2(0,T)}$$

holds with a certain constant C, depending on L and T.

For proofs and further results we refer the reader to the book by Yakhno (1990) and the references therein. Here we would like to stress attention on one of the simplified formulations of inverse problems, which are very convenient for theoretical and numerical study.

The certain conditions have been specified by Yakhno (1990), under which solution to problem (4.1.2), (4.1.4) is such that $U_1 \equiv U_2 \equiv 0$. Thus, in terms of the formal Fourier transform $u = F_t[F_{x_1 x_2}[U_3]]$, problem under study could be reduced to the following problem for Helmholtz equation:

$$\frac{\partial^2 u}{\partial z^2} + \omega^2 v^2(z) u = 0, \quad z \in \mathbb{R}_+,$$
$$u_z(0, \omega) = g(\omega), \quad (4.1.16)$$
$$u_z - i\omega v u \to 0 \quad \text{as} \quad z \to +\infty.$$

Here $z = x_3$ and $\nu_1 = \nu_2 = 0$. The symbols F_t, $F_{x_1 x_2}$ represent the Fourier transform operators with respect to t and x_1, x_2 respectively; ν_1, ν_2 are the variables dual to x_1, x_2; $v = (\rho/(\lambda + 2\varkappa))^{1/2}$ is the reciprocal to the longitudinal wave velocity; and $g = -(\lambda(0) + 2\varkappa(0))^{-1}$.

Concerning the system (4.1.16), one of the typical (spectral) inverse problems could be formulated as follows.

Inverse Problem 4.1.3 (IP 4.1.3). *Given the function $\varphi(\omega)$ for $\omega \in [\omega_1, \omega_2]$, find the coefficient $v(z)$ such that*

$$\varphi(\omega) = i\omega u(0, \omega), \quad \omega \in [\omega_1, \omega_2], \quad (4.1.17)$$

where $u(0, \omega)$ solves equations (4.1.16).

4.2. INVERSE PROBLEMS FOR QUASI-STATIONARY MAXWELL'S EQUATIONS

In this section we discuss electromagnetic field. All the statements and notations are given after Romanov and Kabanikhin (1994). We are grateful to the authors for their kind permission to use here a part of their monograph.

When speaking about applications of Inverse Problems in Mathematical Physics various forms of Lamé system are mentioned first of all. After that, Maxwell's equations could be ranged. These equations describe electromagnetic waves and have the form

$$\text{rot } \boldsymbol{H} = \varepsilon \frac{\partial \boldsymbol{E}}{\partial t} + \sigma \boldsymbol{E} + \boldsymbol{j}, \qquad \text{rot } \boldsymbol{E} = -\mu \frac{\partial \boldsymbol{H}}{\partial t}. \qquad (4.2.1)$$

Here $\boldsymbol{E} = (E_1, E_2, E_3)$ and $\boldsymbol{H} = (H_1, H_2, H_3)$ are vector-functions depending on a point $x \in \mathbb{R}^3$, $x = (x_1, x_2, x_3)$, and on the time variable t; the coefficients ε, μ, σ are functions of x; and $\boldsymbol{j} = \boldsymbol{j}(x,t)$.

Due to geophysical applications here we consider the case then the space $\mathbb{R}^3$ is divided by the plane $x_3 = 0$ into two half-spaces

$$\mathbb{R}^3_- = \{x \in \mathbb{R}^3 \mid x_3 < 0\}, \qquad \mathbb{R}^3_+ = \{x \in \mathbb{R}^3 \mid x_3 > 0\}.$$

The coefficients ε, μ, σ are thought of as constant in $\mathbb{R}^3_-$ and smooth in $\mathbb{R}^3_+$. On the common boundary $x_3 = 0$ of $\mathbb{R}^3_+$, $\mathbb{R}^3_-$ these coefficients may have a finite jump. Equations (4.2.1) are, therefore, assumed to hold in each of the domains $\mathbb{R}^3_-$, $\mathbb{R}^3_+$, while at the plane $x_3 = 0$ the tangential components of $\boldsymbol{E}$ and $\boldsymbol{H}$ should satisfy the continuity "matching" conditions

$$E_i\big|_{x_3=-0} = E_i\big|_{x_3=+0}, \qquad H_i\big|_{x_3=-0} = H_i\big|_{x_3=+0}, \qquad i = 1,2. \qquad (4.2.2)$$

The vectors $\boldsymbol{E}$, $\boldsymbol{H}$ are also supposed to vanish identically up to the moment $t = 0$, whence

$$(\boldsymbol{E}, \boldsymbol{H})_{t<0} \equiv 0, \qquad \boldsymbol{j}\big|_{t<0} \equiv 0. \qquad (4.2.3)$$

The function $\boldsymbol{j}(x,t)$, generally speaking, is a distribution (generalized function in the distributional sense), so the solution to problem (4.2.1)–(4.2.3) can be regarded as a linear functional over the space of infinitely differentiable functions with compact supports.

The problem of finding $\boldsymbol{E}$, $\boldsymbol{H}$ from (4.2.1)–(4.2.3), provided that the coefficients ε, μ, σ, and the function $\boldsymbol{j}(x,t)$ are given, is said to be the *direct* or *Cauchy* problem.

The inverse problem consists in determining the coefficients ε, μ, σ as functions of $x \in \mathbb{R}^3_+$ (in $\mathbb{R}^3_-$ these coefficients, as well as $\boldsymbol{j}(x,t)$, are assumed to be known). For ε, μ, σ to be found in $\mathbb{R}^3_+$, it is required that the tangential components of $\boldsymbol{E}$, $\boldsymbol{H}$ be given at the plane $x_3 = 0$:

$$E_i\big|_{x_3=0} = \varphi_i(x_1, x_2, t), \qquad H_i\big|_{x_3=0} = \psi_i(x_1, x_2, t), \qquad i = 1, 2. \qquad (4.2.4)$$

Thus, we are facing with the following

Inverse Problem 4.2.1 (IP 4.2.1). *To recover the coefficients ε, μ, σ from information (4.2.4) on the solution to the Cauchy problem (4.2.1)–(4.2.3), provided that the function $\boldsymbol{j}(x,t)$ is given.*

It is evident that, together with the tangential components of $\boldsymbol{E}$, $\boldsymbol{H}$, the normal components E_3, H_3 can be also set at the plane $x_3 = 0$. However, it should be kept in mind that they are, generally speaking, discontinuous at $x_3 = 0$, and, therefore, their values at this plane can be treated only as the limits at $x_3 = +0$, $x_3 = -0$. Moreover, E_3, H_3 are uniquely determined at $x_3 = -0$ from (4.2.1), (4.2.4); while their upper limits can be found in the same way after setting the coefficients ε, μ, σ at $x_3 = +0$.

Notice that the functions φ_i, ψ_i, $i = 1, 2$, can not be set arbitrarily since they are connected by two linear relations. Indeed, given φ_i, we can solve the initial boundary value problem (4.2.1)–(4.2.3) in the domain D^- with the boundary conditions $E_i|_{x_3=0} = \varphi_i$, $i = 1, 2$. Thereby, we uniquely determine in D^- all the components of $\boldsymbol{E}$, $\boldsymbol{H}$ and, hence, calculate ψ_i from the functions φ_i, $i = 1, 2$. Thus, for the problem to be well-posed, only two of the four functions in (4.2.4) can be set independently. Any pair of (φ_1, φ_2), (ψ_1, ψ_2), (φ_1, ψ_1), (φ_2, ψ_2) may be taken. For each of them one can define a well-posed initial boundary value problem in D^- and, having solved it, find the rest of $\boldsymbol{E}$, $\boldsymbol{H}$ components at the plane $x_3 = 0$. But we shall not restrict ourselves to a special choice of independent functions, because for the general formulation of inverse problems it is more convenient to use the information on solution to (4.2.1)–(4.2.3) in the form (4.2.4).

The presented inverse problems arise frequently in the mathematical modelling of various physical phenomena. In particular, they are closely connected with the practical needs of exploration geophysics. As an illustration, we mention the following problem of geoelectrics. Let the Earth, with the flat surface $x_3 = 0$, be modeled as the half-space $\mathbb{R}^3_+$, and the air correspond to $\mathbb{R}^3_-$. The function $\boldsymbol{j}(x,t)$ stands for an exterior current generating an electromagnetic field. The functions $\boldsymbol{E}$, $\boldsymbol{H}$ are treated as the electric and magnetic strength vectors of the electromagnetic field described by equations (4.2.1). The coefficients ε, μ, σ denote the permittivity, permeability, and conductivity of a medium, respectively. These coefficients are some known constants in the air $\mathbb{R}^3_-$ and varying unknowns in the earth $\mathbb{R}^3_+$. They are of major interest in geophysical exploration. To determine them, an exterior current is generated and the strength components φ, ψ of the electromagnetic field are measured on the surface.

The mathematical theory presented by Romanov and Kabanikhin (1994) affords ground for the interpretation of the electromagnetic field observed, i.e., for recovering the unknown medium parameters from the measurements.

Let us make now a general terminological note. Usually system (4.2.1) is supplemented by the two equations

$$\operatorname{div} \mu \boldsymbol{H} = 0, \qquad \operatorname{div} \varepsilon \boldsymbol{E} = \rho_e. \qquad (4.2.5)$$

According to the classical theory, it is the system (4.2.1), (4.2.5) which is used to be called a system of Maxwell's equations. However, it is often used to regard system (4.2.1) as an independent object, ignoring equations (4.2.5) (Romanov and Kabanikhin, 1994). This treatment is based on the following reasons. The first equation in (4.2.5) is a direct corollary of (4.2.1), (4.2.3); so it is fulfilled for any solution to problem (4.2.1)–(4.2.3). The second equation in (4.2.5) can be naturally considered as an independent equation for determining the charge density ρ, which problem is beyond our interest here. At the same time, the electric strength vector $\boldsymbol{E}$ can be found from (4.2.1)–(4.2.3). Thus, equations (4.2.1) are the major and quite independent part of Maxwell's equations.

Below we give some results on correspondence between the solutions of quasi-stationary and wave Maxwell's equations.

We consider here the case when the displacement current $\boldsymbol{j}_{\text{dis}} = \varepsilon \, \partial \boldsymbol{E}/\partial t$ is negligibly smaller comparing to the conduction current $\boldsymbol{j}_{\text{con}} = \sigma \boldsymbol{E}$. In this case the wave propagation is usually described by the following electrodynamic equations:

$$\operatorname{rot} \boldsymbol{H} = \sigma \boldsymbol{E} + \boldsymbol{j}, \qquad \operatorname{rot} \boldsymbol{E} = -\mu \frac{\partial \boldsymbol{H}}{\partial t}. \qquad (4.2.6)$$

Equations (4.2.6) are said to be quasi-stationary Maxwell's equations. They can be obtained from the usual Maxwell's equations (4.2.1) by setting $\varepsilon = 0$.

It is natural to assume $\sigma(x) > 0$ when considering the quasi-stationary approximation. In this case, system (4.2.6) is parabolic and characterizes the diffusion of electromagnetic waves. The coefficients of this system are supposed to be given constants, $\sigma = \sigma^- > 0$, $\mu = \mu^- > 0$, in $\mathbb{R}^3_-$ and smooth bounded functions of x in $\mathbb{R}^3_+$. On the interface, the tangential components

Chapter 4. One dimensional inverse problems

of electromagnetic field must satisfy the continuity conditions

$$[H_k]_{x_3=0} = 0, \quad [E_k]_{x_3=0} = 0, \quad k = 1, 2. \tag{4.2.7}$$

It is necessary to add initial conditions. Keeping in mind that the parabolic approximation in (4.2.6) has been obtained from the hyperbolic system (4.2.1), we consider the initial conditions in the form

$$(\boldsymbol{E}, \boldsymbol{H})_{t<0} \equiv 0. \tag{4.2.8}$$

As for behavior of the solution as $|x| \to \infty$, we take the assumption that $\boldsymbol{E}$, $\boldsymbol{H}$ are generalized functions of slow growth (Vladimirov, 1977). This assumption enables us to apply to them the Fourier transform with respect to the space variables.

Inverse Problem 4.2.2 (IP 4.2.2). *Recover the coefficients μ, σ in the domain $\mathbb{R}^3_+$ provided that the solution to problem (4.2.6)–(4.2.8) is given on the plane $x_3 = 0$:*

$$E_k\big|_{x_3=0} = f_k(\bar{x}, t), \quad k = 1, 2, \quad \bar{x} \in \mathbb{R}^2, \quad t \in \mathbb{R}. \tag{4.2.9}$$

It is very convenient to reduce the IP 4.2.2 to an auxiliary inverse problem for hyperbolic equations.

The equivalence between the initial and auxiliary problems is established below, following the papers by Priimenko (1982, 1983) and Romanov, Kabanikhin, and Puchnacheva (1984).

Set $\boldsymbol{j}(x,t) = \boldsymbol{j}^0(x)\varphi(t)$, $\varphi(t) \equiv 0$, $t < 0$, in (4.2.6) and consider the following auxiliary problem for hyperbolic equations:

$$\operatorname{rot} \hat{\boldsymbol{H}} = \sigma \frac{\partial \hat{\boldsymbol{E}}}{\partial t} + \boldsymbol{j}^0(x)\delta(t), \quad \operatorname{rot} \hat{\boldsymbol{E}} = -\mu \frac{\partial \hat{\boldsymbol{H}}}{\partial t},$$

$$[\hat{H}_k]_{x_3=0} = 0, \quad [\hat{E}_k]_{x_3=0} = 0, \quad k = 1, 2,$$

$$(\hat{\boldsymbol{E}}, \hat{\boldsymbol{H}})_{t<0} \equiv 0. \tag{4.2.10}$$

Lemma 4.2.1. *Solutions to problems (4.2.10) and (4.2.6)–(4.2.8) are connected through the following relations:*

$$\boldsymbol{H}(x,t) = \int_{-\infty}^{\infty} \hat{\boldsymbol{H}}(x,\tau) G(t,\tau) \, d\tau,$$

$$\boldsymbol{E}(x,t) = \int_{-\infty}^{\infty} \hat{\boldsymbol{E}}_\tau(x,\tau) G(t,\tau)\, d\tau, \qquad (4.2.11)$$

where

$$G(t,\tau) = \theta(t) \int_0^t \varphi(s) \frac{\tau}{2\sqrt{\pi(t-s)^3}} \exp\left[-\frac{\tau^2}{4(t-s)}\right] ds. \qquad (4.2.12)$$

Prior to prove the lemma, we note that $G(t,\tau)$ solves equations

$$G_t = G_{\tau\tau},$$

$$G\big|_{\tau=0} = \varphi(t), \qquad G\big|_{t<0} \equiv 0. \qquad (4.2.13)$$

The only problem arises then verifying (4.2.13), namely to establish the condition $G|_{\tau=0} = \varphi(t)$. To check it, we consider the integral in (4.2.12) and change the variable s to

$$z = \tau/(2\sqrt{t-s}).$$

Then

$$G(t,\tau) = \theta(t) \frac{2}{\sqrt{\pi}} \int_{-\infty}^{\infty} \varphi\left(t - \frac{\tau^2}{4z^2}\right) \exp(-z^2)\, dz$$

$$\underset{\tau \to 0}{\longrightarrow} \theta(t)\varphi(t) \frac{2}{\sqrt{\pi}} \int_0^{\infty} \exp(-z^2)\, dz = \varphi(t).$$

Using (4.2.13), we find

$$\operatorname{rot} \boldsymbol{H} - \sigma \boldsymbol{E} = \int_{-\infty}^{\infty} (\operatorname{rot} \hat{\boldsymbol{H}} - \sigma \hat{\boldsymbol{E}}_\tau) G(t,\tau)\, d\tau$$

$$= \boldsymbol{j}^0(x) G(t,0) = \boldsymbol{j}^0(x)\varphi(t) = \boldsymbol{j}(x,t),$$

$$\frac{\partial \boldsymbol{H}}{\partial t} = \int_{-\infty}^{\infty} \hat{\boldsymbol{H}}(x,\tau) \frac{\partial}{\partial t} G(t,\tau)\, d\tau$$

$$= \int_{-\infty}^{\infty} \hat{\boldsymbol{H}}(x,\tau) \frac{\partial^2}{\partial \tau^2} G(t,\tau)\, d\tau = -\int_{-\infty}^{\infty} \boldsymbol{H}_\tau(x,\tau) G_\tau(t,\tau)\, d\tau,$$

$$\operatorname{rot} \boldsymbol{E} = \int_{-\infty}^{\infty} (\operatorname{rot} \hat{\boldsymbol{E}}_\tau) G(t,\tau)\, d\tau = -\int_{-\infty}^{\infty} (\operatorname{rot} \hat{\boldsymbol{E}}) G_\tau(t,\tau)\, d\tau,$$

$$\operatorname{rot} \boldsymbol{E} + \mu \frac{\partial \boldsymbol{H}}{\partial t} = -\int_{-\infty}^{\infty} (\operatorname{rot} \hat{\boldsymbol{E}} + \mu \hat{\boldsymbol{H}}_\tau) G_\tau(t,\tau)\, d\tau = 0.$$

The rest of the equalities (4.2.10) are evident. The above calculations need justification, which is rather simple under the assumption that σ, μ, $j^0(x)$, $\varphi(t)$ are bounded and $\varphi(t)$ is finite. In this case the solution to (4.2.10) grows, as $t \to \infty$, not faster than $\exp(Ct)$, with a constant C, while the function $G(t,\tau)$ decreases, for a fixed t, as fast as $\tau \exp(-C_1\tau^2)$, $C_1 > 0$. This ensures the required convergence of the integrals in (4.2.11) and completes the justification.

Relations (4.2.11) can be considered as integral transformations with the kernel $G(t,\tau)$. Their invertibility depends on the properties of $\varphi(t)$. In particular, if $\varphi(t) = \delta(t)$, then the kernel $G(t,\tau)$ is defined by

$$G(t,\tau) = \theta(t) \frac{\tau}{2\sqrt{\pi t^3}} \exp\left(-\frac{\tau^2}{4t}\right), \qquad (4.2.14)$$

and (4.2.11) are, in fact, the Laplace transforms and, thereby, they are uniquely invertible. Note that relations (4.2.11) had been pointed out originally by Priimenko (1982) for exactly the case under consideration. Under the unique invertibility condition, the transformations (4.2.11) determine one-to-one correspondence between the solutions to (4.2.10) and (4.2.6)–(4.2.8). Therefore, it is not difficult to formulate the inverse problem, equivalent to the IP 4.2.2, for the wave system (4.2.10). Indeed, let $\hat{f}_k(\bar{x},t)$, $k = 1,2$, be solutions of the integral equations

$$f_k(\bar{x},t) = \int_{-\infty}^{\infty} \hat{f}_k(\bar{x},\tau) G(t,\tau)\, d\tau, \qquad k = 1,2, \qquad (4.2.15)$$

provided that $f_k|_{t<0} \equiv 0$. Then, assuming

$$\hat{E}_k(\bar{x},0,t) = \hat{F}_k(\bar{x},t) \equiv \int_{-\infty}^{t} \hat{f}_k(\bar{x},\tau)\, d\tau, \qquad k = 1,2, \qquad (4.2.16)$$

we obtain the inverse problem (4.2.10), (4.2.16) equivalent to the IP 4.2.2. Thereby, the inverse problem for parabolic system (4.2.6) is reduced to that for system (4.2.10) which is really the wave system of Maxwell's equations.

Thus, the IP 4.2.2 can be solved in the following way: 1) solve equations (4.2.11); 2) determine the solution to (4.2.10), (4.2.16). As a result, we obtain the coefficients σ, μ which solve the initial problem. If the problems considered in each step of the procedure are uniquely solvable, so is the IP 4.2.2. Therefore, one can formulate uniqueness theorems. In particular,

Theorem 4.2.1. If $\sigma = \sigma(x_3) > 0$, $\mu = \mu^+ > 0$, where μ^+ is a given constant, $\sigma(x_3) \in C^1(\mathbb{R}_+)$, and $\boldsymbol{j} = (1,0,0)\delta(x)\delta(t)$, then the coefficient

$\sigma(x_3)$ is uniquely determined in the domain $x_3 \geq 0$ from the function

$$f(t) = \int_{\mathbb{R}^2} E_1(\bar{x}, 0, t) \exp\left[i(\lambda^0, \bar{x})\right] d\bar{x}$$

for a fixed λ^0.

The uniqueness theorem for the linearized inverse problem of determining $\sigma(x)$ also follows

Theorem 4.2.2. Let $j = (1, 0, 0)\delta(\bar{x} - \bar{x}^0)\delta(x_3)\delta(t)$, $\mu = \mu^+ > 0$, where μ^+ is a fixed constant. Let also $\sigma(x)$ be represented as $\sigma(x) = \sigma_0(x_3) + \sigma_1(x)$, where $\sigma_0(x_3) > 0$ is a known function of the class $C^1(\overline{\mathbb{R}_+})$, $\sigma_1(x)$ is small, finite in $\mathbb{R}^3_+$ and belongs to $C(\overline{\mathbb{R}^3_+})$. Then $\sigma_1(x)$, in linear approximation, is uniquely determined from the function

$$f(t, \nu) = \int_{\mathbb{R}^4} E_1(\bar{x}, 0, t, \bar{x}^0) \exp\left\{i[(\lambda^0, \bar{x} - \bar{x}^0) + (\nu, \bar{x}^0)]\right\} d\bar{x}\, d\bar{x}^0,$$

$$\nu \in \mathbb{R}^2, \quad t \in \mathbb{R},$$

for any fixed λ^0.

Theorem 4.2.3. Assume that $j = (1, 0, 0)g(\bar{x})\delta(x_3)\delta(t)$, $\mu = \mu^+ > 0$, where μ^+ is a given constant. Let also $g(\bar{x})$, $\sigma(x)$ be analytic functions in an origin neighborhood, $g(\bar{x}) \neq 0$, $|\bar{x}| \leq s_0$. Then the data (4.2.9) uniquely determine the function $\sigma(x)$ in a subdomain of $\mathbb{R}^3_+$ adjoining to the piece of the plane $z = 0$ such that $|\bar{x}| \leq s_0$.

With details of the proofs reader can be acquainted by original works (Priimenko, 1983, 1986, 1990; Romanov and Kabanikhin, 1994).

4.3. CONNECTIONS AMONG INVERSE PROBLEMS OF HYPERBOLIC, ELLIPTIC, AND PARABOLIC TYPE

In the previous paragraph we have use special relations between solutions to Maxwell's equations (4.2.1) and their quasi-stationary approximation (4.2.6). Here, following the works of Lavrentiev and Saveliev (1995, 1999) we would like to give more general notes about connections among solutions of Inverse Problems for different kind of governing equations.

It is possible to establish a one-to-one correspondence among solutions to problems of mathematical physics, describing different physical processes.

This fact has been known for years. In case of equations with constant coefficients the aforementioned correspondence is derived in a most simple way after using Fourier and Laplace transforms. Besides, for the first time it has been noted in the work of Lavrentiev, Reznitskaya, and Yakhno (1982) that relations between solutions to equations of different type are useful in view of inverse problems. Indeed, it is possible to use well developed theory, established for hyperbolic problems, then studied equations of elliptic or parabolic type. Reverse approach is also useful. Below we describe general ideas on the simplest model examples of inverse problems.

In the domain $D \subset \mathbb{R}^n$, bounded by the surface S, consider the following problem for parabolic equation

$$u_t = Lu + f(x,t), \qquad (x,t) \in D \times \mathbb{R}_+,$$
$$u\big|_{t=0} = \varphi(x), \qquad x \in D, \qquad (4.3.1)$$
$$\frac{\partial u}{\partial n} = \psi(x,t), \qquad (x,t) \in S \times \mathbb{R}_+.$$

Here L stands for the uniformly elliptic operator with continuous coefficients (depending only on variable x); n is co-normal vector to S.

Let us formulate the "corresponding" hyperbolic problem:

$$\tilde{u}_{tt} = L\tilde{u} + \tilde{f}(x,t), \qquad (x,t) \in D \times \mathbb{R}_+,$$
$$\tilde{u}\big|_{t=0} = \varphi(x), \quad \tilde{u}_t\big|_{t=0} = 0, \qquad x \in D, \qquad (4.3.2)$$
$$\frac{\partial \tilde{u}}{\partial n} = \tilde{\psi}(x,t), \qquad (x,t) \in S \times \mathbb{R}_+,$$

where we give rise an order of differentiation with respect to time variable in the governing equation and add the necessary initial condition.

Let $\tilde{f}$ and $\tilde{\psi}$ be the smooth functions, increasing as $t \to \infty$ not faster then $Ce^{\alpha t}$. Moreover, assume that the following relations hold true:

$$f(x,t) = \int_0^\infty \tilde{f}(x,t) G(t,\tau) \, d\tau, \quad \psi(x,t) = \int_0^\infty \tilde{\psi}(x,t) G(t,\tau) \, d\tau, \quad (4.3.3)$$

where

$$G(t,\tau) = \frac{1}{\sqrt{\pi t}} e^{-\tau^2/4t}$$

is the well-known fundamental solution to the heat equation

$$G_t = G_{\tau\tau}.$$

When, it is easy to check that solutions to problems in (4.3.1) and (4.3.2) are connected through the following convolution type relation

$$u(x,t) = \int_0^\infty \tilde{u}(x,\tau)G(t,\tau)\,d\tau. \tag{4.3.4}$$

Indeed,

$$u_t = \int_0^\infty \tilde{u}(x,\tau)G_t(t,\tau)\,d\tau = \int_0^\infty \tilde{u}(x,\tau)G_{\tau\tau}(t,\tau)\,d\tau$$

$$= (\tilde{u}G_\tau - \tilde{u}_\tau G)\Big|_{\tau=0}^{\tau=\infty} + \int_0^\infty \tilde{u}_{\tau\tau}(x,\tau)G(t,\tau)\,d\tau$$

$$= \int_0^\infty \tilde{u}_{\tau\tau}(x,\tau)G(t,\tau)\,d\tau,$$

$$Lu = \int_0^\infty G(t,\tau)L\tilde{u}(x,\tau)\,d\tau.$$

Thus, for $t \in \mathbb{R}_+$ we have

$$u_t - Lu - f = \int_0^\infty (\tilde{u}_{\tau\tau} - L\tilde{u} - \tilde{f})G(t,\tau)\,d\tau = 0.$$

On the other hand, the following relations are fulfilled

$$u\big|_{t=0} = \lim_{t\to 0}\int_0^\infty \tilde{u}(x,\tau)G(t,\tau)\,d\tau$$

$$= \lim_{t\to 0}\frac{2}{\sqrt{\pi}}\int_0^\infty \tilde{u}(x,2\sqrt{t}\,\tau)e^{-\tau^2}\,d\tau = \tilde{u}(x,0) = \varphi(x),$$

$$\frac{\partial u}{\partial n}\Big|_S = \int_0^\infty \frac{\partial \tilde{u}}{\partial n}\Big|_S G(t,\tau)\,d\tau = \int_0^\infty \tilde{\psi}(x,\tau)G(t,\tau)\,d\tau = \psi(x,t).$$

Thus, representation (4.3.4) gives the solution to problem (4.3.1) through the solution to problem (4.3.2), provided that the problems data are connected via formulae in (4.3.3). As well known, solution to problem (4.3.1) is unique for rather general class of operators L and surfaces S, moreover, this solution has the form (4.3.4). Note that equality in (4.3.4) is invertible for any fixed x as it could be represented in the form of Laplace transform with the dual variable $p = 1/4t$:

$$u(x,1/4p) = \sqrt{\frac{p}{\pi}}\int_0^\infty e^{-zp}\tilde{u}(x,\sqrt{z})\frac{dz}{\sqrt{z}}. \tag{4.3.5}$$

So, formula in (4.3.4) could be treated as one-to-one correspondence between solutions to problems (4.3.1) and (4.3.2).

Now consider the following inverse problem for equation (4.3.1).

Chapter 4. One dimensional inverse problems

Inverse Problem 4.3.1 (IP 4.3.1). *It is necessary to determine $f(x,t)$ (or, possibly, one of the coefficients of the operator L) having known solution to equation (4.3.1) at S:*

$$u\big|_S = g(x,t), \qquad t \in \overline{\mathbb{R}}_+. \tag{4.3.6}$$

For the given function $g(x,t)$, $x \in S$, let us determine the new function $\tilde{g}(x,\tau)$ as the solution to equation

$$g(x,t) = \int_0^\infty \tilde{g}(x,\tau) G(t,\tau)\, d\tau, \qquad x \in S, \ t \in \overline{\mathbb{R}}_+. \tag{4.3.7}$$

Let

$$\tilde{u}\big|_S = \tilde{g}(x,t), \qquad t \in \overline{\mathbb{R}}_+. \tag{4.3.8}$$

Function $\tilde{g}(x,t)$ for $x \in S$ is uniquely determined from (4.3.7). Therefore, the IP 4.3.1 is equivalent to the corresponding inverse problem (4.3.2), (4.3.8).

As is known from the Mathematical Physics, for a wide class of coefficients of elliptic operator L solution to the problem (4.3.2) satisfies the estimate

$$|\tilde{u}(x,t)| \leq C e^{\alpha t}.$$

Formula (4.3.4) (or (4.3.5), which is equivalent) shows that function $u(x,t)$ is analytic with respect to t for $t \in \mathbb{R}_+$. Thus, it is sufficient to determine function $g(x,t)$ in (4.3.6) at an arbitrarily small neighborhood of the value $t = 0$, for instance, at the interval $0 \leq t \leq \delta$, $\delta > 0$.

Due to the correspondence among different inverse problems, established above, a number of results proved for hyperbolic equations could be applied for parabolic problems. It is possible in all the cases, then inversion data are given at the cylinder with the element parallel to time axis.

Consider one of the simplest inverse problems: to determine the coefficient $q(z)$ of parabolic equation

$$u_t = u_{zz} - q(z)u, \qquad (z,t) \in \mathbb{R} \times \mathbb{R}_+, \tag{4.3.9}$$

through the overinformation about solutions to Cauchy problem

$$u\big|_{t=0} = \delta(z), \qquad z \in \mathbb{R}, \tag{4.3.10}$$

given at the point $z = 0$

$$u\big|_{z=0} = f_1(t), \qquad u_z\big|_{z=0} = f_2(t), \qquad t \in \overline{\mathbb{R}}_+. \tag{4.3.11}$$

As it has been explained above, this task is reduced to the following inverse problem:

$$\tilde{u}_{tt} = \tilde{u}_{zz} - q(z)\tilde{u}, \tag{4.3.12}$$

$$\tilde{u}|_{t=0} = \delta(z), \quad \tilde{u}_t|_{t=0} = 0, \quad \tilde{u}|_{z=0} = \tilde{f}_1(t), \quad u_z|_{z=0} = \tilde{f}_2(t).$$

Here functions $\tilde{f}_k$ are determined by f_k according to

$$f_k(t) = \int_0^\infty G(t,\tau)\tilde{f}_k(\tau)\,d\tau, \quad k = 1,2.$$

The coefficient $q(z) \in C(\mathbb{R})$ in inverse problem (4.3.9)–(4.3.11) is uniquely determined by setting the functions f_1 and f_2 at the interval $[0,\delta]$ with any $\delta > 0$.

Now consider elliptic equations. Let domain D_0 of variables x and y ($x \in \mathbb{R}^n$) has the form of semi-infinite cylinder with the element being parallel to y axis:

$$D_0 = D \times \mathbb{R}_+, \qquad x \in D, \quad y \in \mathbb{R}_+.$$

In the domain D_0 bounded by the surface S consider the elliptic equation

$$u_{yy} + Lu + f(x,y) = 0, \qquad (x,y) \in D_0, \tag{4.3.13}$$

with following boundary conditions

$$u|_{y=0} = \varphi(x), \quad \frac{\partial u}{\partial n}\Big|_\Gamma = \psi(x,y), \quad \Gamma = S \times \mathbb{R}_+, \tag{4.3.14}$$

vanishes as $y \to \infty$. Here L is the uniformly elliptic with respect to $x_1,\ldots,x_n$ operator, coefficients of which are independent of y. In addition to the problem (4.3.13)–(4.3.14) consider the problem (4.3.2) in case $\tilde{f}$ and $\tilde{\psi}$ are connected with f and ψ according to

$$f(x,y) = \int_0^\infty \tilde{f}(x,t)H(y,t)\,dt, \qquad \psi(x,y) = \int_0^\infty \tilde{\psi}(x,t)H(y,t)\,dt. \tag{4.3.15}$$

Here the function

$$H(y,t) = \frac{2}{\pi}\frac{y}{y^2 + t^2}$$

stands for solution to Laplace equation

$$H_{tt} + H_{yy} = 0.$$

Suppose that both $\tilde{f}$ and $\tilde{\psi}$ are bounded in order that the integrals in (4.3.15) converge. Under a similar assumption about solution to problem (4.3.12), the unique solution to (4.3.13), (4.3.14) is given by

$$u(x,t) = \int_0^\infty \tilde{u}(x,t) H(y,t)\,\mathrm{d}t. \tag{4.3.16}$$

Let us verify this formula assuming that all the calculations below are correct. Indeed,

$$u_{yy} = \int_0^\infty H_{yy}(y,t)\tilde{u}(x,t)\,\mathrm{d}t = -\int_0^\infty H_{tt}(y,t)\tilde{u}(x,t)\,\mathrm{d}t$$

$$= (-H_t\tilde{u} + H\tilde{u}_t)\big|_{t=0}^{t=\infty} - \int_0^\infty H(y,t)\tilde{u}_{tt}(x,t)\,\mathrm{d}t$$

$$= -\int_0^\infty H(y,t)\tilde{u}_{tt}(x,t)\,\mathrm{d}t,$$

$$Lu = \int_0^\infty H(y,t) L\tilde{u}(x,t)\,\mathrm{d}t.$$

Thus

$$u_{yy} + Lu + f = \int_0^\infty (-\tilde{u}_{tt} + L\tilde{u} + \tilde{f}) H(y,t)\,\mathrm{d}t = 0.$$

At the same time we have that

$$\frac{\partial u}{\partial n}\bigg|_\Gamma = \int_0^\infty \frac{\partial \tilde{u}}{\partial n}\bigg|_\Gamma H(y,t)\,\mathrm{d}y = \int_0^\infty \tilde{\psi}(x,t) H(y,t)\,\mathrm{d}t = \psi(x,y),$$

$$u\big|_{y=0} = \lim_{y\to 0} \int_0^\infty \tilde{u}(x,t) H(y,t)\,\mathrm{d}t$$

$$= \lim_{y\to 0} \frac{2}{\pi} \int_0^\infty \tilde{u}(x,y\tau) \frac{\mathrm{d}\tau}{1+\tau^2} = \tilde{u}(x,0) = \varphi(x).$$

Therefore, the function $u(x,y)$ in (4.3.16) solves the problem (4.3.13), (4.3.14).

Being an integral transform with the Cauchy kernel, equality in (4.3.16) is uniquely invertible for any fixed x. Thus, inverse problem (4.3.2), (4.3.8) is equivalent to the following one.

Inverse Problem 4.3.2 (IP 4.3.2). Determine one of the coefficients of operator L through the information about solution to (4.3.13), (4.3.14) of the type

$$u\big|_\Gamma = g(x,y). \tag{4.3.17}$$

Here the functions g and $\tilde{g}$ are connected by the following "invertible relation":
$$g(x,y) = \int_0^\infty \tilde{g}(x,t) H(y,t)\,dt, \qquad (x,y) \in \Gamma.$$

As it follows the reasons above, replacement of the elliptic equation by the hyperbolic one is correct *only* under two conditions, namely: 1) considered domain should be a cylinder with the element, parallel to y axis; 2) equation coefficients should be independent of y.

These two conditions are really restrictive as in the most of applied problems space variables x and y have similar role.

Let us describe an alternative class of inverse problems for elliptic equations, which could be studied by reduction to corresponding problems for hyperbolic equations. We restrict our self with a single example. For equation
$$\Delta u + \omega^2 c(x) u = f(x), \qquad x \in \mathbb{R}^n, \qquad (4.3.18)$$
consider the problem of determination of function $f(x)$ or coefficient $c(x) \geq c_0 > 0$ by the following information.

For a series of "frequencies" ω_k ($k = 1, 2, \ldots$) solution to equation (4.3.18) (satisfying irradiation condition at infinity) is know at some points of certain surface S:
$$u(x, \omega_k) = g(x, \omega_k), \qquad x \in S, \quad k = 1, 2, \ldots. \qquad (4.3.19)$$

Equation (4.3.18) arise then studying the steady harmonic oscillations. Such oscillations are described by the wave equation
$$c(x) v_{tt} = \Delta v + f(x) e^{i\omega t}.$$

Searching the function v in the form
$$v = u(x) e^{i\omega t},$$
we get equation in (4.3.18). In case $c \equiv 1$ equation (4.3.18) is nothing but Helmholtz equation.

Consider the following Cauchy problem:
$$c(x) w_{tt} = \Delta w + f(x) \delta(t), \qquad (z, t) \in \mathbb{R} \times \mathbb{R}_+, \qquad (4.3.20)$$
$$w\big|_{t<0} \equiv 0, \qquad z \in \mathbb{R}.$$

Under assumption that $c(x)$ and $f(x)$ are bounded, solution to this problem satisfies the estimate $|w| \leq C e^{\alpha t}$. Thus, Laplace transform $\tilde{w}(x, p)$ of the function $w(x,t)$ exists for $\operatorname{Re} p > \alpha$. Laplace image is analytic function of variable p. In some cases (in particular for $c = 1$ and $f(x)$ being compactly supported) the function $\tilde{w}(x,p)$ possesses analytic continuation on imaginary axis $p = i\omega$. Therefore,

$$\tilde{w}(x, i\omega_k) = g(x, \omega_k), \qquad x \in S, \quad k = 1, 2, \ldots.$$

If $\omega_k \to \omega_0$ with ω_0 being an inner point of the analyticity domain for $\tilde{w}(x, i\omega)$, then the values $\tilde{w}(x, i\omega k)$ make it possible to determine $\tilde{w}(x, p)$ (as well as $w(x,t)$) into the whole analyticity domain, and thus to calculate the traces

$$w\big|_S = \tilde{g}(x,t). \tag{4.3.21}$$

As a result we conclude that the inverse problem (4.3.20), (4.3.21) is equivalent to (4.3.18), (4.3.19).

4.4. PROBLEMS WITH A FOCUSED SOURCE OF DISTURBANCES

In these two sections we are giving rather elementary considerations, from time to time accompanying by the complete proofs. The point is to supply the reader with the demonstration of some popular methods in application to simplified problems. This make it possible to follow the text even with not advanced knowledge of mathematical background.

In applied problems the (measured) data often have been represented by compactly supported functions, which are nontrivial only within a very small space area. It is convenient to simulate such situations by considering the problem data as *distributions* (generalized functions) whose supports are concentrated at some fixed points in the space. In this case the solution to the boundary value problem is also a distribution, and fulfillment of differential equalities and boundary conditions is understood in terms of the theory of generalized functions. Corresponding inverse problems, where the data of direct problems are distributions, deserve a separate study. These problems are very important both in theory and applications.

Here below we give an introduction to such complicated area. Consider Cauchy problem for the wave equation with a source term. In the domain $D = \{(z,t) \mid z \in \mathbb{R}, t \in \mathbb{R}_+\}$ we have

$$\frac{\partial^2 u}{\partial t^2} - \frac{\partial^2 u}{\partial z^2} = F(z,t), \qquad (z,t) \in D, \tag{4.4.1}$$

with following initial data

$$u(z,0) = 0, \quad \frac{\partial u}{\partial t}(z,0) = \psi(z), \quad z \in \mathbb{R}. \quad (4.4.2)$$

By the d'Alembert formula the solution to the Cauchy problem (4.4.1), (4.4.2) is given by

$$u(z,t) = \frac{1}{2}\int_{z-t}^{z+t} \psi(\xi)\,d\xi + \frac{1}{2}\iint_{\Delta(z,t)} F(\xi,\tau)\,d\xi\,d\tau, \quad (4.4.3)$$

where $\Delta(z_0, t_0)$ stands for the triangle on the (z,t)-plane bounded by the z-axis and characteristics of equation (4.4.2) passing through the point (z_0, t_0):

$$\Delta(z_0, t_0) = \{(z,t) \mid 0 \leq t \leq t_0 - |z - z_0|\}. \quad (4.4.4)$$

As is well known, solution in (4.4.3) is classical: $u(z,t) \in C^2$, provided that $\psi(z) \in C^1$ and $F(z,t)$, $(\partial F/\partial t)(z,t) \in C$. On the other hand, function $u(z,t)$ of the form (4.4.3) solves equations (4.4.1), (4.4.2) also in the case when $\psi(z)$ and $F(z,t)$ are only continuous or even piecewise continuous functions. In this case we function $u(z,t)$ does not possess second derivatives and, therefore, could not be considered as a classical solution. In such occasion solution is called *generalized solution* and it should be specified in which sense equation is satisfied. For example, if the function $\psi(z)$ is piecewise continuous, then the solution $u(z,t)$ to the Cauchy problem (4.4.1), (4.4.2) remains continuous, while the partial derivatives are merely piecewise continuous.

Using formula (4.4.3), one can construct such generalized solutions to the Cauchy problem (4.4.1), (4.4.3) which are not even continuous. Indeed, let $F(z,t) \equiv 0$. In this case (4.4.3) assumes the form

$$u(z,t) = \frac{1}{2}\int_{z-t}^{z+t} \psi(\xi)\,d\xi. \quad (4.4.5)$$

Now consider a sequence of piecewise continuous and non-negative functions, $\{\psi_n(z)\}_{n\in\mathbb{N}}$, with the following properties

1) functions ψ_n vanish

$$\psi_n(z) \equiv 0 \quad (4.4.6)$$

outside some interval $[z_0 - \alpha_n, z_0 + \alpha_n]$, where $\lim_{n\to\infty} \alpha_n = 0$;

Chapter 4. One dimensional inverse problems

2) integrals of ψ_n converge with unit

$$\lim_{n \to \infty} \int_{z_0-\alpha_n}^{z_0+\alpha_n} \psi_n(\xi) \, d\xi = 1. \tag{4.4.7}$$

A sequence of functions $\{\psi_n(z)\}_{n \in \mathbb{N}}$ possessing properties 1) and 2) is called a *delta-like* (δ-*like*) *sequence*. Such sequence $\{\psi_n(z)\}_{n \in \mathbb{N}}$ converges with the *Dirac's mass*, concentrated at the point $z = z_0$. Dirac's mass is often called δ-function. Remind, that $\delta(z - z_0)$ is not a function in the usual understanding of this term, but generalized function, or *distribution*. The basic property of such distribution is that for any continuous function $f(z) \in C[a, b]$ we have

$$\int_a^b f(z) \delta(z - z_0) \, dz = \begin{cases} f(z_0), & z_0 \in (a, b), \\ 0, & z_0 \notin (a, b). \end{cases} \tag{4.4.8}$$

Now, based on the delta-like sequence $\{\psi_n(z)\}_{n \in \mathbb{N}}$, introduce the following sequence of functions $\{u_n(z, t)\}_{n \in \mathbb{N}}$:

$$u_n(z, t) = \frac{1}{2} \int_{z-t}^{z+t} \psi_n(\xi) \, d\xi. \tag{4.4.9}$$

This is the sequence of solutions to the corresponding Cauchy problems for equation (4.4.1) (in case $F(z, t) \equiv 0$). Passing to the limit as $n \to \infty$, we obtain the function

$$U(z, z_0, t) = \begin{cases} 1/2, & |z - z_0| < t, \\ 0, & |z - z_0| > t. \end{cases} \tag{4.4.10}$$

This function is said to be a *generalized solution* to the wave equation

$$\left(\frac{\partial^2}{\partial t^2} - \frac{\partial^2}{\partial z^2} \right) U(z, z_0, t) = 0 \tag{4.4.11}$$

with the Cauchy data

$$U(z, z_0, 0) = 0, \qquad \frac{\partial U}{\partial t}(z, z_0, 0) = \delta(z - z_0). \tag{4.4.12}$$

This generalized solution can be obtained if we formally apply the d'Alembert formula to the solution of the problem (4.4.11), (4.4.12) and use the property (4.4.8) of the Dirac's mass. Indeed,

$$U(z, z_0, t) = \frac{1}{2} \int_{z-t}^{z+t} \delta(\xi - z_0) \, d\xi = \begin{cases} 1/2, & x_0 \in (z - t, z + t), \\ 0, & z_0 \notin [z - t, z + t]. \end{cases} \tag{4.4.13}$$

Note that formulae in (4.4.10) and (4.4.13) are representations of the same function, these expressions differ only in their form. They do not give the values of the function $U(z, z_0, t)$ on the characteristics $z = z_0 \pm t$ originating from the point $(z_0, 0)$. We will define the values of $U(z, z_0, t)$ at the points of these characteristics as the limit values of $U(z, z_0, t)$ calculated from the direction of the domain $t > |z - z_0|$. In the present case we have $U(z, z_0, |z - z_0|) = 1/2$.

The function $U(z, z_0, t)$, which solves Cauchy problem (4.4.11), (4.4.12), is called *a fundamental solution* to the Cauchy problem for equation (4.4.11). With the aid of this function the solution to the Cauchy problem for equation (4.4.11) (with function $u(z,t)$) with an arbitrary data

$$u(z,0) = \varphi(z), \quad \frac{\partial u}{\partial t}(z,0) = \psi(z), \quad z \in \mathbb{R} \tag{4.4.14}$$

can be written as

$$u(z,t) = \frac{\partial}{\partial t} \int_{-\infty}^{\infty} \varphi(\xi) U(z,\xi,t) \, d\xi + \int_{-\infty}^{\infty} \psi(\xi) U(z,\xi,t) \, d\xi. \tag{4.4.15}$$

Let us check that the function $u(z,t)$ in (4.4.15) coincides with the solution to the problem (4.4.11), (4.4.14) defined by the d'Alembert formula. Indeed,

$$u(z,t) = \frac{\partial}{\partial t} \int_{-\infty}^{\infty} \varphi(\xi) U(z,\xi,t) \, d\xi + \int_{-\infty}^{\infty} \psi(\xi) U(z,\xi,t) \, d\xi$$

$$= \frac{\partial}{\partial t} \frac{1}{2} \int_{z-t}^{z+t} \varphi(\xi) \, d\xi + \frac{1}{2} \int_{z-t}^{z+t} \psi(\xi) \, d\xi$$

$$= \frac{1}{2}[\varphi(z+t) + \varphi(z-t)] + \frac{1}{2} \int_{z-t}^{z+t} \psi(\xi) \, d\xi.$$

Here we have used the representation (4.4.10) of the function $U(z, z_0, t)$. It turns out that formula (4.4.15) holds true not only for equation (4.4.11), but for more general class of differential equations. For example, consider the equation

$$\frac{\partial^2 u}{\partial t^2} = a^2(z,t) \frac{\partial^2 u}{\partial z^2} + b(z,t) \frac{\partial u}{\partial z} + c(z,t) u, \tag{4.4.16}$$

where $a(z,t) \in C^2$, $b(z,t) \in C^1$, $c(z,t) \in C$, and let $U(z, z_0, t)$ be the fundamental solution to the Cauchy problem for equation (4.4.16), i.e., the solution to equation (4.4.16) with initial data (4.4.12).

Then the solution to the Cauchy problem for equation (4.4.16) with data of the general form (4.4.14) is also given by formula (4.4.15).

Chapter 4. One dimensional inverse problems

Now consider equation (4.4.1) with the trivial Cauchy data ($\varphi(z) = \psi(z) \equiv 0$). In this case the solution to the Cauchy problem is

$$u(z,t) = \frac{1}{2} \iint_{\Delta(z,t)} F(\xi, \tau) \, d\xi \, d\tau \qquad (4.4.17)$$

(see (4.4.4)).

Let us construct a two-dimensional version of the above delta-like sequence $\{\psi_n(z)\}_{n \in \mathbb{N}}$. Let $F_n(z,t)$ be piecewise continuous and non-negative functions satisfying the following conditions:

1) $F_n(z,t) \equiv 0$ outside the disk $D_n = \{|z-z_0|^2 + |t-t_0|^2 < \alpha_n^2\}$, where $\lim_{n \to \infty} \alpha_n = 0$;

2) $\lim_{n \to \infty} \iint_{D_n} F_n(z,t) \, dz \, dt = 1$.

The limit of the functional sequence $F_n(z,t)$ with the above properties is called the two-dimensional Dirac's mass concentrated at the point (z_0, t_0) (or, simply, the 2-D δ-function) and is denoted by the symbol $\delta(z-z_0, t-t_0)$. The main property of the function $\delta(z-z_0, t-t_0)$ consists in the equality

$$\iint_D f(z,t) \, \delta(z-z_0, t-t_0) \, dz \, dt = \begin{cases} f(z_0, t_0), & (z_0, t_0) \in D, \\ 0, & (z_0, t_0) \notin \overline{D}, \end{cases} \qquad (4.4.18)$$

where $f(z,t) \in C(D)$ and $\overline{D}$ stands for a closure of open domain D with a piecewise smooth boundary.

Consider the functional sequence $\{u_n(z, t, z_0, t_0)\}_{n \in \mathbb{N}}$

$$u_n(z, t, z_0, t_0) = \frac{1}{2} \iint_{\Delta(z,t)} F_n(\xi, \tau) \, d\xi \, d\tau. \qquad (4.4.19)$$

Formula (4.4.19) describes the sequence of solutions to the Cauchy problem for equation (4.4.1) with the trivial initial data. In view of the properties of the delta-like sequence $\{F_n(z,t)\}_{n \in \mathbb{N}}$ we have

$$U(z, t, z_0, t_0) = \lim_{n \to \infty} u_n(z, t, z_0, t_0) = \begin{cases} 1/2, & t - t_0 \geq |z - z_0|, \\ 0, & t - t_0 < |z - z_0|, \end{cases} \qquad (4.4.20)$$

since $D_n \cap \Delta(z,t) \neq \emptyset$ for all $n \in \mathbb{N}$ only when $t - t_0 \geq |z - z_0|$.

The function $U(z, t, z_0, t_0)$ defined by formula (4.4.20) is called *the fundamental solution* to equation (4.4.1). It is important that fundamental solution is associated with the "action" of the type $\delta(z - z_0, t - t_0)$, concentrated at the point (z_0, t_0), and trivial Cauchy data.

The solution to equation (4.4.1) with zero Cauchy data is represented via the fundamental solution (4.4.20) by the formula

$$u(z,t) = \int_0^\infty d\tau \int_{-\infty}^\infty F(\xi,\tau) U(z,t,\xi,\tau) \, d\xi. \tag{4.4.21}$$

Let us check that the function $u(z,t)$ in (4.4.21) coincides with the solution to the Cauchy problem with trivial initial data. Indeed,

$$u(z,t) = \int_0^\infty d\tau \int_{-\infty}^\infty F(\xi,\tau) U(z,t,\xi,\tau) \, d\xi = \frac{1}{2} \iint_{\Delta(x,t)} F(\xi,\tau) \, d\xi \, d\tau,$$

since the function $U(z,t,\xi,\tau)$ differs from zero only at (ξ,τ) such that $0 \le \tau \le |\xi - z|$, which condition defines the domain $\Delta(z,t)$ in (4.4.4).

Formula (4.4.21) is valid also for more general (compared to (4.4.1)) nonhomogeneous equations, e.g., for the nonhomogeneous version of equation (4.4.16).

The solution to the Cauchy problem with initial data of the general type (4.4.14) for the nonhomogeneous equation (4.4.1) has the form

$$u(z,t) = \frac{\partial}{\partial t} \int_{-\infty}^\infty \varphi(\xi) U(z,\xi,t) \, d\xi + \int_{-\infty}^\infty \psi(\xi) U(z,\xi,t) \, d\xi$$
$$+ \int_0^\infty d\tau \int_{-\infty}^\infty F(\xi,\tau) U(z,t,\xi,\tau) \, d\xi, \tag{4.4.22}$$

where

$$U(z,t,z_0,t_0) = U(z,z_0,t-t_0).$$

Last identity is the corollary from the fact that the coefficients of equation (4.4.1) are independent of the variable t.

Formula (4.4.21) holds true also for equations of the type

$$\frac{\partial^2 u}{\partial t^2} = Lu + F(z,t),$$

where L is a linear differential operator of elliptic type in the n-dimensional space $x = (x_1, \ldots, x_n)$ with, in general, variable coefficients, if the integrals in (4.4.22) are understood as integrals over the n-dimensional space.

Now we proceed with the statement of the inverse problems. In the domain $D = \{(z,t) \mid z \in \mathbb{R}, t > 0\}$, consider the differential equation

$$L_q u := \frac{\partial^2 u}{\partial t^2} - \frac{\partial^2 u}{\partial z^2} + q(z) u = 0, \quad (z,t) \in D, \tag{4.4.23}$$

Chapter 4. One dimensional inverse problems

and the Cauchy data

$$u(z,0) = 0, \qquad \frac{\partial u}{\partial t}(z,0) = \delta(z), \qquad z \in \mathbb{R}. \qquad (4.4.24)$$

Inverse Problem 4.4.1 (IP 4.4.1). *Find a continuous function $q(z)$, if the solution to (4.4.23), (4.4.24) is known at $z = 0$ along with the partial derivative with respect to z*

$$u(0,t) = f_1(t), \qquad \frac{\partial u}{\partial z}(0,t) = f_2(t), \qquad t \in \mathbb{R}_+. \qquad (4.4.25)$$

We will study this problem according to the classical scheme (Romanov, 1969), e.g., starting with the properties of the solution to (4.4.23), (4.4.24). Let us assume $-q(z)u = F(z,t)$ in the problem ((4.4.23), (4.4.24)) and apply formula (4.4.3). Using the basic property of the Dirac's mass, δ, cf. (4.4.6) (with $z_0 = 0$), and the d'Alembert formula, cf. (4.4.3) (with $\psi(z) = \delta(z)$, $F(z,t) = -q(z)u(z,t)$), one concludes that the generalized solution to (4.4.23), (4.4.24) is a piecewise continuous solution to the integral equation

$$u(z,t) = \frac{1}{2}\theta(t-|z|) - \frac{1}{2}\iint_{\Delta(z,t)} q(\xi)u(\xi,\tau)\,d\xi\,d\tau, \quad (z,t) \in D, \quad (4.4.26)$$

where $\theta(t)$ stands for the Heaviside function

$$\theta(t) = \begin{cases} 1, & t \geq 0, \\ 0, & t < 0. \end{cases}$$

Indeed, we have the equality

$$u(z,t) = \frac{1}{2}\int_{z-t}^{z+t} \delta(\xi)\,d\xi - \frac{1}{2}\iint_{\Delta(z,t)} q(\xi)u(\xi,\tau)\,d\xi\,d\tau.$$

But

$$\int_{z-t}^{z+t} \delta(\xi)\,d\xi = \begin{cases} 1, & 0 \in (z-t, z+t) \\ 0, & 0 \notin [z-t, z+t] \end{cases} = \theta(t-|z|),$$

and we come out with equality (4.4.26).

From (4.4.26) one obtains

$$u(z,t) \equiv 0, \qquad t < |z|, \qquad (z,t) \in D. \qquad (4.4.27)$$

Indeed, let us consider the domain $\Delta(z_1, t_1)$ for an arbitrary fixed point $(z_1, t_1) \in D$, $t_1 < |z_1|$. Then for points $(z, t) \in \Delta(z_1, t_1)$ formula (4.4.26) gives the homogeneous equation

$$u(z, t) = -\frac{1}{2} \iint_{\Delta(z,t)} q(\xi) u(\xi, \tau) \, d\xi \, d\tau, \qquad (z, t) \in \Delta(z_1, t_1). \qquad (4.4.28)$$

Let

$$Q = \max_{z_1 - t_1 \leq z \leq z_1 + t_1} |q(z)|,$$

$$U(t) = \max_{z_1 - (t_1 - t) \leq z \leq z_1 + (t_1 - t)} |u(z, t)|.$$

Then from (4.4.28) one gets the inequality

$$U(t) \leq \frac{1}{2} Q \int_0^t d\tau \int_{z+\tau-t}^{-\tau+z+t} |u(\xi, \tau)| \, d\xi \leq Q t_1 \int_0^t U(\tau) \, d\tau, \qquad t \in [0, t_1].$$

This inequality implies that $U(t) \equiv 0$, $t \in [0, t_1]$, and, therefore, $u(z, t) \equiv 0$ in $\Delta(z_1, t_1)$. Since the point (z_1, t_1) has been taken arbitrarily, it is equivalent to (4.4.27).

Equality (4.4.27) demonstrates that the support of the function $u(z, t)$ is concentrated inside the sector formed by the characteristics outgoing from the point $(0, 0)$. Let

$$D' = \{(z, t) \mid t > |z|\},$$

and henceforth consider the function $u(z, t)$ only in the domain D'. From (4.4.26) and (4.4.27) one obtains

$$u(z, t) = \frac{1}{2} - \frac{1}{2} \iint_{\square(z,t)} q(\xi) u(\xi, \tau) \, d\xi \, d\tau, \qquad (z, t) \in D'. \qquad (4.4.29)$$

Here $\square(z, t)$ denotes the rectangular domain, bounded by the characteristics passing through the points $(0, 0)$ and (z, t):

$$\square(z, t) = \{(\xi, \tau) \mid |\xi| \leq \tau \leq t - |z - \xi|\}.$$

Equality (4.4.29) can be rewritten as

$$u(z, t) = \frac{1}{2} - \frac{1}{2} \int_{(z-t)/2}^{(z+t)/2} d\xi \int_{|\xi|}^{t - |z - \xi|} q(\xi) u(\xi, \tau) \, d\xi \, d\tau, \qquad (z, t) \in D'. \qquad (4.4.30)$$

The following lemma is valid:

Lemma 4.4.1. *If for some point $(z_1, t_1) \in D'$ we have*

$$q(z) \in C[(z_1 - t_1)/2, (z_1 + t_1)/2],$$

then the solution to (4.4.29) exists and belongs to the class $C^2(\Box(z_1, t_1))$.

Equation (4.4.29) shows that on the boundary of the domain D' the function $u(z, t)$ is constant:

$$u(z, |z|) = 1/2. \qquad (4.4.31)$$

Now write the formulas for partial derivatives $\partial u/\partial t$, $\partial u/\partial z$, $\partial^2 u/\partial z\,\partial t$, $\partial^2 u/\partial t^2$ in the domain D' which will be useful for investigation of the inverse problem. By differentiating equality (4.4.30) we obtain

$$\frac{\partial u}{\partial t}(z, t) = -\frac{1}{2} \int_{(z-t)/2}^{(z+t)/2} q(\xi) u(\xi, t - |\xi - z|) \, d\xi, \qquad (4.4.32)$$

$$\frac{\partial u}{\partial z}(z, t) = -\frac{1}{2} \int_{(z-t)/2}^{(z+t)/2} q(\xi) u(\xi, t - |\xi - z|) \operatorname{sign}(\xi - z) \, d\xi, \qquad (4.4.33)$$

$$\frac{\partial^2 u}{\partial z\,\partial t}(z, t) = -\frac{1}{8}\left[q\!\left(\frac{z+t}{2}\right) - q\!\left(\frac{z-t}{2}\right)\right]$$
$$- \frac{1}{2}\int_{(z-t)/2}^{(z+t)/2} q(\xi)\,\frac{\partial u}{\partial t}(\xi, t - |\xi - z|)\operatorname{sign}(\xi - z)\,d\xi, \qquad (4.4.34)$$

$$\frac{\partial^2 u}{\partial t^2}(z, t) = -\frac{1}{8}\left[q\!\left(\frac{z+t}{2}\right) + q\!\left(\frac{z-t}{2}\right)\right]$$
$$- \frac{1}{2}\int_{(z-t)/2}^{(z+t)/2} q(\xi)\,\frac{\partial u}{\partial t}(\xi, t - |\xi - z|)\,d\xi. \qquad (4.4.35)$$

Lemma 4.1 and equalities (4.4.31)–(4.4.35) result in the following lemma, specifying the properties for the data of the inverse problem.

Lemma 4.4.2. *If $q(z) \in C[-T/2, T/2]$, $T > 0$, then $f_1 \in C^2[0, T]$, $f_2 \in C^1[0, T]$, and*

$$f_1(0) = 1/2, \qquad f_1'(0) = f_2(0) = f_2'(0) = 0, \qquad (4.4.36)$$

where the values of the functions f_1 and f_2 and their derivatives at $t = 0$ are understood as the right-hand side limits.

Considering equalities (4.4.34), (4.4.35) at the point $z = 0$, it is possible to obtain the integral equation of the second kind with respect to the function $q(z)$

$$q(z) = q_0(z) - 4\operatorname{sign}(z) \int_0^z q(\xi) \frac{\partial u}{\partial t}(\xi, 2|z| - \xi)\,d\xi, \qquad (4.4.37)$$

$$q_0(z) = -4[f_1''(2|z|) + f_1'(2|z|)\operatorname{sign}(z)].$$

Equations (4.4.37), (4.4.29), and (4.4.32) are the closed system of integral equations, in D', with respect to three functions u, $\partial u/\partial t$, and q. Using the system obtained, we can easily prove the following three theorems:

Theorem 4.4.1. *If at $t_0 > 0$ we have $f_1 \in C^2[0, t_0]$, $f_2 \in C^1[0, t_0]$, and conditions (4.4.36) are satisfied, then there exists $h \in (0, t_0/2)$ such that the solution $q(z)$ to the IP 4.4.1 in the functional space $C[-h, h]$ exists and is unique.*

Theorem 4.4.2. *Under conditions of Theorem 4.2.1 the function $q(z) \in C[-t_0/2, t_0/2]$ is uniquely defined by the information (4.4.25) for $t \in (0, t_0]$.*

Theorem 4.4.3. *Let $q(z)$, $\bar{q}(z)$ be two solutions to the IP 4.4.1 with the data f_1, f_2 and $\bar{f}_1$, $\bar{f}_2$, respectively. Then the estimate*

$$|q(z) - \bar{q}(z)| \le C(\|f_1'' - \bar{f}_1''\|_{C[0,t_0]} + \|f_2' - \bar{f}_2'\|_{C[0,t_0]}),$$

$$z \in [-t_0/2, t_0/2],$$

is valid where C depends only on the norms of the functions q, $\bar{q}$ in the space $C[-t_0/2, t_0/2]$ and the parameter t_0.

These theorems are rather well known and thus we address the reader to Romanov (1987) for proofs.

Now consider the case when the additional information on the solution to the Cauchy problem (4.4.23), (4.4.24) is given at the point $z_1 \ne 0$.

Inverse Problem 4.4.2 (IP 4.4.2). Find a continuous function $q(z)$, if the solution to (4.4.23), (4.4.24) is known at $z = z_1$ along with the partial derivative with respect to z

$$u(z_1, t) = f_1(t), \qquad \frac{\partial u}{\partial z}(z_1, t) = f_2(t), \qquad t \in \mathbb{R}_+. \qquad (4.4.38)$$

In this case the uniqueness property of solution to the inverse problem is essentially different. Let $z_1 > 0$. Then the following theorem characterizing the uniqueness of the solution to the inverse problem is valid:

Chapter 4. One dimensional inverse problems

$0 < z < |t|$ as the solution to a homogeneous integral equation. For points $(z, t) \in D''$, where
$$D'' = \{(z, t) \mid z \geq |t|\},$$
the domain of nontrivial integration in (4.5.7) is a rectangle
$$\Box'(z, t) := \{(\xi, \tau) \mid |\tau| \leq \xi \leq z - |t - \tau|\},$$
bounded by the characteristics outgoing from the points $(0, 0)$ and (z, t).

Let us introduce the notation
$$\tilde{w}(z, t) = w(z, t) - [\delta(t - z) + \delta(t + z)]/2. \tag{4.5.9}$$

The function $\tilde{w}(z, t)$ is piecewise continuous function (not the distribution) satisfying the equation
$$\tilde{w}(z, t) = \frac{1}{4} \theta(z - |t|) \left[\int_0^{(z+t)/2} q(\xi) \, d\xi + \int_0^{(z-t)/2} q(\xi) \, d\xi \right]$$
$$+ \frac{1}{2} \iint_{\Box'(z,t)} q(\xi) \tilde{w}(\xi, \tau) \, d\xi \, d\tau, \qquad z \in \mathbb{R}_+, \tag{4.5.10}$$

which follows from (4.5.7).

Equation (4.5.10) shows that the function $\tilde{w}(z, t)$ equals zero outside D'', is even with respect to the argument t, has continuous first derivatives with respect to its arguments inside D'', and, besides,
$$\tilde{w}(z, z) = \frac{1}{4} \int_0^z q(\xi) \, d\xi, \qquad z \in \mathbb{R}_+. \tag{4.5.11}$$

Using the function $w(z, t)$, we can represent the solution to (4.5.1) under conditions (4.5.3) as
$$u(z, t) = \int_{-\infty}^\infty f(\tau) w(z, t - \tau) \, d\tau = \int_{-\infty}^\infty f(t - \tau) w(z, \tau) \, d\tau. \tag{4.5.12}$$

Indeed, from (4.5.4) and (4.5.5), using the property (4.4.7) of the delta-function, one obtains
$$L_q u = \int_{-\infty}^\infty f(\tau) L_q w(z, t - \tau) \, d\tau = 0,$$
$$u(0, t) = \int_{-\infty}^\infty f(\tau) \delta(t - \tau) \, d\tau = f(t), \qquad \frac{\partial u}{\partial z}(0, t) = 0.$$

Using the definition (4.5.9) and identity (4.5.8), we can rewrite formula (4.5.12) as follows:

$$u(z,t) = \frac{1}{2}[f(t-z) + f(t+z)] + \int_{-z}^{z} f(t-\tau)\tilde{w}(z,\tau)\,d\tau. \qquad (4.5.13)$$

If we consider now those points (z,t), where identity (4.5.8) is valid, then

$$\frac{1}{2}[f(t-z) + f(t+z)] + \int_{-z}^{z} f(t-\tau)\tilde{w}(z,\tau)\,d\tau = 0, \qquad z > |t|. \qquad (4.5.14)$$

At any fixed point $z \in \mathbb{R}_+$ relation (4.5.14) can be regarded as an integral equation of the first kind with respect to the function $\tilde{w}(z,t)$, $t \in (-z,z)$. Since at $t = \tau$ the kernel of this equation has a jump: $f(+0) = 1/2$, $f(-0) = -1/2$, therefore, its differentiation with respect to t results in the Fredholm's equation of the second kind

$$\tilde{w}(z,t) + \int_{-z}^{z} f'(t-\tau)\tilde{w}(z,\tau)\,d\tau = -\frac{1}{2}[f'(t-z) + f'(t+z)],$$
$$z \in \mathbb{R}_+, \qquad t \in (-z,z). \qquad (4.5.15)$$

Note that the kernel $f'(t-\tau)$ of this equation is continuous and symmetric (latter is due to the fact that function $f'(t)$ is even). Taking into account the evenness of the function $\tilde{w}(z,t)$ with respect to the argument t, we can replace equation in (4.5.15) by the following one

$$\tilde{w}(z,t) + \int_{0}^{z} [f'(t-\tau) + f'(t+\tau)]\tilde{w}(z,\tau)\,d\tau = -\frac{1}{2}[f'(t-z) + f'(t+z)],$$
$$z \in \mathbb{R}_+, \qquad t \in [0,z), \qquad (4.5.16)$$

which is more convenient for investigation.

Equation (4.5.16) is equivalent to (4.5.14) under the additional condition that the function $\tilde{w}(z,t)$ is even with respect to the argument t. Indeed, any solution to (4.5.14), which is even with respect to t, evidently satisfies (4.5.16). On the other hand, if we extend any solution to (4.5.16) in an even way (with respect to t) to the interval $t \in (-z,0)$, then the function obtained will satisfy (4.5.15) (obtained by differentiating (4.5.14) with respect to t) and the condition

$$\int_{-z}^{z} f(-\tau)\tilde{w}(z,\tau)\,d\tau = 0,$$

Chapter 4. One dimensional inverse problems

which coincide with equality (4.5.14) at $t = 0$ (remember that $f(t)$ is an odd function). Therefore, any solution to (4.5.16), when extended in an even way with respect to t, is a solution to (4.5.14). Thus, the equivalence of (4.5.14) and (4.5.16) is established.

Assume that for any $z \in \mathbb{R}_+$ equation (4.5.15) possesses a unique solution in the class of continuous functions. In this case such solution defines the function $\tilde{w}(z, t)$ which is continuous in the domain D'' (with the even extension of $\tilde{w}$ to that part of the domain D'' where $t \in \mathbb{R}_-$ is taken into account). Since the right-hand side of (4.5.16) is a function continuously differentiable in D'', while the kernel is piecewise continuously differentiable, it follows that the solution to (4.5.16) is continuously differentiable in D''. In particular, equation (4.5.16) gives $(\tilde{w}/\partial t)(z, +0) = 0$ (remember that for $t \neq 0$ the function $f(t) \in C^2$ is an odd function, hence $f''(t)$ is an odd function also, and in this case $f''(t) + f''(-t) = 0$). Therefore, the even extension to the domain $t \in \mathbb{R}_-$ is carried out with preserving the continuity of partial derivatives.

From equality (4.5.11) one obtains

$$q(z) = 4 \frac{d}{dz} \tilde{w}(z, z), \qquad z \in \mathbb{R}_+. \qquad (4.5.17)$$

In the same way one can obtain the equation for negative values of z, which differs from (4.5.16) only by the upper limit of integration, where z is replaced by $|z|$. Thus, we have $\tilde{w}(-z, t) = \tilde{w}(z, t)$ and, as a result, $q(-z) = q(z)$.

Thus, the following algorithm for solving the inverse problem arises: (4.5.16) is solved for every z, and then $q(z)$ is found by formula (4.5.17). It follows from the theory of Fredholm's equations that for small values of z equation (4.5.16) is uniquely solvable. In this case the following statement holds true:

Lemma 4.5.1. *If a solution to the IP 4.4.1 exists, then (4.5.16) is uniquely solvable for any $z \in \mathbb{R}_+$.*

One of the corollaries of Lemma 4.5.1 is the theorem on the uniqueness of the solution to the inverse problem. Also the following lemma, which is converse to Lemma 4.5.1 is valid:

Lemma 4.5.2. *If for some $T > 0$ we have $f(t) \in C^2(0, T]$, $f(+0) = 1/2$, $f'(+0) = 0$; and equation (4.5.16) with $f(t)$ extended onto the interval $t \in [-T, 0)$ in the odd way is uniquely solvable for $z \in (0, T/2]$, then a solution to the IP 4.4.1 exists on the segment $[-T/2, T/2]$.*

Lemmas 4.5.1 and 4.5.2 result in the following theorem:

Theorem 4.5.1. *The IP 4.4.1 is uniquely solvable on the segment $[-T/2, T/2]$ in the class of continuous functions if and only if the function $f(t)$ satisfies the conditions*

1) $f(t) \in C^2[0,T]$, $f(+0) = 1/2$, $f'(0) = 0$;
2) *the integral equation (4.5.16) in which $f(-t) = -f(t)$, $t \in [0,T]$, is uniquely solvable at any $z \in (0, T/2)$.*

4.6. DETERMINATION OF THE PIECE-WISE CONSTANT COEFFICIENT FOR WAVE EQUATION

Here we study a simple version of model inverse problem for Lamé system. Suppose (see the previous paragraphs) that only one component of the displacement vector is nontrivial and this component depends only on time and depth. Moreover, the only parameter to be determined is the longitudinal wave velocity, which is a piecewise-constant function of the depth. Such assumption is often used in applications. Indeed, layered structure occurs in the Earth crust then the sediment structures are considered. On the other hand, any given function (from the reasonable functional space, could be approximated with the help of piecewise-constant function. Finally, the so-called "semi-analytic" approach is used to develop cost effective inversion algorithm in such case (see Fatianov and Mikhailenko, 1988; Fatianov, 1990; and the references therein).

Consider the second initial boundary value problem for the wave equation. We have:

$$\frac{\partial^2 u}{\partial t^2} = v^{-2}(z) \frac{\partial^2 u}{\partial z^2}, \qquad (z,t) \in \mathbb{R}_+ \times \mathbb{R}_+, \qquad (4.6.1)$$

with initial data

$$u\big|_{t<0} = 0, \qquad z \in \mathbb{R}_+, \qquad (4.6.2)$$

and boundary condition

$$u_z\big|_{z=0} = H(t), \qquad z \in \overline{\mathbb{R}}_+. \qquad (4.6.3)$$

Here the function $v(z)$ is supposed to be piecewise-constant and can have discontinuities at the points $a_1 < a_2 < \ldots < a_k$, i.e., assuming $a_0 = 0$, we

Chapter 4. One dimensional inverse problems

can write the equality

$$v(z) = \begin{cases} v_m, & z \in (a_{m-1}, a_m), \ m = 1, \ldots, k, \\ v_{k+1}, & z > a_k, \end{cases} \qquad (4.6.4)$$

where $v_m = $ const. In such approximation the coefficient $v(z)$ is reciprocal to the longitudinal wave velocity (Yakhno, 1990).

Remark 4.6.1. The model above could be regarded as the simulation of the layered media, where each interval (a_{m-1}, a_m) corresponds to the material layer with certain properties.

At points of the discontinuity a_m of the coefficient $v^{-2}(z)$ we add to (4.6.1)–(4.6.3) the following matching conditions:

$$[u]_{z=a_m} = [v^{-2}(z)u_z]_{z=a_m} = 0, \qquad m = 1, \ldots, k, \qquad (4.6.5)$$

where $[\cdot]$ stands for the jumps; i.e., $[f]_{z=z^*} = f(z^* + 0) - f(z^* - 0)$.

Note that for application under study the matching conditions above are nothing but physical requirement that the displacement vector and material flux remain continuous then passing the layer boundary.

The additional information to determine the coefficient $v(z)$, often used in applications, is as follows

$$\Phi(\omega) = -\int_{-\infty}^{\infty} u_t(0, t) e^{-i\omega t} \, dt, \qquad \omega \in \mathbb{R}. \qquad (4.6.6)$$

After using the formal Fourier transform, from (4.6.1)–(4.6.5) we obtain the problem: for $z \in \mathbb{R} \setminus \{a_1, \ldots, a_k\}$ and $\omega \in \mathbb{R}$ solve equations:

$$\frac{\partial^2 u}{\partial z^2} + \omega^2 v^2(z) u = 0,$$

$$u_z(0, \omega) = h(\omega), \qquad u_z - i\omega v u \to 0 \quad \text{as} \quad z \to +\infty, \qquad (4.6.7)$$

$$[u]_{z=a_m} = [v^{-2} u_z]_{z=a_m} = 0, \qquad m = 1, \ldots, k.$$

Prior to formulate the inverse problem we stress attention on the following points.

1) In the sequel we suppose that the quantity

$$\tau = v_m(a_m - a_{m-1}), \qquad m = 1, \ldots, k \qquad (4.6.8)$$

is known and is such that $\tau > \pi/(\omega_2 - \omega_1)$. Note that the fact the number τ is the same for each m is the part of assumption.

2) The dual variable ω of the Fourier transform has clear physical meaning of the oscillation frequency (Aki and Richards, 1980).

Remark 4.6.2. From the point of view of applications the above value τ is nothing more but the travelling time of a seismic wave through each of the media layers. Note that the assumption (4.6.8) corresponds to the so-called Goupillaud hypothesis and is widely used in geophysics (Aki and Richards, 1980).

Remark 4.6.3. Physical measurements are far nontrivial for the low frequencies (that is for small values of ω). Therefore, it is important to use the overinformation in (4.6.9) only at the bounded interval of variation of ω, chosen away from zero. This point differs our approach from the most available algorithms, which are derived under assumption that additional information is given for all frequencies, $\omega \in \mathbb{R}$ (see, e.g., Yakhno, 1990).

So, the inverse problem we are interested in is:

Inverse Problem 4.6.1 (IP 4.6.1). *Given the function $\Phi(\omega)$ for $\omega \in [\omega_1, \omega_2]$, find the coefficient $v(z)$ of the form (4.6.5) such that*

$$\Phi(\omega) = i\omega u(0, \omega), \qquad \omega \in [\omega_1, \omega_2] \tag{4.6.9}$$

where $u(0, \omega)$ is the solution to the problem (4.6.7).

Below we are going to describe mostly the general ideas omitting, therefore, all the technical details. For the complete proofs we refer the reader to the paper by Lavrentiev, Jr. (1992).

Equation in (4.6.7) is ODE with *constant* coefficient at each interval $a_{m-1} < z < a_m$. Thus, using well know form of general solution to such equation along with the boundary and matching conditions it is not difficult to conclude that

$$\Phi(\omega) = \frac{h(\omega)}{c_1} \frac{P_1^{k-1} e^{-i\omega\tau} + P_2^{k-1} e^{i\omega\tau}}{P_1^{k-1} e^{-i\omega\tau} - P_2^{k-1} e^{i\omega\tau}}, \tag{4.6.10}$$

where $P_j^{k-1}(e^{-i\omega\tau}, e^{i\omega\tau})$ are homogeneous polynomials of degree $k-1$.

Assume that $h(\omega) \neq 0$ for $\omega \in [\omega_0, \omega_0 + \pi/\tau] \subset [\omega_1, \omega_2]$ and consider the following Fourier coefficients of the ratio $\Phi(\omega)/h(\omega)$:

$$\varphi_m = \int_{\omega_0}^{\omega_0 + \pi/\tau} \frac{\Phi(\omega)}{h(\omega)} e^{-2im\omega\tau} d\omega, \qquad m = 0, \ldots, k. \tag{4.6.11}$$

A number of rather simple (that is algebraic instead of the ones in terms of differential equations) connections among the numbers φ_m and the parameters v_m of inverse problem have been discovered by Lavrentiev, Jr. (1992). These relations are based on calculation of the above integrals.

Chapter 4. One dimensional inverse problems 143

To calculate integrals in (4.6.11) we use the substitution of variables $\zeta = e^{2i\omega\tau}$ ($d\zeta = 2i\tau\zeta\, d\omega$), an one-to-one mapping of semi-interval $[\omega_0, \omega_0 + \pi/\tau)$ of the ω axis onto the unit circle $|\zeta| = 1$ of the complex ζ plane (integration is carried out in the direction of increasing $\arg\zeta$).

Multiplying the numerator and denominator in the right hand side of (4.6.10) by the function $e^{ki\omega\tau} = \zeta^{k/2}$, we obtain that the function $c_1\Phi(\omega)h^{-1}(\omega)$, which in terms of the variable ζ we denote by $F_k(\zeta)$, is the ratio of two polynomials of degree k:

$$F_k(\zeta) = \frac{f_0^{(k)} + f_1^{(k)}\zeta + \ldots + f_k^{(k)}\zeta^k}{g_0^{(k)} + g_1^{(k)}\zeta + \ldots + g_k^{(k)}\zeta^k}. \quad (4.6.12)$$

Here the subscript of the function $F_k(\zeta)$ and the superscripts of the coefficients $f_j^{(k)}$ and $g_j^{(k)}$ denote the number of layers k in the problem under study.

In this notation, definition (4.6.11) takes the form

$$\varphi_m = \frac{1}{2i\tau c_1} \int_{|\zeta|=1} F_k(\zeta) \frac{d\zeta}{\zeta^{m+1}}. \quad (4.6.13)$$

Integrals of the form (4.6.13) (remember that the integrand is rational function) can be directly calculated with the help of the residues theory (cf. the theory of complex valued functions). The simplest formulae occur in the case that all the roots in the denominator of the function $F_k(\zeta)$ lie outside the unit circle of the (complex) ζ plane.

All further formulae are valid under the assumptions indicated above that the polynomial $g_0^{(k)} + g_1^{(k)}\zeta + \ldots + g_k^{(k)}\zeta^k$ does not have roots in the circle $|\zeta| \leq 1$. A sufficient condition for the validity of this assumption by virtue of the Rouche theorem (Lavrentiev and Shabat, 1977), is the inequality

$$|g_0^{(k)}| > \sum_{j=1}^{k} |g_j^{(k)}|.$$

Thus we assume that the functions in the integrand in (4.6.13) within the bounded contour of integration of the disk $|\zeta| < 1$ have the only singularity – a pole of order $m + 1$ at the point $\zeta = 0$, respectively. This enables us to use residues theorem (Lavrentiev and Shabat, 1977), on the basis of which

$$\varphi_m = \frac{\pi}{\tau c_1} \operatorname*{Res}_{\zeta=0} \frac{F_k(\zeta)}{\zeta^{m+1}} = \frac{1}{m!} \frac{\pi}{\tau c_1} F_k^{(m)}(0). \quad (4.6.14)$$

Values of derivatives $F_k^{(m)}(0)$ are closely related with coefficients v_m (Lavrentiev, Jr., 1992), which results in the following

Theorem 4.6.1. *Suppose that*

i) *equalities (4.6.8) are satisfied;*

ii) *the problem data in (4.6.3) are such that Fourier transform $h(\omega)$ of the function $H(t)$ does not vanish $h(\omega) \neq 0$ for $\omega \in [\omega_0, \omega_0 + \pi/\tau] \subset [\omega_1, \omega_2]$;*

iii) *the denominator in (4.6.12) as a polynomial of $\zeta = e^{2i\omega\tau}$ does not have roots in the disk $|\zeta| \leq 1$.*

Then for the function $F_k(\zeta)$ from (4.6.12), constructed from the solution to the problem (4.6.7), the following formulae hold:

$$F_k^{(m)}(0) = m! \left[2\gamma_m \prod_{p=1}^{m-1}(1 - \gamma_p^2) - \sum_{p=1}^{m-1} \frac{h_p^m}{(m-p)!} F_k^{(m-p)}(0) \right]. \quad (4.6.15)$$

Here the numbers γ_m and h_p^m are determined by

$$\gamma_m = \frac{v_{m+1} - v_m}{v_{m+1} + v_m}, \quad m = 1, \ldots, k;$$

$$h_p^m = h_{m-1}^m h_{m-p-1}^{m-1} + h_p^{m-1}, \quad m = 2, \ldots, k, \quad p = 0, \ldots, m-2,$$

$$h_{m-1}^m = -\gamma_{m-1}, \quad m = 2, \ldots, k, \quad h_0^1 = h_m^m = 0. \quad (4.6.16)$$

Proof of the theorem is given by Lavrentiev, Jr. (1992). Basic tools are: general solution of ODE with *constant* coefficients and mathematical induction. Here we only indicate the basis formulae and list the sequence of lemmas, which lead to the complete proof.

First we note, that by direct calculations it is easy to obtain the following representations for "one-layer" (that is the case of homogeneous media with *constant* coefficient $v(z)$) and "two-layered" media:

$$F_1(\zeta) = \frac{1 + \gamma_1 \zeta}{1 - \gamma_1 \zeta}, \quad F_2(\zeta) = \frac{1 + \gamma_1 \zeta + \gamma_1 \gamma_2 \zeta + \gamma_2 \zeta^2}{1 - \gamma_1 \zeta + \gamma_1 \gamma_2 \zeta - \gamma_2 \zeta^2}. \quad (4.6.17)$$

Then, by induction, it is easy to establish that

$$F_k(\zeta) = \frac{Q_k^1(\zeta) + \gamma_k Q_k^2(\zeta)}{Q_k^3(\zeta) + \gamma_k Q_k^4(\zeta)}, \quad (4.6.18)$$

Chapter 4. One dimensional inverse problems

where the coefficients of the polynomials $Q_k^j(\zeta)$ depend on $\gamma_1, \ldots, \gamma_{k-1}$ and are independent of γ_k.

The following statements could be established one by one with the help of induction proposed by Lavrentiev, Jr. (1992):

Lemma 4.6.1. *If the function $F_k(\zeta)$ from (4.6.18) is known, then F_{k+1} (remind that the subscript indicates the number of layers in the considered problem) is obtained from it after replacement of the coefficient γ_k by the function*
$$\frac{\gamma_k + \gamma_{k+1}\zeta}{1 + \gamma_k \gamma_{k+1}\zeta}.$$

Lemma 4.6.2. *For calculation of the derivatives of the rational function $F_k(\zeta)$ in (4.6.12) for $\zeta = 0$ the following formulas hold:*
$$F_k^{(m)}(0) = m! \left[\frac{f_m^{(k)}}{g_0^{(k)}} - \sum_{p=1}^{m} \frac{g_p^{(k)}}{g_0^{(k)}} \frac{1}{(m-p)!} F_k^{(m-p)}(0) \right]. \tag{4.6.19}$$

Lemma 4.6.3. *For arbitrary, natural $m \leq k$ the following equation hold:*
$$F_k^{(m)}(0) = F_{k+1}^{(m)}(0). \tag{4.6.20}$$

Lemma 4.6.4. *Coefficients h_p^m in (4.6.15) satisfy relations (4.6.16).*

Lemma 4.6.5. *The following formula holds:*
$$F_k^{(k)}(0) = k! \left(2\gamma_k \prod_{p=1}^{k-1} (1 - \gamma_p^2) + G_k \right), \tag{4.6.21}$$

where the G_k depend on $\gamma_1, \ldots, \gamma_{k-1}$ and do not depend on the γ_k.

The above Theorem can be reformulated in terms of *recursive algorithm*, solving IP 4.6.1:

Algorithm of solution to IP 4.6.1

1. Calculate $k+1$ integrals
$$\varphi_m = \int_{\omega_0}^{\omega_0 + \pi/\tau} \frac{\Phi(\omega)}{h(\omega)} e^{-2im\omega\tau} \, d\omega, \qquad m = 0, \ldots, k.$$

2. Find the numbers γ_m, $m = 1, \ldots, k$, recursively

$$\gamma_1 = \frac{\varphi_1}{2\varphi_0}, \qquad \gamma_m = \frac{1}{2\varphi_0} \frac{\varphi_m + \sum_{p=1}^{m-1} h_p^m \varphi_{m-p}}{\prod_{p=1}^{m-1}(1 - \gamma_p^2)}, \qquad (4.6.22)$$

where the coefficients h_p^m are calculated according to (4.6.16).

3. Solve the IP 4.6.1 by

$$v_1 = \frac{\pi}{\tau \varphi_0}, \qquad v_{m+1} = \frac{1 + \gamma_m}{1 - \gamma_m} v_m, \qquad m = 1, \ldots, k. \qquad (4.6.23)$$

Chapter 5.

Inverse problems for the coupled Maxwell and Lamé systems

The interaction of electromagnetic fields with deformable media is a subject of many theoretical and experimental investigations in the field of continuum mechanics and geophysics in the recent decades. For description of simple enough interactions, the theories of magnetohydrodynamics (Alfvén, 1950, pp. 76–91; Landau and Lifshitz, 1984), electroelasticity (Balakirev and Gilinskii, 1982; Parton and Kudrjavtsev, 1988), and magnetoelasticity (Knopoff, 1955; Tamm, 1976) were developed. These theories are, basically, a combination (without introducing new conceptions) of objects and phenomena considered in continuum mechanics and electrodynamics.

Investigation of more complex electromagnetoelastic interactions in a continuous medium requires to consider more complex models. For a more profound acquaintance with the modern state of the theory of electromagnetoelastic interactions the reader is referred to, e.g., Eringen and Maugin (1990), Maugin (1988), Pride (1994).

The aim of this chapter is to study some inverse problems of reconstructing the electromagnetic and elastic characteristics of a medium connected with electromagnetoelastic interactions. The models considered here are based on the simplest variants of combination of the Lamé and Maxwell equations.

Let us give a brief characteristic of basic types of electromagnetoelastic interactions. It is well known that when an electrical-conducting elastic body oscillates in an electromagnetic field, variations of the electrical and magnetic fields are observed as a result of this motion. Similar processes are also observed when seismic waves propagate in the Earth's crust. Variations of elastic and electromagnetic fields arising in this case are called *electromagnetoelastic waves*. Such waves contain a certain information about electromagnetic and elastic parameters of the medium. In this case, as a rule, the following types of electromagnetoelastic interactions are distinguished:

a) the interaction based on the *electrokinetic properties of a medium*;

b) the interaction based on the *piezoelectric properties of a medium*;

c) the interaction based on the *velocity effect*.

Whereas, for example, the electrokinetic effect is connected with local interactions of elastic waves with the flow of the pore liquid, the velocity effect is based on moving of particles in external electromagnetic field. In the case of geophysical or seismological problems the velocity effect leads to so-called *seismomagnetic effect*.

We need to make several remarks about the defining equations of electromagnetoelasticity. We consider an electromagnetoelastic medium from the viewpoint of linear elasticity connected with electrodynamics through the motion of particles. For our purposes, it is more convenient to define the equations separately in each particular case.

The first result on inverse problems of electromagnetoelasticity was obtained, apparently, by Burdakova and Yakhno (1989). A systematic study of inverse problems for the system of electromagnetoelasticity equations was begun in the nineties of the last century. In this connection, we note the works by Avdeev, Goryunov, and Priimenko (1996a, 1996b, 1997); Avdeev, Goryunov, Soboleva, and Priimenko (1999); Imomnazarov (1998); Kabanikhin and Lorenzi (1999); Klimenko (1995); Lavrentiev, Jr. and Priimenko (1995); Lorenzi and Priimenko (1996); Lorenzi and Romanov (1993); Merazhov and Yakhno (1995); Priimenko and Vishnevsky (2002, 2003a, 2003b); Romanov (1995a, 1995b); Yakhno (1998); and Yakhno and Merazhov (2000).

This chapter is not a complete exposition of this field. Its main task is to give a conception of the range of problems and methods for their solution.

5.1. ONE-DIMENSIONAL INVERSE PROBLEM OF ELECTROMAGNETOELASTICITY IN THE CASE OF THE SEISMOMAGNETIC EFFECT

In this section, we give some results of solution of inverse problems for the equations of electromagnetoelasticity in the case of seismomagnetic inter-

Chapter 5. Coupled Maxwell and Lamé systems

action. In our exposition, we follow the work by Lavrentiev, Jr. and Priimenko (1995).

Let $\mathbb{R}^3$ and $\mathbb{R}^3_\pm$ be the three-dimensional Euclidean space of points $x = (x_1, x_2, x_3)$ and the half-spaces $\{x \in \mathbb{R}^3 \mid \pm x_3 > 0\}$, respectively. We assume that the half-space $\mathbb{R}^3_-$ corresponds to the Earth's atmosphere with constant electromagnetic parameters $\varepsilon_0 > 0$, $\mu_0 > 0$, and $\sigma_0 = 0$. We shall describe the process of propagation of electromagnetic waves in $\mathbb{R}^3_-$ with the help of the following Maxwell system:

$$\varepsilon_0 \frac{\partial \boldsymbol{E}}{\partial t} = \operatorname{rot} \boldsymbol{H},$$
$$\mu_0 \frac{\partial \boldsymbol{H}}{\partial t} = -\operatorname{rot} \boldsymbol{E}, \qquad \operatorname{div} \boldsymbol{H} = 0. \tag{5.1.1}$$

At the same time, in the half-space $\mathbb{R}^3_+$ corresponding to the Earth's crust, we observe the interaction of electromagnetic and elastic waves described by the following system of electromagnetoelasticity:

$$\boldsymbol{J} = \operatorname{rot} \boldsymbol{H},$$
$$\frac{\partial \boldsymbol{B}}{\partial t} = -\operatorname{rot} \boldsymbol{E}, \qquad \operatorname{div} \boldsymbol{B} = 0, \tag{5.1.2}$$

$$\rho \frac{\partial^2 \boldsymbol{U}}{\partial t^2} = \operatorname{Div} T, \tag{5.1.3}$$

where

$$\operatorname{Div} T = \Big(\sum_{j=1}^{3} \frac{\partial}{\partial x_j} T_{ij} \Big)_{i=1}^{3}. \tag{5.1.4}$$

For the stress tensor T, the vectors of electric and magnetic induction $\boldsymbol{D}$ and $\boldsymbol{B}$, and the vector of the electric current density $\boldsymbol{J}$, we have the following defining relations:

$$T = \lambda \operatorname{tr} S \cdot I + 2\varkappa S,$$
$$\boldsymbol{B} = \mu \boldsymbol{H}, \qquad \boldsymbol{J} = \sigma \Big(\boldsymbol{E} + \mu \frac{\partial \boldsymbol{U}}{\partial t} \times \boldsymbol{H}^0 \Big), \tag{5.1.5}$$

where S is the strain tensor defined by the formula

$$S_{ij} = \frac{1}{2} \Big(\frac{\partial U_i}{\partial x_j} + \frac{\partial U_j}{\partial x_i} \Big), \qquad i, j = 1, 2, 3, \tag{5.1.6}$$

and I is the unit matrix of order 3×3. In the above formulas, $\rho, \lambda, \varkappa$: $\mathbb{R}^3_+ \to \mathbb{R}_+$ are the density of the inhomogeneous medium (the Earth's crust) and the Lamé coefficients, respectively; $\varepsilon, \mu, \sigma : \mathbb{R}^3_+ \to \mathbb{R}_+$ are the dielectric and magnetic permeabilities and electrical conductivity of the crust, respectively; and $\boldsymbol{H}^0$ is a constant vector characterizing the Earth's magnetic field. Equations (5.1.1)–(5.1.6) were derived from the general model of equations of electromagnetoelasticity (Eringen and Maugin, 1990) for the case of an isotropic inhomogeneous earth by means of linearization in a neighborhood of the constant solution $(\boldsymbol{U}^0, \boldsymbol{E}^0, \boldsymbol{H}^0) = (\boldsymbol{0}, \boldsymbol{0}, \boldsymbol{H}^0)$ and discarding the nonlinearities of order higher than the first. We also assume that the propagation of electromagnetic oscillations in the Earth's crust $\mathbb{R}^3_+$ is described by the quasi-stationary approximation of the Maxwell equations. Knopoff has shown (Knopoff, 1955) that in this case it is natural to neglect the reverse effect of the electromagnetic field on the process of propagation of elastic waves. We assume that the following matching conditions are fulfilled on the Earth's surface $\Gamma = \{x \in \mathbb{R}^3 \mid x_3 = 0\}$:

$$[\mu H_k] = [E_k] = 0, \qquad k = 1, 2, \tag{5.1.7}$$

where the symbol $[\,\cdot\,]$ denotes the jump of a function across the surface where the coefficients of the problem have breaks. As a direct problem for the system (5.1.1)–(5.1.7), we consider the Cauchy problem

$$(\boldsymbol{U}, \boldsymbol{E}, \boldsymbol{H})_{t<0} \equiv 0 \tag{5.1.8}$$

in the case where the electromagnetic oscillations arise under the action of a vertical elastic source concentrated on the Earth's surface Γ:

$$T_{k3}\big|_\Gamma = \delta_{k3}\delta(t, x_1, x_2), \qquad k = 1, 2, 3. \tag{5.1.9}$$

Here $\delta(\,\cdot\,)$ is the delta-function concentrated at the point $(t, x_1, x_2) = (0, 0, 0)$ and δ_{k3} is the Kronecker symbol. Since the propagation of electromagnetic waves in the Earth's crust $\mathbb{R}^3_+$ is described by the quasi-stationary approximation of the Maxwell equations, therefore, to single out the unique solution it is natural to assume that the following radiation conditions at infinity are fulfilled:

$$|\boldsymbol{E}| \to 0, \quad |\boldsymbol{H}| \to 0 \quad \text{for} \quad |x| \to \infty. \tag{5.1.10}$$

Thus, the direct problem consists in finding the vectors $(\boldsymbol{U}, \boldsymbol{E}, \boldsymbol{H})$ satisfying relations (5.1.1)–(5.1.10) under the condition that the properties of the medium described by the coefficients ε_0, μ_0, ε, μ, σ, ρ, λ, and $\varkappa$ are known. The inverse problem is understood as the problem of finding the electromagnetic and elastic parameters from equations (5.1.1)–(5.1.10) if we know some additional information about the behavior of certain components of the vectors $(\boldsymbol{U}, \boldsymbol{E}, \boldsymbol{H})$ on the Earth's surface Γ. As an illustration of the proposed approach we shall consider one of the simplest variants of the above inverse problem.

Statement of a simplified problem

We shall focus our attention on one of the simplest versions of equations (5.1.1)–(5.1.10). In this version, however, the main properties of more general models are kept. Assume that all the coefficients in the problem depend only on one variable x_3; in the sequel, we shall denote it by z. For this case, consider the system

$$\frac{\partial^2 u}{\partial t^2} = v^2(z) \frac{\partial^2 u}{\partial z^2}, \qquad (z,t) \in \mathbb{R}_+ \times \mathbb{R}, \qquad (5.1.11a)$$

$$\frac{\partial e}{\partial t} = c^2(z) \frac{\partial^2 e}{\partial z^2} + \mu h^0 \frac{\partial^2 u}{\partial t^2}, \qquad (z,t) \in \mathbb{R}_+ \times \mathbb{R}, \qquad (5.1.11b)$$

$$\left.\frac{\partial u}{\partial z}\right|_{z=0} = b\delta(t), \qquad t \in \mathbb{R}, \qquad (5.1.11c)$$

$$\left.\left(\frac{\partial}{\partial z} - c_0 \left(\frac{c_0}{c}\right)^2 \frac{\partial}{\partial t}\right) e\right|_{z=0} = 0, \qquad t \in \mathbb{R}, \qquad (5.1.11d)$$

$$\lim_{z \to \infty} e = 0, \qquad t \in \mathbb{R}, \qquad (5.1.11e)$$

$$(e, u)|_{t<0} \equiv 0, \qquad z \in \mathbb{R}_+. \qquad (5.1.11f)$$

Here

$$u = \operatorname{Re} F_{x_1 x_2}(U_3)\big|_{\nu_1 = \nu_2 = 0}, \qquad e = \operatorname{Re} F_{x_1 x_2}(E_1)\big|_{\nu_1 = \nu_2 = 0}$$

are the values of the generalized Fourier transforms with respect to the variables x_1, x_2 of the functions U_3 and E_1, respectively, at $\nu_1 = \nu_2 = 0$; ν_1, ν_2 are the variables dual to x_1, x_2; $v(z) = \sqrt{(\lambda + 2\varkappa)/\rho}$ is the velocity of longitudinal elastic waves; $c(z) = \sqrt{\sigma\mu}$ is the characteristic of the process of diffusion of electromagnetic waves in the Earth's crust; $c_0 = (\varepsilon_0\mu_0)^{-1/2}$ is the velocity of electromagnetic waves in the Earth's atmosphere; $b = (\lambda(0) + 2\varkappa(0))^{-1}$; and h^0 is the number characterizing the constant magnetic field of the Earth.

System (5.1.11) is obtained by applying the Fourier transform to the system (5.1.1)–(5.1.10). Since the coefficients of the problem are known functions in $\mathbb{R}^3_-$, the upper half-space was excluded from consideration. For this, we replaced its effect by the corresponding boundary condition (5.1.11d).

Let us introduce the following class of functions:

Definition 5.1.1. We shall say that the functions $c(z)$ and $v(z)$ belong to the class $\mathfrak{M}$ if

a) there exist positive numbers c_m, v_m, z_n, and z'_n, $m = 1, 2, \ldots, k+1$, $n = 1, 2, \ldots, k$, such that

$$c(z) = \begin{cases} c_m, & z \in (z_{m-1}, z_m), \quad m = 1, 2, \ldots, k, \\ c_{k+1}, & z > z_k, \end{cases}$$

$$v(z) = \begin{cases} v_m, & z \in (z'_{m-1}, z'_m), \quad m = 1, 2, \ldots, k, \\ v_{k+1}, & z > z'_k, \end{cases}$$

where $z'_0 = z_0 = 0$;

b) there exist positive constants τ_v and τ_c such that

$$\begin{aligned} \tau_v &= \frac{z'_1 - z'_0}{v_1} = \frac{z'_2 - z'_1}{v_2} = \cdots = \frac{z'_k - z'_{k-1}}{v_k}, \\ \tau_c &= \frac{z_1 - z_0}{c_1} = \frac{z_2 - z_1}{c_2} = \cdots = \frac{z_k - z_{k-1}}{c_k}. \end{aligned} \quad (5.1.12)$$

We can now formulate the inverse problem.

Inverse Problem 5.1.1 (IP 5.1.1). We know the functions $u_0(t)$ and $e_0(t)$ and the constants τ_v and τ_c. It is required to find the functions $c(z), v(z) \in \mathfrak{M}$, i.e., the set $\{c_m, v_m, z_n, z'_n; m = 1, \ldots, k+1; n = 1, \ldots, k\}$, such that

$$u|_{z=0} = u_0(t), \qquad e|_{z=0} = e_0(t), \quad t \in \mathbb{R}_+, \qquad (5.1.13)$$

where u, e is the solution of the problem (5.1.11).

Inverse problem for a system of ordinary differential equations

To illustrate a possible approach to solution of Inverse Problem 5.1.1 we consider the following system. In equations (5.1.11), we formally replace the derivative $\partial e/\partial t$ by the derivative $\partial^2 e/\partial t^2$, i.e., instead of equation (5.1.11b) we consider the equation

$$\frac{\partial^2 e}{\partial t^2} = c^2(z) \frac{\partial^2 e}{\partial z^2} + \mu h^0 \frac{\partial^2 u}{\partial t^2}.$$

To construct an algorithm for solving the so-modified problem we use the generalized Fourier transform with respect to the variable t. For conve-

Chapter 5. Coupled Maxwell and Lamé systems

nience, we represent our problem in the form of two subproblems:

$$\frac{d^2 u}{dz^2} + \omega^2 v^{-2} u = 0, \qquad z \in \mathbb{R}_+, \tag{5.1.14a}$$

$$\left.\frac{du}{dz}\right|_{z=0} = h_u(\omega), \tag{5.1.14b}$$

$$\frac{du}{dz} - i\omega v^{-1} u \xrightarrow{z \to \infty} 0, \tag{5.1.14c}$$

$$[u]\big|_{z=z'_m} = \left[v^2 \frac{du}{dz}\right]\bigg|_{z=z'_m} = 0, \qquad m = 1, 2, \ldots, k, \tag{5.1.14d}$$

$$\frac{d^2 e}{dz^2} + \omega^2 c^{-2} e = \omega^2 c^{-2} \mu h^0 u, \qquad z \in \mathbb{R}_+, \tag{5.1.15a}$$

$$\left.\frac{de}{dz}\right|_{z=0} = h_e(\omega), \tag{5.1.15b}$$

$$\frac{de}{dz} - i\omega c^{-1} e \xrightarrow{z \to \infty} 0, \tag{5.1.15c}$$

$$[e]\big|_{z=z_m} = \left[c^2 \frac{de}{dz}\right]\bigg|_{z=z_m} = 0, \qquad m = 1, 2, \ldots, k. \tag{5.1.15d}$$

Here $h_u(\omega)$ is the Fourier transform with respect to the variable t of the corresponding boundary condition on the function u in equations (5.1.11) and the function $h_e(\omega)$ will be defined later. Note that the problems (5.1.14) and (5.1.15) can be solved successively.

Inverse Problem 5.1.2 (IP 5.1.2). *It is required to find a function* $v(z) \in \mathfrak{M}$ *such that*

$$u(0, \omega) = \Phi_u(\omega), \tag{5.1.16}$$

where $\Phi_u(\omega)$ *is a given function and* $u(z, \omega)$ *is the solution of the problem* (5.1.14).

If we know the solution of the IP 5.1.2, we can also determine the function $c(z)$ because in this case the right-hand side of equation (5.1.15a) will be a known function. Thus we can consider the following inverse problem.

Inverse Problem 5.1.3 (IP 5.1.3). *We know the functions* $\Phi_e(\omega)$ *and* $u(z, \omega)$. *It is required to find a function* $c(z) \in \mathfrak{M}$ *such that*

$$e(0, \omega) = \Phi_e(\omega), \tag{5.1.17}$$

where $e(z, \omega)$ *is the solution of the problem* (5.1.15).

Remark 5.1.1. Since the function $e(0,\omega)$ is assumed to be known, therefore, we can also calculate the value of the derivative $\partial e/\partial t$ for $z = 0$. Thus, under the assumption that we know the constant c_1, the function $h_e(\omega)$ in equations (5.1.15) is the Fourier transform with respect to the variable t of the expression

$$c_0 \left(\frac{c_0}{c_1}\right)^2 \frac{\partial e(0,t)}{\partial t}$$

and thus can be calculated.

Summarizing the above considerations we can say that the original IP 5.1.1 was decomposed in the IP 5.1.2 and the IP 5.1.3. Using the results of solving the corresponding inverse problems (see Section 4.6) we can propose the following recursive algorithm for solution of the IP 5.1.2:

a) determine the numbers

$$\varphi_m = \int_{\omega_0}^{\omega_0 + \pi/\tau_v} h_u^{-1}(\omega)\, \Phi_u(\omega) \exp(-2im\omega\tau_v)\, d\omega, \qquad (5.1.18)$$

$$h_n^m = h_{m-1}^m \cdot h_{m-n-1}^{m-1} + h_n^{m-1}, \quad n = 0, 1, \ldots, m-2, \quad m = 2, 3, \ldots, k,$$

$$h_{m-1}^m = -\gamma_{m-1}, \qquad h_0^1 = h_m^m = 0, \qquad \gamma_1 = \varphi_1/2\varphi_0,$$

$$\gamma_m = \frac{1}{2\varphi_0}\left(\varphi_m + \sum_{n=1}^{m-1} h_n^m \varphi_{m-n}\right) \Big/ \prod_{n=1}^{m-1}(1 - \gamma_n^2), \qquad m = 2, 3, \ldots, k,$$

where ω_0 is an arbitrary positive number;

b) determine the numbers

$$v_1 = \frac{\pi}{\tau_v \varphi_0}, \qquad v_{m+1} = \frac{1+\gamma_m}{1-\gamma_m} v_m, \qquad m = 1, 2, \ldots, k. \qquad (5.1.19)$$

Using formulas (5.1.12), find the numbers z'_m, $m = 1, 2, \ldots, k$. This completes solution of the IP 5.1.2.

To construct a solution of the IP 5.1.3 we consider the sets

$$(z_{m-1}, z_m) \cap (z'_{n-1}, z'_n), \qquad m = 1, 2, \ldots, k, \quad n = 1, 2, \ldots, k.$$

Since the function $u_n(z,\omega)$, $n = 1, 2, \ldots, k$, on the right-hand side of the differential equation (5.1.15a) has the form

$$u_n(z,\omega) = u_n^1 + u_n^2 = B_1^n \exp\left(\frac{i\omega z}{v_n}\right) + B_2^n \exp\left(-\frac{i\omega z}{v_n}\right),$$

therefore, we can use the following representation for the solution of the problem (5.1.15):
$$e_m(z,\omega) = e_m^0 + e_m^1 + e_m^2, \qquad (5.1.20)$$

where
$$e_m^0 = A_1^m \exp\left(\frac{i\omega z}{c_m}\right) + A_2^m \exp\left(-\frac{i\omega z}{c_m}\right)$$

is the general solution to the homogeneous differential equation (5.1.15a) and e_m^j, $j = 1, 2$, are its particular solutions in the cases where the function u on the right-hand side is replaced by u_m^j, $j = 1, 2$, respectively. These particular solutions can be represented in the form
$$e_m^j = r_m^j \exp\left\{(-1)^{j+1}\frac{i\omega z}{v_n}\right\}, \quad j = 1, 2, \qquad (5.1.21)$$

where
$$r_m^j = -\frac{\omega^2 v_n^2 h^0 \mu B_j^m}{\omega^2(v_n^2 - c_m^2)} = -\mu h^0 B_j^m \left(1 - \left(\frac{c_m}{v_n}\right)^2\right)^{-1}, \quad j = 1, 2. \qquad (5.1.22)$$

It is known that for the most of physical materials the ratio c_m/v_n is less than 10^{-1}, therefore, we can neglect the term $(c_m/v_n)^2$ in (5.1.22) and set
$$r_m^j = -\mu h^0 B_j^m, \qquad j = 1, 2. \qquad (5.1.23)$$

Taking into account the fact that we can represent the boundary condition (5.1.15b) in the form of recursive relationships with B_j^m (Section 4.6), analogous formulas can be derived from relations (5.1.15), (5.1.18)–(5.1.21), and (5.1.23). Thus, the algorithm of solution of the IP 5.1.2 can also be used for solution of the IP 5.1.3. There is only one difference in the use of the additional information (5.1.17). We have

$$e_1(0,\omega) = e_1^0(0,\omega) + e_1^1(0,\omega) + e_1^2(0,\omega)$$
$$= A_1^1 + A_2^1 - \mu h^0(B_1^1 + B_2^1) = \Phi_e(\omega),$$

$$\left.\frac{de_1}{dz}\right|_{z=0} = i\omega c_1^{-1}(A_1^1 - A_2^1) - i\omega v_1^{-1}\mu h^0(B_1^1 - B_2^1) = h_e(\omega).$$

Since
$$B_1^1 + B_2^1 = \Phi_u(\omega), \qquad i\omega v_1^{-1}(B_1^1 - B_2^1) = h_u(\omega),$$

therefore, the algorithm of solution of the IP 5.1.3 can be formulated as follows:

a) find the numbers

$$\varphi'_m = \int_{\omega_1}^{\omega_1+\pi/\tau_c} (h')^{-1}(\omega)\, \Phi'(\omega) \exp(-2i m\omega\tau_c)\, d\omega, \quad (5.1.24)$$

where ω_1 is some positive number and

$$h'(\omega) = h_e(\omega) + \mu h^0 h_u(\omega), \qquad \Phi'(\omega) = \Phi_e(\omega) + \mu h^0 \Phi_u(\omega);$$

b) find the numbers

$$c_1 = \frac{\pi}{\tau_c \varphi'_0}, \qquad c_{m+1} = c_m \frac{1+\gamma_m}{1-\gamma_m}, \qquad m = 1, 2, \ldots, k, \quad (5.1.25)$$

where γ_m, $m = 1, 2, \ldots, k$, are determined by formulas (5.1.18) with φ_m replaced by φ'_m. The numbers z_m are found by formulas (5.1.12). Thus, the IP 5.1.1 is solved by formulas (5.1.18), (5.1.19), (5.1.24), and (5.1.25).

Remark 5.1.2. In fact, we have formulated an algorithm for solving the IP 5.1.2 and the IP 5.1.3 of reconstructing the parameters of a medium from additional information of the form (5.1.13) about the solution of the corresponding direct problems (5.1.14) and (5.1.15). Using the relationship between the solutions of differential equations of parabolic and hyperbolic types (see Chapter IV, Section 4.4), we can formulate a similar algorithm for solving the original IP 5.1.1. Since the corresponding formulas are very bulky, we dwelt on demonstrating this approach in a more simple case where we have formally changed the type of the differential equation for $e(z, t)$ from parabolic to hyperbolic.

5.2. INVERSE PROBLEMS OF ELECTROMAGNETOELASTICITY IN THE CASE OF PIEZOELECTRIC INTERACTION

In this section, following the work (Yakhno and Merazhov, 2000), we give some results of solution of inverse problems (and the corresponding direct ones) for the system of electromagnetoelasticity in the case of piezoelectric mechanism of interaction between electromagnetic and elastic waves.

In this case, the defining relations for the stress tensor T and the components of electric and magnetic inductions are as follows (Parton and Kudrjavtsev, 1988):

$$T_{ij} = \sum_{k,l=1}^{3} c_{ijkl} \frac{\partial U_k}{\partial x_l} - \sum_{k=1}^{3} e_{kij} E_k, \qquad i,j = 1,2,3,$$

$$D_j = \sum_{k=1}^{3} \varepsilon_{jk} E_k - \sum_{k,l=1}^{3} e_{jkl} \frac{\partial U_k}{\partial x_l}, \qquad j = 1,2,3, \tag{5.2.1}$$

$$B_j = \mu H_j, \qquad j = 1,2,3,$$

where $c_{ijkl} = c_{ijkl}(x)$ are the elastic moduli, $e_{ijk} = e_{ijk}(x)$ are the piezoelectric moduli, $\varepsilon_{ij} = \varepsilon_{ij}(x)$ are the dielectric moduli, and $\mu = \mu(x)$ is the magnetic permeability.

It is convenient to write the system of electromagnetoelasticity in the Gauss system of units:

$$\rho \frac{\partial^2 \boldsymbol{U}}{\partial t^2} = \operatorname{Div} T, \qquad \frac{1}{c} \frac{\partial \boldsymbol{D}}{\partial t} = \operatorname{rot} \boldsymbol{H},$$

$$\frac{1}{c} \frac{\partial \boldsymbol{B}}{\partial t} = -\operatorname{rot} \boldsymbol{E}, \qquad \operatorname{div} \boldsymbol{B} = 0. \tag{5.2.2}$$

Further, by analogy with Dieulesaint and Royer (1974), we set

$$c_{ijkl} = c_{jikl} = c_{ijlk} = c_{klij}, \qquad e_{ijk} = e_{jik}, \qquad \varepsilon_{ij} = \varepsilon_{ji}.$$

The relationship between elastic and electromagnetic processes is defined by the piezoelectric moduli of the medium.

In the general case, equations (5.2.1) and (5.2.2) connect three types of "slow" elastic waves with two types of "fast" electromagnetic waves. But if we study the propagation of "slow" waves, we can neglect the quantities of the order of v/c and v/c^2 (here v is the velocity of propagation of elastic waves and c is that of electromagnetic waves). In this case, we can formally set that the velocity of light c is equal to infinity, and system (5.2.2) splits into two groups. The equations

$$\rho \frac{\partial^2 \boldsymbol{U}}{\partial t^2} = \operatorname{Div} T,$$

$$\operatorname{div} \boldsymbol{D} = 0, \qquad \operatorname{rot} \boldsymbol{E} = 0, \qquad \boldsymbol{E} = -\operatorname{grad} \Phi \tag{5.2.3}$$

form the first group, and the second group consists of the equations

$$\operatorname{rot} \boldsymbol{H} = 0, \qquad \operatorname{div} \boldsymbol{B} = 0. \tag{5.2.4}$$

The symmetry of the stress tensor reduces the number of independent elastic moduli from 81 to 21. If we write $c_{\alpha\beta} = c_{ijkl}$, where $\alpha = (i,j)$ and $\beta = (k,l)$ according to the rule $(1,1) \to 1$, $(2,2) \to 2$, $(3,3) \to 3$, $(2,3) = (3,2) \to 4$, $(3,1) = (1,3) \to 5$, and $(1,2) = (2,1) \to 6$, then the matrix of independent elastic moduli can be written in the form of a symmetric 6×6 matrix since the order in the pair of indices (i,j) plays no role and there exist only six different pair combinations. The complete set of characteristics of an electromagnetoelastic medium is given in the literature (Parton and Kudrjavtsev, 1988), in the form of a matrix

$$\begin{pmatrix} c_{\alpha\beta} \; (6 \times 6) & e_{\alpha k} \; (6 \times 3) \\ e_{k\alpha} \; (3 \times 6) & \varepsilon_{ij} \; (3 \times 3) \end{pmatrix}.$$

In this section, we give the results of solution of some direct and inverse problems for anisotropic media with cubic structure and the matrix of characteristics in the form

$$\begin{pmatrix} \begin{bmatrix} c_{11} & c_{12} & c_{12} \\ c_{12} & c_{22} & c_{12} \\ c_{12} & c_{12} & c_{33} \end{bmatrix} & \mathbb{O} & \mathbb{O} \\ \mathbb{O} & \begin{bmatrix} c_{44} & 0 & 0 \\ 0 & c_{44} & 0 \\ 0 & 0 & c_{44} \end{bmatrix} & \begin{bmatrix} e_{14} & 0 & 0 \\ 0 & e_{14} & 0 \\ 0 & 0 & e_{14} \end{bmatrix} \\ \mathbb{O} & \begin{bmatrix} e_{14} & 0 & 0 \\ 0 & e_{14} & 0 \\ 0 & 0 & e_{14} \end{bmatrix} & \begin{bmatrix} \varepsilon & 0 & 0 \\ 0 & \varepsilon & 0 \\ 0 & 0 & \varepsilon \end{bmatrix} \end{pmatrix}.$$

Formulation of direct problems and the theorem on existence and uniqueness of their solutions

In the domain $\mathbb{R}_+^3 \times \mathbb{R}$, consider the system of differential equations (5.2.3) with the following initial and boundary conditions:

$$\boldsymbol{U}\big|_{t<0} \equiv 0, \tag{5.2.5}$$

$$\boldsymbol{E}\big|_{t<0} \equiv 0, \tag{5.2.6}$$

$$T_{i3}\big|_{x_3=+0} = f_i(x_1, x_2, t), \quad i = 1, 2, 3, \tag{5.2.7}$$

$$\Phi\big|_{x_3=+0} = g(x_1, x_2, t), \quad \frac{\partial \Phi}{\partial x_3}\bigg|_{x_3=+0} = 0. \tag{5.2.8}$$

Chapter 5. Coupled Maxwell and Lamé systems

Equalities (5.2.3) and (5.2.5)–(5.2.8) for anisotropic media of cubic structure can be written as

$$\rho \frac{\partial^2 U_1}{\partial t^2} = \frac{\partial}{\partial x_3}\left(c_{44}\frac{\partial U_1}{\partial x_3}\right) + c_{12}\frac{\partial^2 U_3}{\partial x_1 \partial x_3} + \frac{\partial}{\partial x_3}\left(c_{44}\frac{\partial U_3}{\partial x_1}\right) + c_{44}\frac{\partial^2 U_1}{\partial x_2^2}$$
$$+ c_{11}\frac{\partial^2 U_1}{\partial x_1^2} + (c_{12}+c_{44})\frac{\partial^2 U_2}{\partial x_1 \partial x_2} + \frac{\partial}{\partial x_3}\left(e_{14}\frac{\partial \Phi}{\partial x_2}\right) + e_{14}\frac{\partial^2 \Phi}{\partial x_2 \partial x_3}, \quad (5.2.9)$$

$$\rho \frac{\partial^2 U_2}{\partial t^2} = \frac{\partial}{\partial x_3}\left(c_{44}\frac{\partial U_2}{\partial x_3}\right) + c_{12}\frac{\partial^2 U_3}{\partial x_2 \partial x_3} + \frac{\partial}{\partial x_3}\left(c_{44}\frac{\partial U_3}{\partial x_2}\right) + c_{44}\frac{\partial^2 U_2}{\partial x_1^2}$$
$$+ c_{11}\frac{\partial^2 U_2}{\partial x_2^2} + (c_{12}+c_{44})\frac{\partial^2 U_1}{\partial x_1 \partial x_2} + \frac{\partial}{\partial x_3}\left(e_{14}\frac{\partial \Phi}{\partial x_1}\right) + e_{14}\frac{\partial^2 \Phi}{\partial x_1 \partial x_3}, \quad (5.2.10)$$

$$\rho \frac{\partial^2 U_3}{\partial t^2} = \frac{\partial}{\partial x_3}\left(c_{11}\frac{\partial U_3}{\partial x_3}\right) + c_{44}\frac{\partial^2 U_1}{\partial x_1 \partial x_3} + \frac{\partial}{\partial x_3}\left(c_{12}\frac{\partial U_1}{\partial x_1}\right) + c_{44}\frac{\partial^2 U_2}{\partial x_2 \partial x_3}$$
$$+ \frac{\partial}{\partial x_3}\left(c_{12}\frac{\partial U_2}{\partial x_2}\right) + c_{44}\frac{\partial^2 U_3}{\partial x_2^2} + e_{14}\frac{\partial^2 \Phi}{\partial x_1 \partial x_2} + e_{14}\frac{\partial^2 \Phi}{\partial x_2 \partial x_1}, \quad (5.2.11)$$

$$\frac{\partial}{\partial x_1}\left(e_{14}\frac{\partial U_2}{\partial x_3}\right) - \frac{\partial}{\partial x_1}\left(\varepsilon\frac{\partial \Phi}{\partial x_1}\right) + \frac{\partial}{\partial x_1}\left(e_{14}\frac{\partial U_3}{\partial x_2}\right) + \frac{\partial}{\partial x_2}\left(e_{14}\frac{\partial U_1}{\partial x_3}\right)$$
$$+ \frac{\partial}{\partial x_2}\left(e_{14}\frac{\partial U_3}{\partial x_1}\right) - \frac{\partial}{\partial x_2}\left(\varepsilon\frac{\partial \Phi}{\partial x_2}\right) + \frac{\partial}{\partial x_3}\left(e_{14}\frac{\partial U_1}{\partial x_2}\right)$$
$$+ \frac{\partial}{\partial x_3}\left(e_{14}\frac{\partial U_2}{\partial x_1}\right) - \frac{\partial}{\partial x_3}\left(\varepsilon\frac{\partial \Phi}{\partial x_3}\right) = 0, \quad (5.2.12)$$

$$\left(c_{44}\frac{\partial U_3}{\partial x_1} + c_{44}\frac{\partial U_1}{\partial x_3} + e_{14}\frac{\partial \Phi}{\partial x_2}\right)\bigg|_{x_3=+0} = f_1(x_1, x_2, t), \quad (5.2.13)$$

$$\left(c_{44}\frac{\partial U_3}{\partial x_2} + c_{44}\frac{\partial U_2}{\partial x_3} + e_{14}\frac{\partial \Phi}{\partial x_1}\right)\bigg|_{x_3=+0} = f_2(x_1, x_2, t), \quad (5.2.14)$$

$$\left(c_{12}\frac{\partial U_1}{\partial x_1} + c_{12}\frac{\partial U_2}{\partial x_2} + c_{11}\frac{\partial U_3}{\partial x_3}\right)\bigg|_{x_3=+0} = f_3(x_1, x_2, t), \quad (5.2.15)$$

$$U_i\big|_{t<0} \equiv 0, \quad \frac{\partial \Phi}{\partial x_i}\bigg|_{t<0} \equiv 0, \quad i=1,2,3, \quad (5.2.16)$$

$$\Phi\big|_{x_3=+0} = g(x_1, x_2, t), \quad \frac{\partial \Phi}{\partial x_3}\bigg|_{x_3=+0} = 0. \quad (5.2.17)$$

Assume that the vector-function $q = (c_{11}, c_{12}, c_{44}, e_{14})$ belongs to the class

$$\mathfrak{M} = \{c_{11}(x_3),\ c_{12}(x_3),\ c_{44}(x_3),\ e_{14}(x_3) \mid c_{44} > 0,\quad c_{11} > c_{12},$$
$$c_{11} + 2c_{12} > 0,\quad c'_{11}(+0) = 0,\quad c'_{44}(+0) = 0;$$
$$c_{11}, c_{44} \in C^2(\overline{\mathbb{R}}_+),\quad c_{12} \in C(\overline{\mathbb{R}}_+),\quad e_{14} \in C^1(\overline{\mathbb{R}}_+)\}$$

and that $\rho = \rho(x_3)$ and $\varepsilon = \varepsilon(x_3)$ are positive functions from the class $C^2(\overline{\mathbb{R}}_+)$; we also assume that $\rho'(+0) = 0$. As in the previous section, in this case it is also convenient to single out the variable x_3 and denote it by z. For solution of inverse problems we shall need the properties of solutions of the corresponding direct problems.

Direct Problem 5.2.1. *Let $f_j = -\theta(t)\mathrm{e}^{\mathrm{i}\nu x_1}/2$, $j = 1,2,3$, where ν is a parameter; and let $g \equiv 0$. It is required to find the vector-function $(\boldsymbol{U}, \Phi)$ which satisfies equalities (5.2.9)–(5.2.17). The solution of this problem will be denoted by $(\boldsymbol{U}^\mathrm{I}, \Phi^\mathrm{I})$.*

Direct Problem 5.2.2. *Let $f_j \equiv 0$, $j = 1,2,3$, and $g = \theta(t)\mathrm{e}^{\mathrm{i}\nu x_1}$, where ν is a parameter. It is required to find the vector-function $(\boldsymbol{U}, \Phi)$ which satisfies equalities (5.2.9)–(5.2.17). The solution of this problem will be denoted by $(\boldsymbol{U}^\mathrm{II}, \Phi^\mathrm{II})$.*

For Direct Problems 5.2.1 and 5.2.2, the following existence and uniqueness theorem holds.

Theorem 5.2.1. *Let T be a fixed positive number. Then under the above assumptions the solutions of Direct Problems 5.2.1 and 5.2.2 exist and are unique in the class of vector-functions $\Pi(T)$:*

$$\Pi(T) = \{(\boldsymbol{U}, \Phi) \mid U_j = \mathrm{e}^{\mathrm{i}\nu x_1} u_j(z,t,\nu),\ \Phi = \mathrm{e}^{\mathrm{i}\nu x_1}\varphi(z,t,\nu),$$
$$u_j, \varphi \in C^1(\mathbb{R}; C(\Omega(T))),\quad j = 1,2,3\},$$

$$\Omega(T) = \{(z,t) \mid z \in \overline{\mathbb{R}}_+,\ t \in [0,T]\}.$$

Moreover, for the solutions $\boldsymbol{U}^\mathrm{I}$ and $\boldsymbol{U}^\mathrm{II}$, the traces

$$\boldsymbol{U}^\mathrm{I}(0,0,t,0),\quad \boldsymbol{U}^\mathrm{II}(0,0,t,0),\quad \frac{\partial \boldsymbol{U}^\mathrm{I}}{\partial \nu}(0,0,t,\nu)\bigg|_{\nu=0},\quad \frac{\partial \boldsymbol{U}^\mathrm{II}}{\partial \nu}(0,0,t,\nu)\bigg|_{\nu=0}$$

in $C^1(\mathbb{R})$ are defined correctly.

Chapter 5. Coupled Maxwell and Lamé systems

The complete proof of Theorem 5.2.1 can be found in the original paper by Yakhno and Merazhov (2000). Here we only outline briefly the basic idea of the proof. We shall seek the solutions of the direct problems in terms of the functions u_j, $j = 1, 2, 3$, and φ. It is easy to see that the system (5.2.9)–(5.2.17) written in terms of these functions splits into two systems. The first system consists of the equalities

$$\rho \frac{\partial^2 u_1}{\partial t^2} = \frac{\partial}{\partial z}\left(c_{44} \frac{\partial u_1}{\partial z}\right) + i\nu c_{12} \frac{\partial u_3}{\partial z} + i\nu \frac{\partial}{\partial z}(c_{44} u_3) - \nu^2 c_{11} u_1,$$

$$\rho \frac{\partial^2 u_3}{\partial t^2} = \frac{\partial}{\partial z}\left(c_{11} \frac{\partial u_3}{\partial z}\right) + i\nu c_{44} \frac{\partial u_1}{\partial z} + i\nu \frac{\partial}{\partial z}(c_{12} u_1) - \nu^2 c_{44} u_3,$$

$$\left(i\nu c_{44} u_3 + c_{44} \frac{\partial u_1}{\partial z}\right)\bigg|_{z=+0} = -\frac{1}{2}\theta(t), \qquad (5.2.18)$$

$$\left(i\nu c_{12} u_1 + c_{11} \frac{\partial u_3}{\partial z}\right)\bigg|_{z=+0} = -\frac{1}{2}\theta(t),$$

$$u_j\big|_{t<0} \equiv 0, \quad j = 1, 3.$$

The second system is

$$\rho \frac{\partial^2 u_2}{\partial t^2} = \frac{\partial}{\partial z}\left(c_{44} \frac{\partial u_2}{\partial z}\right) - \nu^2 c_{44} u_2 + i\nu \frac{\partial}{\partial z}(\varepsilon\varphi) + i\nu e_{14} \frac{\partial \varphi}{\partial z},$$

$$-\frac{\partial}{\partial z}\left(\varepsilon \frac{\partial \varphi}{\partial z}\right) + \varepsilon \nu^2 \varphi + i\nu e_{14} \frac{\partial u_2}{\partial z} + i\nu \frac{\partial}{\partial z}(e_{14} u_2) = 0,$$

$$\left(c_{44} \frac{\partial u_2}{\partial z} + i\nu e_{14} \varphi\right)\bigg|_{z=+0} = -\frac{1}{2}\theta(t), \qquad (5.2.19)$$

$$u_2\big|_{t<0} \equiv 0, \quad \varphi\big|_{t<0} \equiv 0, \quad \frac{\partial \varphi}{\partial z}\bigg|_{t<0} \equiv 0,$$

$$\varphi\big|_{z=+0} = 0, \quad \frac{\partial \varphi}{\partial z}\bigg|_{z=+0} = 0.$$

One can show that the system (5.2.18), (5.2.19) can be written in the form of a system of linear Volterra integral equations of second kind which is solvable uniquely in the functional classes specified above.

The inverse problem and results of its investigation

The subject of our further research is the following problem.

Inverse Problem 5.2.1 (IP 5.2.1). *Let ρ and ε be given positive constants. Find the vector-function $q \in \mathfrak{M}$ such that the components of the* solutions of Direct Problems 5.2.1 and 5.2.2 satisfy the equalities

$$U_1^I(x_1, z, t, \nu)\big|_{x_1=0, z=+0, \nu=+0} = h_1(t),$$
$$U_3^I(x_1, z, t, \nu)\big|_{x_1=0, z=+0, \nu=+0} = h_2(t),$$
$$\frac{\partial U_1^I}{\partial \nu}(x_1, z, t, \nu)\big|_{x_1=0, z=+0, \nu=+0} = h_3(t), \quad (5.2.20)$$
$$\frac{\partial U_2^{II}}{\partial \nu}(x_1, z, t, \nu)\big|_{x_1=0, z=+0, \nu=+0} = h_4(t),$$

where $h_j(t)$, $j = 1, 2, 3, 4$, are known functions for $t \in [0, T]$.

The following theorems hold true for the IP 5.2.1.

Theorem 5.2.2. *Let T and Z be fixed positive numbers. Assume that the data of the inverse problem (5.2.20) satisfy the conditions*

$$h_j(t) \in C^3[0, T], \quad h_j(+0) > 0, \quad j = 1, 2,$$
$$h_j(t) \in C^2[0, T], \quad h_j(+0) = 0, \quad j = 3, 4, \quad (5.2.21)$$
$$\max\{-h_3'(+0), 2h_3'(+0)\} < \frac{h_1(0)}{2\rho(0)[h_1(0) + h_2(0)]}.$$

Then there exists a vector-function $q \in \mathfrak{M}$ which is a unique solution to the IP 5.2.1 for $z \in [0, Z]$.

Theorem 5.2.3. *Let T be a fixed positive number and let the functions $h_j(t)$, $j = 1, 2, 3, 4$, satisfy conditions (5.2.21). Then there exists a number $Z^* > 0$ such that there exists a vector-function $q \in \mathfrak{M}$ which is a unique solution to the IP 5.2.1 for $z \in [0, Z^*]$ and which complies with the information $h_j(t)$, $t \in [0, T]$, $j = 1, 2, 3, 4$.*

Let us define the class of functions $\mathfrak{M}_0(m, M, Z) \subset \mathfrak{M}$ as follows:

$$\mathfrak{M}_0(m, M, Z) = \{q \in \mathfrak{M} \mid \|c_{11}\|_2(Z) \leq M, \ c_{11} \geq m, \ \|c_{44}\|_2(Z) \leq M,$$
$$c_{44} \geq m, \ \|c_{12}\|(Z) \leq M, \ \|e_{14}\|_1(Z) \leq M\},$$

where m, M, T, and Z are fixed positive numbers and

$$\|\cdot\|(Z) = \|\cdot\|_{C[0,Z]}, \quad \|\cdot\|_s(Z) = \|\cdot\|_{C^s[0,Z]}.$$

Chapter 5. Coupled Maxwell and Lamé systems

Theorem 5.2.4. *Let the vector-functions $q, q^* \in \mathfrak{M}_0(m, M, Z)$ be the solutions to the IP 5.2.1 corresponding to the informations $h_j(t)$ and $h_j^*(t)$, $j = 1, 2, 3, 4$, respectively, for $t \in [0, T]$. Then the following estimate is valid:*

$$\|q - q^*\|(Z_0) \leq C\Big[\sum_{k=1}^{2} \|h_k - h_k^*\|_3(T) + \|h_3 - h_3^*\|_2(T_1) + \|h_4 - h_4^*\|_1(T)\Big], \quad (5.2.22)$$

where C is a certain constant depending on m, M, and T; $m \leq M$;

$$T_1 = \frac{M}{m}T; \quad Z = \frac{T}{2}\sqrt{\frac{M}{m}}; \quad Z_0 = \frac{T}{2}\sqrt{\frac{m}{M}}.$$

The scheme of solution of the inverse problem

Let us show that solution of the IP 5.2.1 can be reduced to successive solution of certain auxiliary inverse problems for scalar hyperbolic equations. From relations (5.2.18) for the function $u_1^0(z, t) \equiv u_1^I(z, t, \nu)|_{\nu=+0}$ we find

$$\rho \frac{\partial^2 u_1^0}{\partial t^2} = \frac{\partial}{\partial z}\Big(c_{44} \frac{\partial u_1^0}{\partial z}\Big), \quad (z, t) \in \mathbb{R}_+ \times \mathbb{R},$$

$$u_1^0\big|_{t<0} \equiv 0, \quad c_{44} \frac{\partial u_1^0}{\partial z}\Big|_{z=+0} = -\frac{1}{2}\theta(t). \quad (5.2.23)$$

Equalities (5.2.20) and (5.2.23) allow us to consider the following inverse problem.

Inverse Problem 5.2.2 (IP 5.2.2). *Find the function $c_{44} \in \mathfrak{M}_1$, where*

$$\mathfrak{M}_1 = \{c(z) \in C^2(\mathbb{R}_+), \ c > 0, \ c'(+0) = 0\},$$

such that the solution of the problem (5.2.23) satisfies the condition

$$u_1^0(z, t)\big|_{z=+0} = h_1(t), \quad (5.2.24)$$

where $h_1(t)$, $t \in [0, T]$, is a known function and ρ is a given positive constant.

Relations (5.2.18) for the function $u_3^0(z, t) \equiv u_3^I(z, t, \nu)|_{\nu=+0}$ imply

$$\rho \frac{\partial^2 u_3^0}{\partial t^2} = \frac{\partial}{\partial z}\Big(c_{11} \frac{\partial u_3^0}{\partial z}\Big), \quad (z, t) \in \mathbb{R}_+ \times \mathbb{R},$$

$$u_3^0\big|_{t<0} \equiv 0, \quad c_{11} \frac{\partial u_3^0}{\partial z}\Big|_{z=+0} = -\frac{1}{2}\theta(t). \quad (5.2.25)$$

Equalities (5.2.20) and (5.2.25) allow us to consider the following inverse problem.

Inverse Problem 5.2.3 (IP 5.2.3). *Find the function $c_{11} \in \mathfrak{M}_1$ such that the solution of the problem (5.2.25) satisfies the condition*

$$u_3^0(z,t)\big|_{z=+0} = h_2(t), \qquad (5.2.26)$$

where $h_2(t)$, $t \in [0,T]$, is a known function and ρ is a given positive constant.

Relations (5.2.18) for the function $u_1^1(z,t) \equiv \dfrac{\partial u_1^{\mathrm{I}}(z,t,\nu)}{\partial \nu}\bigg|_{\nu=+0}$ imply

$$\rho \frac{\partial^2 u_1^1}{\partial t^2} = \frac{\partial}{\partial z}\left(c_{44}\frac{\partial u_1^1}{\partial z}\right) + ic_{12}\frac{\partial u_3^0}{\partial z} + i\frac{\partial}{\partial z}(c_{44}u_3^0), \quad (z,t) \in \mathbb{R}_+ \times \mathbb{R},$$

$$u_1^1\big|_{t<0} \equiv 0, \quad \left(c_{44}\frac{\partial u_1^1}{\partial z} + ic_{44}u_3^0\right)\bigg|_{z=+0} = 0. \qquad (5.2.27)$$

Equalities (5.2.27) allow us to consider the following inverse problem.

Inverse Problem 5.2.4 (IP 5.2.4). *Find the function $c_{12} \in C(\overline{\mathbb{R}_+})$ such that the solution of the problem (5.2.27) satisfies the condition*

$$u_1^1(z,t)\big|_{z=+0} = h_3(t), \qquad (5.2.28)$$

where $h_3(t)$, $t \in [0,T]$, is a known function; ρ is a given positive constant; and $u_3^0(z,t)$ is the solution of the problem (5.2.25).

Relations (5.2.19) for the function $u_2^1(z,t) \equiv \dfrac{\partial u_2^{\mathrm{II}}(z,t,\nu)}{\partial \nu}\bigg|_{\nu=+0}$ imply

$$\rho \frac{\partial^2 u_2^1}{\partial t^2} = \frac{\partial}{\partial z}\left(c_{44}\frac{\partial u_2^1}{\partial z}\right) + i\frac{\partial}{\partial z}(e_{14}\varphi^0) + ie_{14}\frac{\partial \varphi^0}{\partial z}, \quad (z,t) \in \mathbb{R}_+ \times \mathbb{R},$$

$$u_2^1\big|_{t<0} \equiv 0, \quad \left(c_{44}\frac{\partial u_2^1}{\partial z} + ie_{14}\varphi^0\right)\bigg|_{z=+0} = 0. \qquad (5.2.29)$$

Here the function $\varphi^0(z,t)$ is the solution of the problem

$$\frac{\partial^2 \varphi^0}{\partial z^2} = 0, \quad (z,t) \in \mathbb{R}_+ \times \mathbb{R},$$

$$\varphi^0\big|_{z=+0} = \theta(t), \quad \frac{\partial \varphi^0}{\partial z}\bigg|_{z=+0} = 0. \qquad (5.2.30)$$

Equalities (5.2.29) allow us to consider the following inverse problem.

Inverse Problem 5.2.5 (IP 5.2.5). *Find the function $e_{14} \in C(\overline{\mathbb{R}_+})$ such that the solution of the problem (5.2.29) satisfies the condition*

$$u_2^1(z,t)\big|_{z=+0} = h_4(t), \qquad (5.2.31)$$

where $h_4(t)$, $t \in [0,T]$, is a known function; ρ is a given positive constant; c_{44} and c_{11} are the functions found by solving the IP 5.2.1 and the IP 5.2.2; and the function $\varphi^0(z,t)$ is the solution of the problem (5.2.30).

Applying the methods of solution of one-dimensional inverse problems for hyperbolic equations developed in Chapter IV, Sections 4.4 and 4.5, to solution of the Inverse Problems 5.2.2–5.2.5, we can prove the basic Theorems 5.2.2–5.2.4 on solvability of the original IP 5.2.1. For the complete proofs of the above-mentioned theorems the reader is referred to the original paper by Yakhno and Merazhov (2000).

5.3. INVERSE PROBLEMS OF ELECTROMAGNETOELASTICITY FOR WEAKLY CONDUCTING MEDIA

In this section, following the work (Romanov, 1995b), we give some results of solution of inverse problems for a system of equations describing the linear process of interaction of electromagnetic and elastic waves in a weakly conducting elastic medium. The system of equations of electromagnetoelasticity is as follows:

$$\operatorname{rot} \boldsymbol{H} = \frac{\partial \boldsymbol{D}}{\partial t} + \boldsymbol{J} + \boldsymbol{j},$$

$$\operatorname{rot} \boldsymbol{E} = -\frac{\partial \boldsymbol{B}}{\partial t}, \quad \operatorname{div} \boldsymbol{B} = 0, \qquad (5.3.1)$$

$$\rho \frac{\partial^2 \boldsymbol{U}}{\partial t^2} = \operatorname{Div} T + \mu \boldsymbol{J} \times \boldsymbol{h}^0 + \boldsymbol{f},$$

where the vector-functions $\boldsymbol{j}$ and $\boldsymbol{f}$ characterize the external source of currents and the external source of elastic oscillations, and are distributions with finite supports and equaled to zero for $t \in \mathbb{R}_-$ in our case. The defining relations for the stress tensor T and the components of the electric induction $\boldsymbol{D}$ and magnetic induction $\boldsymbol{B}$ for an electromagnetoelastic isotropic inhomogeneous space in the Galilean approximation, see (Eringen and Maugin, 1990; Maugin, 1988), are as follows:

$$\boldsymbol{D} = \varepsilon \boldsymbol{E} + \alpha \frac{\partial \boldsymbol{U}}{\partial t} \times \boldsymbol{h}^0, \quad \boldsymbol{J} = \sigma \boldsymbol{E} + \sigma \mu \frac{\partial \boldsymbol{U}}{\partial t} \times \boldsymbol{h}^0, \quad \boldsymbol{B} = \mu \boldsymbol{H},$$

$$T = \lambda \cdot \operatorname{tr} S \cdot I + 2\varkappa S, \qquad (5.3.2)$$

$$S: \quad S_{ij} = \frac{1}{2}\left(\frac{\partial U_i}{\partial x_j} + \frac{\partial U_j}{\partial x_i}\right), \quad i,j = 1,2,3.$$

Here $\alpha = \varepsilon\mu - \varepsilon_0\mu_0$; ε_0 and μ_0 are the values of the parameters ε and μ in vacuum; and $\boldsymbol{h}^0$ is the magnetic intensity vector characterizing the external constant magnetic field. In this case we assume that μ is a positive constant and $\boldsymbol{h}^0$ is a constant nonzero vector. The material properties of a medium are described by smooth bounded functions

$$\varepsilon, \rho, \lambda, \varkappa : \mathbb{R}^3 \to \mathbb{R}_+, \qquad \sigma : \mathbb{R}^3 \to \overline{\mathbb{R}}_+.$$

For the system (5.3.1), (5.3.2), we consider the Cauchy problem with the initial data

$$(\boldsymbol{H}, \boldsymbol{E}, \boldsymbol{U})_{t<0} \equiv 0. \tag{5.3.3}$$

We shall treat a solution of the Cauchy problem (5.3.1)–(5.3.3) as a generalized function defined over the space of infinitely differentiable compactly supported functions. These equations show that the interaction of an elastic medium with the electromagnetic field is a one-way interaction for a nonconductive medium ($\sigma \equiv 0$): the displacement vector $\boldsymbol{U}$ is independent of the electromagnetic field, but at same time the components of $\boldsymbol{E}$ and $\boldsymbol{H}$ depend on $\boldsymbol{U}$. For weakly conducting media, we have $\sigma \sim 0$; by this reason we can linearize our system (5.3.1), (5.3.2) by calculating the Fréchet derivative on $\sigma \equiv 0$. The system obtained is the basic subject of our investigation. Thus, we arrive at the following system in the domain $(x, t) \in \mathbb{R}^3 \times \mathbb{R}$:

$$\operatorname{rot} \boldsymbol{h} - \varepsilon \frac{\partial \boldsymbol{e}}{\partial t} - \alpha \frac{\partial^2 \boldsymbol{u}}{\partial t^2} \times \boldsymbol{h}^0 + \sigma \boldsymbol{E}^0 + \mu \sigma \frac{\partial \boldsymbol{U}^0}{\partial t} \times \boldsymbol{h}^0 = \boldsymbol{j}, \tag{5.3.4}$$

$$\operatorname{rot} \boldsymbol{e} = -\mu \frac{\partial \boldsymbol{h}}{\partial t}, \qquad \operatorname{div} \mu \boldsymbol{h} = 0 \tag{5.3.5}$$

$$\rho \frac{\partial^2 \boldsymbol{u}}{\partial t^2} - \operatorname{Div} T - \sigma \mu \left(\boldsymbol{E}^0 + \mu \frac{\partial \boldsymbol{U}^0}{\partial t} \times \boldsymbol{h}^0 \right) \times \boldsymbol{h}^0 = \boldsymbol{f}, \tag{5.3.6}$$

$$(\boldsymbol{h}, \boldsymbol{e}, \boldsymbol{u})_{t<0} \equiv 0, \tag{5.3.7}$$

where $\boldsymbol{H}^0, \boldsymbol{E}^0, \boldsymbol{U}^0$ solves the Cauchy problem (5.3.1)–(5.3.3) under the condition $\sigma = 0$.

The structure of the solution of the Cauchy problem (5.3.4)–(5.3.7) was studied in the paper (Romanov, 1995a) under the condition that the vector-functions $\boldsymbol{j}$ and $\boldsymbol{f}$ have the form

$$\boldsymbol{j} = \boldsymbol{j}^0 \, \delta(x - x^0, t), \qquad \boldsymbol{f} = \boldsymbol{f}^0 \, \delta(x - x^0, t), \tag{5.3.8}$$

where $\boldsymbol{j}^0$ and $\boldsymbol{f}^0$ are constant vectors and $x^0 \in \mathbb{R}^3$ is a fixed point.

Let $c_1, c_2,$ and c_3 be the velocities of electromagnetic and longitudinal and transverse elastic waves, respectively:

$$c_1 = \sqrt{\frac{1}{\varepsilon \mu}}, \qquad c_2 = \sqrt{\frac{\lambda + 2\varkappa}{\rho}}, \qquad c_3 = \sqrt{\frac{\varkappa}{\rho}}. \tag{5.3.9}$$

Chapter 5. Coupled Maxwell and Lamé systems

For each of them we introduce the Riemannian metric with the length element $\mathrm{d}\tau_k$ defined by the formula

$$\mathrm{d}\tau_k = c_k(x)\,\mathrm{d}s, \quad k = 1, 2, 3, \tag{5.3.10}$$

where $\mathrm{d}s$ is the length element in the Euclidean metric. Let us denote by $\Gamma_k(x^0, x)$ the geodesic of the metric (5.3.10) connecting the points x^0 and x and by $\tau_k(x^0, x)$ its length. It is well-known that $\tau_k(x^0, x)$ as a function of the variable x satisfies the relations

$$|\mathrm{grad}_x\, \tau_k(x^0, x)| = c_k^{-2}(x),$$
$$\tau_k(x^0, x) = O(|x - x^0|) \quad \text{for} \quad x \to x^0. \tag{5.3.11}$$

In what follows we shall assume that each of the metrics (5.3.10) is simple, i.e., each pair of points x^0, x is connected by one and only one geodesic $\Gamma_k(x^0, x)$. In addition, we assume that $c_k(x)$, $k = 1, 2, 3$, satisfy the conditions

$$0 < m_3 \leq c_3(x) < c_2(x) < c_1(x) < M_1 < \infty, \tag{5.3.12}$$

where m_3 and M_1 are some constants.

Let $\theta_0(t)$ be the Heaviside function:

$$\theta_0(t) = 1 \quad \text{for} \quad t \geq 0, \qquad \theta_0(t) = 0 \quad \text{for} \quad t < 0.$$

Introduce the functions

$$\theta_n(t) = \frac{t^n}{n!}\theta_0(t), \quad \theta_{-n}(t) = \frac{\mathrm{d}^n}{\mathrm{d}t^n}\theta_0(t), \quad n = 1, 2, 3, \ldots, \tag{5.3.13}$$

$$S_k = S_k(x, t, x^0) \equiv t - \tau_k(x^0, x), \quad k = 1, 2, 3. \tag{5.3.14}$$

The differentiation in (5.3.13) is understood in the sense of the theory of distributions. The equalities $S_k = 0$, $k = 1, 2, 3$, define the characteristic cones corresponding to the velocities c_k, $k = 1, 2, 3$.

We now introduce new vector functions $\boldsymbol{V}^i$, $i = 1, 2, 3$, defining them by the formulas $\boldsymbol{V}^1 = \boldsymbol{h}$, $\boldsymbol{V}^2 = \boldsymbol{e}$, $\boldsymbol{V}^3 = \boldsymbol{u}$. In the paper (Romanov, 1995a), the following theorem was proved.

Theorem 5.3.1. *Let there be a certain number $\delta_0 > 0$ such that the coefficients $\varepsilon, \sigma, \rho, \lambda,$ and $\varkappa$ are constants in the domain*

$$D_0 = \{x \in \mathbb{R}^3 \mid |x - x^0| < \delta_0\}$$

and belong to $C^m(\mathbb{R}^3)$, $m > 3(N+10) + 10$, for a certain integer $N \geq 5$. Then the solution to the Cauchy problem (5.3.4)–(5.3.8) can be represented in the form

$$V^i(x,t) = \sum_{n=-2}^{N} \{\alpha_n^i(x)\theta_n(S_1) + \beta_n^i(x)\theta_n(S_2) + \gamma_n^i(x)\theta_n(S_3)\}$$
$$+ V_N^i(x,t), \quad i = 1, 2, 3, \quad (5.3.15)$$

where $V_N^i(x,t) \in C^N(\mathbb{R}^3 \times \mathbb{R})$ and the coefficients $\alpha_n^i(x)$, $\beta_n^i(x)$, and $\gamma_n^i(x)$ have the following properties:

a) $\alpha_{-2}^3(x) = \beta_{-2}^3(x) = \gamma_{-2}^3(x) \equiv 0$;

b) $\alpha_n^i(x)$, $\beta_n^i(x)$, and $\gamma_n^i(x)$ are analytic real-valued functions in the domain $D_0 \setminus \{x^0\}$ and smooth outside D_0 (more precisely, functions of the class $C^{m-2n-8}(\mathbb{R}^3 \setminus D_0)$). Moreover, there exists a positive constant $C > 0$ depending only on the values of the coefficients $\varepsilon, \mu, \sigma, \rho$, λ, and $\varkappa$ in the domain D_0 and on the values of $|j^0|$, $|f^0|$, and $|h^0|$ such that

$$(|\alpha_n^i|, |\beta_n^i|, |\gamma_n^i|) \leq C \cdot \begin{cases} |x-x^0|^{-(3+n)}, & i=1,2, \\ |x-x^0|^{-(2+n)}, & i=3; \end{cases} \quad (5.3.16)$$

c) in the domain $\{(x,t) \mid S_3 > 0, \, x \in D_0\}$, the functions

$$\widetilde{V}_N^i(x,t) = \sum_{n=-2}^{N} \{\alpha_n^i(x)\theta_n(S_1) + \beta_n^i(x)\theta_n(S_2) + \gamma_n^i(x)\theta_n(S_3)\}$$

satisfy the estimates

$$|\widetilde{V}_N^1| \leq C |x-x^0|^{-2}, \quad |\widetilde{V}_N^2| \leq C |x-x^0|^{-3}, \quad |\widetilde{V}_N^3| \leq C |x-x^0|^{-1} \quad (5.3.17)$$

with the same constant C as in (5.3.16).

Remark 5.3.1. For a homogeneous medium, the solution to the Cauchy problem (5.3.4)–(5.3.8) is representable in the form (5.3.15) with $N = 5$ and $V_N^i \equiv 0$.

The above theorem is a basis for formulating an inverse problem for weakly conducting media.

Formulation of an inverse problem and main results

We consider the case where the coefficients of equations (5.3.4)–(5.3.8) are known constants outside a finite domain D and are unknown functions of x in D. We also assume that the magnetic permeability μ is constant everywhere in $\mathbb{R}^3$. For simplicity we assume that all the coefficients are functions of the class $C^m(\mathbb{R}^3)$ with a sufficiently large m and are continuous together with all their derivatives of order up to m on the boundary of the domain D. Let S be some closed sufficiently smooth surface enclosing the domain D. Assume that the function $\boldsymbol{V} = (\boldsymbol{h}, \boldsymbol{e}, \boldsymbol{u})$, the solution to the Cauchy problem (5.3.4)–(5.3.8), is known on the set $S \times [0, T]$, where T is a sufficiently large positive number, and x^0 is an arbitrary point of the surface S, i.e.,

$$\boldsymbol{V} = \boldsymbol{F}(x, t, x^0), \qquad (x, t, x^0) \in S \times [0, T] \times S \equiv G. \tag{5.3.18}$$

Remark 5.3.2. In what follows it will be convenient for us to mark explicitly the dependence of functions on the variable x^0 because in the inverse problems concerned the source position is a variable parameter which will be used.

Inverse Problem 5.3.1 (IP 5.3.1). *Find the coefficients $\varepsilon, \sigma, \lambda, \rho$, and $\varkappa$ in the domain D if the function $\boldsymbol{F}(x, t, x^0)$ is given.*

In the sequel, we shall assume that $T \geq d/m_3$, where d is the diameter of the domain D and m_3 is the constant from formula (5.3.12). Using the representation (5.3.15), from the function $\boldsymbol{F}(x, t, x^0)$ we can uniquely determine the following quantities:

a) $\tau_k(x^0, x)$, $(x^0, x) \in S \times S$, $k = 1, 2, 3$;

b) $\alpha_n^i(x^0, x)$, $\beta_n^i(x^0, x)$, $\gamma_n^i(x^0, x)$, $(x^0, x) \in S \times S$, $i = 1, 2, 3$, $n \geq -2$.

The IP 5.3.1 is reduced to successive solving the inverse kinematic problem of reconstructing the velocities $c_k(x)$, $k = 1, 2, 3$, inside the domain D from the functions $\tau_k(x^0, x)$, $(x^0, x) \in S \times S$, $k = 1, 2, 3$, and subsequent determining the still unknown combinations of the sought functions from the functions $\alpha_n^i(x^0, x)$, $\beta_n^i(x^0, x)$, and $\gamma_n^i(x^0, x)$ given on $S \times S$. The theory of solving the inverse kinematic problem in the case of isotropic media is developed sufficiently well (see, e.g., Romanov, 1987). In particular, it was established that in the case of simple metrics, the values of the functions $\tau_k(x^0, x)$, $k = 1, 2, 3$, on the set $S \times S$ determine uniquely the functions $c_k(x)$, $k = 1, 2, 3$, inside the domain D. Thus, we can think that the velocities $c_k(x)$, $k = 1, 2, 3$, were determined inside the domain D as a result of solving the corresponding inverse kinematic problems. If we know the functions $c_k(x)$, $k = 1, 2, 3$, then we know three nonlinear combinations of the parameters of the medium, see formulas (5.3.9). Since the coefficient μ is assumed to be known and constant everywhere, therefore, the coefficient ε is determined uniquely from these relations, and the elastic moduli can be expressed in terms of the density of the medium ρ and known velocities c_2 and c_3 only. Thus, among the sought parameters only the electrical conductivity σ and the density of the medium ρ remain unknown. To find them, we use the functions $\alpha_n^i(x^0, x)$, $\beta_n^i(x^0, x)$, and $\gamma_n^i(x^0, x)$.

It should be noted that this method of reducing an inverse problem to solution of the inverse kinematic problem and subsequent use of the amplitudes for determining the rest of the relations between the desired coefficients was used earlier in (Romanov, 1987) and (Yakhno, 1998) for investigation of more simple inverse problems for hyperbolic equations of second order and for equations arising in the theory of elasticity.

Consider one possible statement of the inverse problem of reconstructing the parameters σ and ρ; we shall use the function $\alpha_{-2}^1(x^0, x)$ for its solution. Assume that $\boldsymbol{f}^0 = 0$ and $\sigma(x) \equiv \sigma_0$ for $x \in \mathbb{R}^3 \setminus D$ and $\alpha \neq 0$ for $x \in D$. We shall show that the problem under consideration can be reduced to the following integral geometry problem: Consider the integrals

$$\int_{\Gamma_1(x^0,x)} (a\eta_1 + b)\, ds = p_1(x^0, x), \quad (x^0, x) \in S \times S, \tag{5.3.19}$$

$$\int_{\Gamma_1(x^0,x)} a\eta_2\, ds = p_2(x^0, x), \quad (x^0, x) \in S \times S. \tag{5.3.20}$$

Assume that we know the functions $p_k(x^0, x)$, $k = 1, 2$, with the weights η_k, $k = 1, 2$, also known. It is required to determine the coefficients a and b inside the domain D if they are known outside of D. The coefficients a and b are related to the parameters of the medium σ and ρ by the formulas

$$a = \frac{\sigma\alpha}{2\rho\,(c_1^2 - c_2^2)\,\epsilon^2\, c_1}, \quad b = -\frac{\sigma}{2c_1\,\epsilon}.$$

The weight functions η_k, $k = 1, 2$, are defined by the relations

$$\begin{aligned}\eta_1 &= (\boldsymbol{h}^0 \cdot \boldsymbol{w}^0)^2\,|\boldsymbol{w}^0|^{-2} + \xi\,(\boldsymbol{h}^0 \cdot \boldsymbol{\nu})^2, \\ \eta_2 &= (\boldsymbol{h}^0 \cdot \boldsymbol{w}^0)\,(\boldsymbol{h}^0 \cdot (\boldsymbol{\nu} \times \boldsymbol{w}^0))\,|\boldsymbol{w}^0|^{-2},\end{aligned} \tag{5.3.21}$$

where $\xi = (c_1^2 - c_2^2)/(c_1^2 - c_3^2)$, $\xi \in (0,1)$; $\boldsymbol{\nu}$ is the unit vector tangent to $\Gamma_1(x^0, x)$ at the point x; and $\boldsymbol{w}^0$ is the solution to the problem (5.3.43)–(5.3.45); this solution is completely determined by specifying the quantities $\boldsymbol{j}^0$ and $c_1(x)$.

Remark 5.3.3. The functions a and b uniquely determine only σ and σ/ρ. Thus, the density of the medium can be uniquely determined only at those points of the domain D where $\sigma \neq 0$. This characteristic feature of the problem concerned is a consequence of the fact that we use the information about the coefficient α_{-2}^1. One can show that in the case where $\alpha_0^3(x^0, x)$, $(x^0, x) \in S \times S$, is used as the additional information (under the condition $f^0 \neq 0$), the density of the medium is determined uniquely in the domain D.

The fact that equalities (5.3.19) and (5.3.20) contain weight functions η_k, $k = 1, 2$, complicates the solution of the corresponding integral geometry problem in the general case. So we give only three special formulations of the original problem.

1. Let $c_1 = \text{const} = c_{10}$, $\boldsymbol{j}^0 = (1, 0, 0) \cdot 4\pi c_{10}^2$, $\boldsymbol{h}^0 = (1, 0, 1)$. Let $S(z)$ and $D(z)$ denote the cross-sections of the surface S and domain D by the plane $x_3 = z$. In this case, the integral geometry problem splits into one-parameter family of planar problems for each cross-section of the domain D by the plane $x_3 = z$. The vector $\boldsymbol{\nu}$ of the corresponding line $\Gamma_1(x^0, x)$, $(x^0, x) \in S(z) \times S(z)$, can be characterized in this case by the angular coordinate φ: $\boldsymbol{\nu} = (\cos\varphi, \sin\varphi, 0)$. This coordinate is constant along $\Gamma_1(x^0, x)$ and, consequently, depends only on the boundary points $(x^0, x) \in S(z) \times S(z)$. Therefore, the weight functions η_k, $k = 1, 2$, can be written as $\eta_1 = 1 + \xi \cos^2\varphi$, $\eta_2 = \sin\varphi$. Since the weight function η_2 is constant along $\Gamma_1(x^0, x)$, equation (5.3.20) can be written as

$$\int_{\Gamma_1(x^0,x)} a\, ds = p_2'(x^0, x), \quad (x^0, x) \in S \times S, \tag{5.3.22}$$

where $p_2' = p_2/\sin\varphi$. The problem of determining the function a from equation (5.3.22) is a standard integral geometry problem. Having found the function a, we can calculate the integral of the product $a\eta_1$ along $\Gamma_1(x^0, x)$ in formula (5.3.19). This yields a similar integral geometry problem for determining the coefficient b. The uniqueness of solution of these problems is evident and follows from the results of Chapter III. A conditional stability estimate is less evident because in the general case the weight function η_1 is not constant on $\Gamma_1(x^0, x)$.

The following theorem holds.

Theorem 5.3.2. *Let (a, b) and $(\bar{a}, \bar{b})$ be two solutions of equations (5.3.19), (5.3.20) corresponding to the data (p_1, p_2) and $(\bar{p}_1, \bar{p}_2)$, respectively, and coinciding outside the domain D. Then the following estimates hold for* the differences $\tilde{a} = a - \bar{a}$, $\tilde{b} = b - \bar{b}$, $\tilde{p}_1 = p_1 - \bar{p}_1$, and $\tilde{p}_2 = p_2 - \bar{p}_2$:

$$\int_{D(z)} \tilde{a}^2 \, d\bar{x} \leq \frac{1}{2\pi} \int_{S(z)} dl' \int_{S(z)} \left| \frac{\partial \tilde{p}_2}{\partial l} \cdot \frac{\partial \tilde{p}_2}{\partial l'} \right| dl, \qquad (5.3.23)$$

$$\int_{D(z)} \tilde{b}^2 \, d\bar{x} \leq \frac{1}{\pi} \int_{S(z)} dl' \int_{S(z)} \left\{ \frac{23}{8} \left| \frac{\partial \tilde{p}_2}{\partial l} \cdot \frac{\partial \tilde{p}_2}{\partial l'} \right| + \left| \frac{\partial \tilde{p}_1}{\partial l} \cdot \frac{\partial \tilde{p}_1}{\partial l'} \right| \right\} dl, \qquad (5.3.24)$$

where $d\bar{x} = dx_1 \, dx_2$; $\partial/\partial l$ and $\partial/\partial l'$ denote the derivatives in the directions tangent to $S(z)$ at the points x^0 and x, respectively; and dl and dl' are the length elements calculated at these points.

It is more convenient to give the proof of Theorem 5.3.2 later.

2. Consider a slightly modified 2D version of the problem. Namely, let D be an infinite cylindrical domain with the ruling parallel to the axis x_3: $D = D_0 \times \mathbb{R}$, where D_0 is the cross-section of the domain D by the plane $x_3 = 0$. Assume that the coefficients of the problem are independent of the variable x_3. Let S be a simple smooth closed curve enclosing the domain D_0, and let the vectors $\boldsymbol{j}^0$ and $\boldsymbol{h}^0$ be the same as in the previous case. In this case, the geodesics $\Gamma_1(x^0, x)$, $(x^0, x) \in S \times S$, are planar curves. Denoting the unit vector tangent to $\Gamma_1(x^0, x)$ at the point x by $\boldsymbol{\nu} = (\cos \varphi, \sin \varphi, 0)$, we conclude that in this case we also have $\eta_2 = \sin \varphi$ and $\eta_1 = 1 + \xi \cos^2 \varphi$. There is an essential distinction between this and previous cases; namely, the weight function is not constant along a geodesic line. In this case, equation (5.3.20) can be written as

$$\int_{\Gamma_1(x^0, x)} a \, dx_2 = p_2(x^0, x), \qquad (x^0, x) \in S \times S. \qquad (5.3.25)$$

Thus, p_2 is the integral along $\Gamma_1(x^0, x)$ of the simplest form of first degree. The problems of finding a form of first degree from its integrals along a family of geodesics of a Riemannian metric were considered by Anikonov (1978), Romanov (1987), Sharafutdinov (1994).

Assume that there are two solutions a and $\bar{a}$ of equation (5.3.25) corresponding to the function p_2 and coinciding outside the domain D_0. Then their difference $\tilde{a} = a - \bar{a}$ solves a homogeneous equation. It follows from Anikonov (1978) that $\partial \tilde{a}/\partial x_1 = 0$, $x \in D_0$. Since $\tilde{a} = 0$ outside the domain D_0, we conclude that $\tilde{a} = 0$, $x \in D_0$, i.e., $a = \bar{a}$. Thus, the following uniqueness theorem holds:

Theorem 5.3.3. *Specifying the functions p_1 and p_2 for $(x^0, x) \in S \times S$ defines the functions a and b in the domain D_0 uniquely.*

Chapter 5. Coupled Maxwell and Lamé systems

It should be noted that certain estimates of the conditional stability of the solution to equations (5.3.19), (5.3.25) can be obtained from the estimates given by Sharafutdinov (1994).

3. Let us return to the original formulation of the problem. We consider the case where the velocities c_1 depends essentially on all three variables x_1, x_2, and x_3. Let $\boldsymbol{j}^0$ be a constant nonzero vector. We also suppose that the vector $\boldsymbol{h}^0$ may have two different values $\boldsymbol{h}^{0k}$, $k = 1, 2$, where $\boldsymbol{h}^{02} = r\boldsymbol{h}^{01}$, $r^2 \neq 1$, and that for each vector $\boldsymbol{h}^{0k}$ the function $\alpha_{-2}^1(x^0, x)$ and, accordingly, the functions $p_k(x^0, x)$, $(x^0, x) \in S \times S$, are known. We shall denote by p_{1k}, $k = 1, 2$, the function p_1 corresponding to the vectors $\boldsymbol{h}^{0k}$, $k = 1, 2$. From equality (5.3.19) for $\boldsymbol{h}^0 = \boldsymbol{h}^{0k}$, $k = 1, 2$, we readily obtain

$$\int_{\Gamma_1(x^0,x)} a\eta_0 \, ds = \bar{p}_1(x^0, x), \quad (x^0, x) \in S \times S, \tag{5.3.26}$$

$$\int_{\Gamma_1(x^0,x)} b \, ds = \bar{p}_2(x^0, x), \quad (x^0, x) \in S \times S, \tag{5.3.27}$$

where

$$\eta_0 = (\boldsymbol{h}^{01} \cdot \boldsymbol{w}^0)^2 |\boldsymbol{w}^0|^{-2} + \xi \, (\boldsymbol{h}^{01} \cdot \boldsymbol{\nu})^2,$$

$$\bar{p}_1 = \frac{p_{11} - p_{12}}{1 - r^2}, \quad \bar{p}_2 = \frac{p_{12} - p_{11} r^2}{1 - r^2}.$$

Thus, using the data that correspond to two different values of the vector $\boldsymbol{h}^0$, we can split the original problem into two problems independent of each other. The problem of determining the function b from equation (5.3.27) is an integral geometry problem which has been studied in full (Romanov, 1987).

The next conditional stability theorem (an analog of the estimate (3.54) from Romanov (1987)) holds true for this problem.

Theorem 5.3.4. *Let b and b' be two solutions of equation (5.3.27) corresponding to the right-hand sides $\bar{p}_2$ and $\bar{p}'_2$. Then the following inequality holds for the differences $\tilde{b} = b - b'$ and $\tilde{p}_2 = \bar{p}_2 - \bar{p}'_2$:*

$$\int_D c_1^{-1} \tilde{b}^2 \, dx \leq \frac{1}{8\pi} \int_S \int_S |\mathrm{grad}_x \, \tilde{p}_2 \times \boldsymbol{n}(x)| \cdot |\mathrm{grad}_{x^0} \, \tilde{p}_2 \times \boldsymbol{n}(x^0)|$$

$$\times \left(\sum_{k,j=1}^3 \left(\frac{\partial^2 \tau_1}{\partial x_k \, \partial x_j^0} \right)^2 \right)^{1/2} dS_x \, dS_{x^0}, \tag{5.3.28}$$

where $\boldsymbol{n}(x)$ is the normal vector to the surface S at the point x.

As concerns the properties of equation (5.3.26), their investigation requires additional efforts.

Reduction to an integral geometry problem

The differential and integral equations considered here correspond to the IP 5.3.1, which was formulated earlier. As the first step, let us derive a differential equation for finding the function $\boldsymbol{\alpha}^1_{-2}(x)$. Let $\boldsymbol{f}^0 = 0$. Then $\boldsymbol{U}^0 \equiv 0$ and, as a consequence, $\alpha^3_k = \beta^3_k = \gamma^3_k \equiv 0$ for $k < 0$. It means that the function $\boldsymbol{u}$ has no singular part. Introduce the following representation for the function $\boldsymbol{V}^0 = (\boldsymbol{H}^0, \boldsymbol{E}^0, \boldsymbol{U}^0) \equiv (\boldsymbol{V}^{10}, \boldsymbol{V}^{20}, \boldsymbol{V}^{30})$:

$$V^{i0}(x,t) = \sum_{n=-2}^{N} \{\alpha_n^{i0}(x)\theta_n(S_1) + \beta_n^{i0}(x)\theta_n(S_2) + \gamma_n^{i0}(x)\theta_n(S_3)\}$$
$$+ V_N^{i0}(x,t), \quad i = 1, 2, 3.$$

Let us substitute the above representation into equations (5.3.1) and (5.3.2) with $\sigma = 0$ and representation (5.3.15) into equations (5.3.4)–(5.3.6). Equating the coefficients at $\theta_n(S_1)$, $n = -3, -2$, we obtain the following relationships between $\boldsymbol{\alpha}^{10}_{-2}, \boldsymbol{\alpha}^{20}_{-2}, \boldsymbol{\alpha}^{10}_{-1}, \boldsymbol{\alpha}^{20}_{-1}$ and $\boldsymbol{\alpha}^1_{-2}, \boldsymbol{\alpha}^2_{-2}, \boldsymbol{\alpha}^1_{-1}, \boldsymbol{\alpha}^2_{-1}, \boldsymbol{\alpha}^3_0$:

$$\operatorname{grad} \tau_1 \times \boldsymbol{\alpha}^{10}_{-2} + \varepsilon \boldsymbol{\alpha}^{20}_{-2} = 0, \qquad \operatorname{grad} \tau_1 \times \boldsymbol{\alpha}^{20}_{-2} - \mu \boldsymbol{\alpha}^{10}_{-2} = 0, \qquad (5.3.29)$$

$$\operatorname{grad} \tau_1 \times \boldsymbol{\alpha}^{10}_{-1} + \varepsilon \boldsymbol{\alpha}^{20}_{-1} - \operatorname{rot} \boldsymbol{\alpha}^{10}_{-2} = 0,$$
$$\operatorname{grad} \tau_1 \times \boldsymbol{\alpha}^{20}_{-1} - \mu \boldsymbol{\alpha}^{10}_{-1} - \operatorname{rot} \boldsymbol{\alpha}^{20}_{-2} = 0, \qquad (5.3.30)$$

$$\operatorname{grad} \tau_1 \times \boldsymbol{\alpha}^1_{-2} + \varepsilon \boldsymbol{\alpha}^2_{-2} = 0, \qquad \operatorname{grad} \tau_1 \times \boldsymbol{\alpha}^2_{-2} - \mu \boldsymbol{\alpha}^1_{-2} = 0, \qquad (5.3.31)$$

$$\operatorname{grad} \tau_1 \times \boldsymbol{\alpha}^1_{-1} + \varepsilon \boldsymbol{\alpha}^2_{-1} - \operatorname{rot} \boldsymbol{\alpha}^1_{-2} + \alpha \boldsymbol{\alpha}^3_0 \times \boldsymbol{h}^0 + \sigma \boldsymbol{\alpha}^{20}_{-2} = 0,$$
$$\operatorname{grad} \tau_1 \times \boldsymbol{\alpha}^2_{-1} - \mu \boldsymbol{\alpha}^1_{-1} - \operatorname{rot} \boldsymbol{\alpha}^2_{-2} = 0, \qquad (5.3.32)$$

$$\rho \boldsymbol{\alpha}^3_0 - (\lambda + 2\varkappa)(\boldsymbol{\alpha}^3_0 \cdot \operatorname{grad} \tau_1)\operatorname{grad} \tau_1 + \varkappa\left(\operatorname{grad} \tau_1 \times (\operatorname{grad} \tau_1 \times \boldsymbol{\alpha}^3_0)\right)$$
$$- \sigma \mu \boldsymbol{\alpha}^{20}_{-2} \times \boldsymbol{h}^0 = 0. \qquad (5.3.33)$$

Using formulas (5.3.11), we can transform relations (5.3.29) and (5.3.31) into the equalities

$$\operatorname{grad} \tau_1 \cdot \boldsymbol{\alpha}^{10}_{-2} = 0, \qquad \operatorname{grad} \tau_1 \cdot \boldsymbol{\alpha}^{20}_{-2} = 0,$$
$$\operatorname{grad} \tau_1 \cdot \boldsymbol{\alpha}^1_{-2} = 0, \qquad \operatorname{grad} \tau_1 \cdot \boldsymbol{\alpha}^2_{-2} = 0, \qquad (5.3.34)$$

which express the principle of orthogonality of the electromagnetic field to the direction of propagation of its main singularity. To derive equations

Chapter 5. Coupled Maxwell and Lamé systems

for the coefficient $\boldsymbol{\alpha}_{-2}^{10}$ we proceed as follows. We take the vector product of the first relation from (5.3.30) by $\operatorname{grad} \tau_1$ and then eliminate the term $\operatorname{grad} \tau_1 \times \boldsymbol{\alpha}_{-1}^{20}$ with the help of the second relation from (5.3.30). Using the first equality from (5.3.29), we eliminate $\boldsymbol{\alpha}_{-2}^{20}$ and finally obtain

$$\operatorname{grad} \tau_1 \times (\operatorname{grad} \tau_1 \times \boldsymbol{\alpha}_{-1}^{10}) + \varepsilon\mu\boldsymbol{\alpha}_{-1}^{10} - \varepsilon \operatorname{rot}\left(\varepsilon^{-1}(\operatorname{grad} \tau_1 \times \boldsymbol{\alpha}_{-2}^{10})\right)$$
$$- \operatorname{grad} \tau_1 \times \operatorname{rot} \boldsymbol{\alpha}_{-2}^{10} = 0.$$

For further simplification, we use the following equalities, which are well-known from vector analysis:

$$\boldsymbol{c} \times (\boldsymbol{a} \times \boldsymbol{b}) = \boldsymbol{a}\,(\boldsymbol{c} \cdot \boldsymbol{b}) - \boldsymbol{b}\,(\boldsymbol{c} \cdot \boldsymbol{a}),$$
$$\operatorname{rot}(\boldsymbol{a} \times \boldsymbol{b}) = \boldsymbol{a} \cdot \operatorname{div} \boldsymbol{b} - \boldsymbol{b} \cdot \operatorname{div} \boldsymbol{a} + (\boldsymbol{b} \cdot \nabla)\boldsymbol{a} - (\boldsymbol{a} \cdot \nabla)\boldsymbol{b},$$
$$\boldsymbol{a} \times \operatorname{rot} \boldsymbol{b} = \operatorname{grad}(\boldsymbol{a} \cdot \boldsymbol{b}) - (\boldsymbol{b} \cdot \nabla)\boldsymbol{a} - (\boldsymbol{a} \cdot \nabla)\boldsymbol{b} - \boldsymbol{b} \times \operatorname{rot} \boldsymbol{a}.$$

Taking into account formulas (5.3.11) for τ_1 and the equality $\operatorname{grad} \tau_1 \cdot \boldsymbol{\alpha}_{-1}^{10} = \operatorname{div} \boldsymbol{\alpha}_{-2}^{10}$, which follows from the second relation from (5.3.30), we obtain

$$L\boldsymbol{\alpha}_{-2}^{10} \equiv 2 \sum_{k=1}^{3} \frac{\partial \boldsymbol{\alpha}_{-2}^{10}}{\partial x_k} \frac{\partial \tau_1}{\partial x_k} + \boldsymbol{\alpha}_{-2}^{10}(\Delta \tau_1 - \operatorname{grad} \ln \varepsilon \cdot \operatorname{grad} \tau_1)$$
$$+ (\operatorname{grad} \ln \varepsilon \cdot \boldsymbol{\alpha}_{-2}^{10}) \operatorname{grad} \tau_1 = 0. \quad (5.3.35)$$

The relations for $\boldsymbol{\alpha}_{-2}^{1}$ are derived in a similar way:

$$L\boldsymbol{\alpha}_{-2}^{1} = 2c_1^{-1}\boldsymbol{F}, \quad (5.3.36)$$

where

$$\boldsymbol{F} \equiv -\frac{c_1}{2} \operatorname{grad} \tau_1 \times (\alpha \boldsymbol{\alpha}_0^3 \times \boldsymbol{h}^0 + \sigma \boldsymbol{\alpha}_{-2}^{20}).$$

Using formula (5.3.33), we can find $\boldsymbol{\alpha}_0^3$:

$$\boldsymbol{\alpha}_0^3 = \frac{\sigma}{\rho}\left[\frac{\boldsymbol{h}^0 \cdot \boldsymbol{\alpha}_{-2}^{10}}{c_1^2 - c_2^2} \operatorname{grad} \tau_1 - \frac{\boldsymbol{h}^0 \cdot \operatorname{grad} \tau_1}{c_1^2 - c_3^2}\boldsymbol{\alpha}_{-2}^{10}\right]. \quad (5.3.37)$$

Thus, the final expression for the function $\boldsymbol{F}$ takes the form

$$\boldsymbol{F} = a\{(\boldsymbol{h}^0 \cdot \boldsymbol{\alpha}_{-2}^{10})\left[\boldsymbol{h}^0 - c_1^2(\boldsymbol{h}^0 \cdot \operatorname{grad} \tau_1) \operatorname{grad} \tau_1\right]$$
$$+ \xi c_1^2(\boldsymbol{h}^0 \cdot \operatorname{grad} \tau_1)^2 \boldsymbol{\alpha}_{-2}^{10}\} + b\boldsymbol{\alpha}_{-2}^{10}, \quad (5.3.38)$$

where

$$a = \frac{\sigma \alpha}{2\rho \left(c_1^2 - c_2^2\right) \varepsilon^2 c_1}, \qquad b = -\frac{\sigma}{2c_1 \varepsilon}, \qquad \xi = \frac{c_1^2 - c_2^2}{c_1^2 - c_3^2}, \qquad \xi \in (0,1). \tag{5.3.39}$$

Equations (5.3.38) and (5.3.39) imply that the coefficient α^1_{-2} depends linearly on σ and σ/ρ. Consequently, in this case the inverse problem can be treated as a problem of determining σ and σ/ρ or the coefficients a and b, which is the same.

The equations of the geodesics $\Gamma_1(x^0, x)$ can be written as

$$\frac{dx}{ds} = \nu, \qquad \frac{d\nu}{ds} = -\operatorname{grad} \ln c_1 + (\operatorname{grad} \ln c_1 \cdot \nu)\nu, \tag{5.3.40}$$

where s is the Euclidean length of the curve passing through the point x^0 and $\nu = c_1 \operatorname{grad} \tau_1$ is the unit vector tangent to $\Gamma_1(x^0, x)$ at the point x and oriented in the direction of increasing of s.

Let us show that along the geodesic $\Gamma_1(x^0, x)$ equations (5.3.35) and (5.3.36) are ordinary differential equations. Indeed,

$$2 \sum_{k=1}^{3} \frac{\partial \alpha^{10}_{-2}}{\partial x_k} \frac{\partial \tau_1}{\partial x_k} = 2 c_1^{-1} \frac{d \alpha^1_{-2}}{ds}.$$

Since the magnetic permeability μ is constant, we have $\operatorname{grad} \ln \varepsilon = -\operatorname{grad} \ln c_1^2$. In further treatment we use the equality

$$\Delta \tau_1 + \operatorname{grad} \ln c_1^2 \cdot \operatorname{grad} \tau_1 = 2 c_1^{-1} \frac{d}{ds} \ln \left[\tau_1 \left(\det \frac{\partial g(x^0, x)}{\partial x} \right)^{-1/2} \right],$$

which follows immediately from (Romanov, 1987, p. 116, formula (4.18)). The function $g(x^0, x)$ defines the Riemannian coordinates of the point x in the metric $d\tau_1 = c_1^{-1} ds$ that corresponds to the point x^0 and in our case can be found by the formula

$$g(x^0, x) = -(2c_1^2(x^0))^{-1} \operatorname{grad}_{x^0} \tau_1^2(x^0, x).$$

Under the above assumptions this function is smooth in both variables, namely, $g(x^0, x) \in C^{m-2}(\mathbb{R}^6)$. In a homogeneous space we have $g(x^0, x) \equiv x - x^0$.

We introduce new functions

$$w^0(x^0, x) = \alpha^{10}_{-2}(x^0, x) \tau_1(x^0, x) \left(\det \frac{\partial g(x^0, x)}{\partial x} \right)^{-1/2}, \tag{5.3.41}$$

$$w(x^0, x) = \alpha^1_{-2}(x^0, x) \tau_1(x^0, x) \left(\det \frac{\partial g(x^0, x)}{\partial x} \right)^{-1/2}. \tag{5.3.42}$$

Chapter 5. Coupled Maxwell and Lamé systems

Then the differential equations for the functions $\boldsymbol{w}^0(x^0, x)$ and $\boldsymbol{w}(x^0, x)$ along the geodesic $\Gamma_1(x^0, x)$ become

$$\frac{d\boldsymbol{w}^0}{ds} - (\boldsymbol{w}^0 \cdot \operatorname{grad} \ln c_1)\boldsymbol{\nu} = 0, \qquad \frac{d\boldsymbol{w}}{ds} - (\boldsymbol{w} \cdot \operatorname{grad} \ln c_1)\boldsymbol{\nu} = \boldsymbol{F}, \quad (5.3.43)$$

where

$$\boldsymbol{F} = a\left\{(\boldsymbol{h}^0 \cdot \boldsymbol{w}^0)\left[\boldsymbol{h}^0 - \boldsymbol{\nu}(\boldsymbol{h}^0 \cdot \boldsymbol{\nu})\right] + \xi(\boldsymbol{h}^0 \cdot \boldsymbol{\nu})^2 \boldsymbol{w}^0\right\} + b\boldsymbol{w}^0. \quad (5.3.44)$$

The initial conditions at the point $s = 0$ can be found by comparing the singularities of the solutions $(\boldsymbol{H}^0, \boldsymbol{E}^0, \boldsymbol{U}^0)$ and $(\boldsymbol{H}, \boldsymbol{E}, \boldsymbol{U})$ in a homogeneous medium. Since the function $\boldsymbol{H}^0$ in a homogeneous medium is calculated by the formula

$$\boldsymbol{H}^0 = \operatorname{rot}\left[\frac{\boldsymbol{j}^0}{4\pi|x - x^0|}\theta_{-1}(S_1)\right],$$

it follows that

$$\boldsymbol{w}^0 = (4\pi c_{10})^{-1}\boldsymbol{j}^0 \times \operatorname{grad} \tau_1\big|_{x=x^0} = (4\pi c_{10}^2)^{-1}(\boldsymbol{j}^0 \times \boldsymbol{\nu}^0), \quad (5.3.45)$$

where $\boldsymbol{\nu}^0$ is the unit vector tangent to $\Gamma_1(x^0, x)$ at the point x^0 and c_{10} is the value of the velocity c_1 outside the domain D. Analyzing the solution to the problem (5.3.4)–(5.3.8) corresponding to the homogeneous medium (see Romanov, 1995a, Section 4), we derive the following initial data for $\boldsymbol{w}$:

$$\boldsymbol{w}\big|_{s=0} = (4\pi c_{10}^2)^{-1}(\boldsymbol{j}^0 \times \boldsymbol{\nu}^0). \quad (5.3.46)$$

It should be noted that the vector-functions $\boldsymbol{w}^0$ and $\boldsymbol{w}$ are orthogonal to the vector $\boldsymbol{\nu}(x^0, x) = c_1 \operatorname{grad} \tau_1(x^0, x)$. The first equation from (5.3.43) implies that $d|\boldsymbol{w}^0|^2/ds = 0$, i.e., the function $|\boldsymbol{w}^0|$ is constant along the geodesic $\Gamma_1(x^0, x)$:

$$|\boldsymbol{w}^0| = (4\pi c_{10}^2)^{-1}|\boldsymbol{j}^0 \times \boldsymbol{\nu}^0|. \quad (5.3.47)$$

Represent the function $\boldsymbol{w}$ in the form

$$\boldsymbol{w} = (1 + p_1)\boldsymbol{w}^0 + p_2(\boldsymbol{\nu} \times \boldsymbol{w}^0). \quad (5.3.48)$$

Simple calculations lead to the following equations for determining the functions p_1 and p_2:

$$\frac{dp_1}{ds} = a\eta_1 + b, \qquad p_1\big|_{s=0} = 0, \quad (5.3.49)$$

$$\frac{dp_2}{ds} = a\eta_2, \qquad p_2\big|_{s=0} = 0, \quad (5.3.50)$$

where
$$\eta_1 = (\boldsymbol{h}^0 \cdot \boldsymbol{w}^0)^2 |\boldsymbol{w}^0|^{-2} + \xi (\boldsymbol{h}^0 \cdot \boldsymbol{w}^0)^2,$$
$$\eta_2 = (\boldsymbol{h}^0 \cdot \boldsymbol{w}^0)(\boldsymbol{h}^0 \cdot (\boldsymbol{\nu} \times \boldsymbol{w}^0)) |\boldsymbol{w}^0|^{-2}. \tag{5.3.51}$$

Integrating equations (5.3.49) and (5.3.50) along $\Gamma_1(x^0, x)$, $(x^0, x) \in S \times S$, we get

$$\int_{\Gamma_1(x^0, x)} (a\eta_1 + b) \, ds = p_1(x^0, x), \quad (x^0, x) \in S \times S, \tag{5.3.52}$$

$$\int_{\Gamma_1(x^0, x)} a\eta_2 \, ds = p_2(x^0, x), \quad (x^0, x) \in S \times S. \tag{5.3.53}$$

Since the function α_{-2}^1 is known for $(x^0, x) \in S \times S$, therefore, the functions $p_k(x^0, x)$, $k = 1, 2$, are also known for $(x^0, x) \in S \times S$. Consequently, the problem of determining the functions a and b from equations (5.3.52) and (5.3.53) is an integral geometry problem. In the special case where $c_1 = \text{const} = c_{10}$ and the geodesics $\Gamma_1(x^0, x)$ are straight lines, this problem is a problem of tomography.

Proof of Theorem 5.3.2. We note that the estimate (5.3.23) was obtained in (Mukhometov, 1977), so we need only prove the estimate (5.3.24). Following the technique developed in (Mukhometov, 1977), we can write the inequality

$$\int_{D(z)} \int_0^{2\pi} \left[(\tilde{a}\eta_1 + \tilde{b})^2 - \left(\tilde{a} \frac{\partial \eta_1}{\partial \varphi} \right)^2 \right] d\varphi \, d\bar{x} \leq \int_{S(z)} \int_{S(z)} \left| \frac{\partial \tilde{p}_1}{\partial l} \cdot \frac{\partial \tilde{p}_1}{\partial l'} \right| dl \, dl', \tag{5.3.54}$$

which is an analog of inequality (5.3.23). Since

$$(\tilde{a}\eta_1 + \tilde{b})^2 - \left(\tilde{a} \frac{\partial \eta_1}{\partial \varphi} \right)^2 \geq \frac{1}{2} \tilde{b}^2 - \tilde{a}^2 (\eta_1^2 + \eta_{1\varphi}^2)$$

$$\geq \frac{1}{2} \tilde{b}^2 - \tilde{a}^2 \left[(1 + \cos^2 \varphi)^2 + \sin^2 2\varphi \right],$$

we have

$$\int_0^{2\pi} \left[(\tilde{a}\eta_1 + \tilde{b})^2 - \left(\tilde{a} \frac{\partial \eta_1}{\partial \varphi} \right)^2 \right] d\varphi \geq \pi \left(\tilde{b}^2 - \frac{23}{4} \tilde{a}^2 \right).$$

Consequently,

$$\int_{D(z)} \left(\tilde{b}^2 - \frac{23}{4} \tilde{a}^2 \right) d\bar{x} \leq \frac{1}{\pi} \int_{S(z)} \int_{S(z)} \left| \frac{\partial \tilde{p}_1}{\partial l} \cdot \frac{\partial \tilde{p}_1}{\partial l'} \right| dl \, dl'. \tag{5.3.55}$$

The estimate (5.3.24) follows immediately from inequalities (5.3.23) and (5.3.55). This proves the theorem. □

5.4. AN INVERSE PROBLEM OF ELECTROMAGNETOELASTICITY IN THE CASE OF NONLINEAR INTERACTION

In this section, following the original work (Lorenzi and Priimenko, 1996), we give some results of solution of direct and inverse problems for the system of electromagnetoelasticity in the case of nonlinear interaction between electromagnetic and elastic fields.

We shall consider one possible statement of the problem which arises in the theory of electromagnetoelasticity under the following assumptions:

1. Ω_1, Ω_2, and Ω are three bounded connected open sets in $\mathbb{R}^3$ such that Ω_2 and Ω belong to the classes C^3 and C^2, respectively, and the following conditions are fulfilled:

$$\overline{\Omega_2} \subset\subset \Omega, \qquad \Omega_1 = \Omega \setminus \overline{\Omega_2}. \tag{5.4.1}$$

2. An oscillating inhomogeneous electrical-conducting elastic body B occupying the domain $\Omega_2 \subset \mathbb{R}^3$ is placed into the domain Ω where the process of propagation of electromagnetic waves occurs.

3. The electromagnetic field arises as a result of propagation of elastic oscillations. Moreover, we neglect the reverse effect of the electromagnetic field on the process of propagation of elastic waves.

4. We neglect the transport currents in the domain Ω.

5. The motion of the medium occurs with velocities that are much smaller than the velocity of electromagnetic waves in the elastic medium.

In contrast to the previous problems, in this section we give some considerations that justify our choice of the defining relations between the vector-functions $\boldsymbol{D}$, $\boldsymbol{B}$, $\boldsymbol{E}$, $\boldsymbol{H}$, $\boldsymbol{U}$, and $\boldsymbol{J}$.

Let us consider the relativistic variant of the theory of electromagnetism (Tamm, 1976). Using the MKS system of units, we obtain the following relations between $\boldsymbol{D}$, $\boldsymbol{B}$ and $\boldsymbol{E}$, $\boldsymbol{H}$, $\partial \widetilde{\boldsymbol{U}}/\partial t$, where $\widetilde{\boldsymbol{U}}$ is the continuation of the function $\boldsymbol{U}$ by zero over the whole of the domain $(0, T) \times \Omega$, $T > 0$:

$$\boldsymbol{D} = \frac{1}{1 - \varepsilon\mu \left|\frac{\partial \widetilde{\boldsymbol{U}}}{\partial t}\right|^2} \left\{ \varepsilon\left(1 - c^{-2}\left|\frac{\partial \widetilde{\boldsymbol{U}}}{\partial t}\right|^2\right) \boldsymbol{E} \right.$$
$$\left. + (\varepsilon\mu - c^{-2}) \left[\frac{\partial \widetilde{\boldsymbol{U}}}{\partial t} \times \boldsymbol{H} - \varepsilon\left(\frac{\partial \widetilde{\boldsymbol{U}}}{\partial t} \cdot \boldsymbol{E}\right)\frac{\partial \widetilde{\boldsymbol{U}}}{\partial t}\right] \right\}, \tag{5.4.2}$$

$$B = \frac{1}{1-\varepsilon\mu\left|\frac{\partial \widetilde{U}}{\partial t}\right|^2}\left\{\mu\left(1-c^{-2}\left|\frac{\partial \widetilde{U}}{\partial t}\right|^2\right)H\right.$$

$$\left. -(\varepsilon\mu - c^{-2})\left[\frac{\partial \widetilde{U}}{\partial t} \times E + \mu\left(\frac{\partial \widetilde{U}}{\partial t} \cdot H\right)\frac{\partial \widetilde{U}}{\partial t}\right]\right\}. \quad (5.4.3)$$

According to assumption 4 the corresponding Ohm law is as follows:

$$J = \sigma\left(1 - c^{-2}\left|\frac{\partial \widetilde{U}}{\partial t}\right|^2\right)^{-1/2}\left[E + \frac{\partial \widetilde{U}}{\partial t} \times B - c^{-2}\left(\frac{\partial \widetilde{U}}{\partial t} \cdot E\right)\frac{\partial \widetilde{U}}{\partial t}\right]. \quad (5.4.4)$$

In virtue of assumption 5, which can be formulated as

$$(\varepsilon\mu)^{1/2}\left|\frac{\partial \widetilde{U}}{\partial t}\right| \ll 1 \quad (\varepsilon\mu \geq c^{-2}), \quad (5.4.5)$$

relations (5.4.2)–(5.4.4) take the form

$$D = \varepsilon E, \quad B = \mu H, \quad J = \sigma\left[E + \mu\frac{\partial \widetilde{U}}{\partial t} \times H\right]. \quad (5.4.6)$$

Remark 5.4.1. The defining relations (5.3.2) in the previous section can be obtained from relations (5.4.2)–(5.4.4) by linearizing them with respect to the vector-functions H, E, and $\widetilde{U}$.

Using the relations obtained, we can write the Maxwell system in the domain $(0, T) \times [\Omega_1 \cup \Omega_2]$ in the form

$$\begin{aligned}\varepsilon\frac{\partial E}{\partial t} &= \operatorname{rot} H - \sigma E - \sigma\mu\frac{\partial \widetilde{U}}{\partial t} \times H, \\ \mu\frac{\partial H}{\partial t} &= -\operatorname{rot} E, \quad \operatorname{div}\mu H = 0.\end{aligned} \quad (5.4.7)$$

By assumptions 2 and 3, the propagation of elastic waves in the body B is governed by the ordinary system of Lamé's equations

$$\rho\frac{\partial^2 U}{\partial t^2} = \operatorname{Div} T + F, \quad (t, x) \in (0, T) \times \Omega_2, \quad (5.4.8)$$

where $\rho : \Omega_2 \to \mathbb{R}_+$ and $F, U : (0, T) \times \Omega_2 \to \mathbb{R}^3$. The stress tensor T is defined by the formula

$$T = \lambda \operatorname{tr} S \cdot I + 2\varkappa S, \quad (5.4.9)$$

where S is the strain tensor with the components

$$S_{ij} = \frac{1}{2}\left(\frac{\partial U_i}{\partial x_j} + \frac{\partial U_j}{\partial x_i}\right), \quad i,j = 1,2,3,$$

and $\lambda, \varkappa : \Omega_2 \to \mathbb{R}_+$. In this section, we assume that the function $\boldsymbol{F}$ has the following representation:

$$\boldsymbol{F}(t,x) = f(t)\,\boldsymbol{g}(t,x), \tag{5.4.10}$$

where $\boldsymbol{g} : [0,T] \times \Omega_2 \to \mathbb{R}^3$ is a known vector-function and the function $f : [0,T] \to \mathbb{R}$ is unknown.

Our main problem consists in determining the function f. For this, we need to supplement the differential equations (5.4.7) and (5.4.8) with the appropriate initial and boundary conditions and the gluing conditions for the solution of the problem on the surfaces where the coefficients of the differential equations have breaks. To reconstruct the function f itself we need some additional information. As usual, we shall denote by $[\boldsymbol{w}]_{\partial\Omega}$ the jump of a vector-function $\boldsymbol{w} : \Omega \to \mathbb{R}^3$ across the oriented surface $\partial\Omega$; this jump is calculated in the direction of the internal normal to $\partial\Omega$:

$$[\boldsymbol{w}]_{\partial\Omega} = \boldsymbol{w}|_{\partial\Omega_+} - \boldsymbol{w}|_{\partial\Omega_-}. \tag{5.4.11}$$

Now we can formulate the inverse problem.

Inverse Problem 5.4.1 (IP 5.4.1). *Determine the set of functions*

$$\boldsymbol{U} : [0,T] \times \Omega_2 \to \mathbb{R}^3, \quad \boldsymbol{E}, \boldsymbol{H} : [0,T] \times \Omega \to \mathbb{R}^3, \quad f : [0,T] \to \mathbb{R}$$

such that

$$\rho\frac{\partial^2 \boldsymbol{U}}{\partial t^2} = \operatorname{Div} T + f(t)\,\boldsymbol{g}(t,x), \quad (t,x) \in (0,T) \times \Omega_2, \tag{5.4.12}$$

$$\boldsymbol{U}(0,x) = \boldsymbol{U}_0(x), \quad x \in \Omega_2, \tag{5.4.13}$$

$$\frac{\partial \boldsymbol{U}}{\partial t}(0,x) = \boldsymbol{U}_1(x), \quad x \in \Omega_2, \tag{5.4.14}$$

$$\boldsymbol{U}(t,x) = 0, \quad (t,x) \in (0,T) \times \partial\Omega_2, \tag{5.4.15}$$

$$\varepsilon\frac{\partial \boldsymbol{E}}{\partial t} = \operatorname{rot} \boldsymbol{H} - \sigma\boldsymbol{E} - \sigma\mu\frac{\partial \widetilde{\boldsymbol{U}}}{\partial t} \times \boldsymbol{H}, \quad (t,x) \in (0,T) \times [\Omega_1 \cup \Omega_2], \tag{5.4.16}$$

$$\mu\frac{\partial \boldsymbol{H}}{\partial t} = -\operatorname{rot} \boldsymbol{E}, \quad \operatorname{div} \mu\boldsymbol{H} = 0, \quad (t,x) \in (0,T) \times [\Omega_1 \cup \Omega_2], \tag{5.4.17}$$

$$\boldsymbol{E}(0,x) = \boldsymbol{E}_0(x), \quad x \in \Omega, \tag{5.4.18}$$

$$\boldsymbol{H}(0,x) = \boldsymbol{H}_0(x), \quad x \in \Omega, \tag{5.4.19}$$

$$\mathbf{n} \times \mathbf{E} = 0, \quad (t,x) \in (0,T) \times \partial\Omega, \qquad (5.4.20)$$
$$[\mathbf{E} \times \mathbf{n}]_{\partial\Omega_2} = [\mathbf{H} \times \mathbf{n}]_{\partial\Omega_2} = 0, \quad (t,x) \in (0,T) \times \partial\Omega_2, \qquad (5.4.21)$$
$$\Phi[\mathbf{E}] = \phi(t), \quad t \in [0,T]. \qquad (5.4.22)$$

It is assumed that the functions $\varepsilon, \mu : \overline{\Omega} \to \mathbb{R}_+$, $\sigma : \overline{\Omega} \to \overline{\mathbb{R}_+}$, and $\mathbf{E}_0, \mathbf{H}_0 : \Omega \to \mathbb{R}^3$ are continuous in the domain $\overline{\Omega} \setminus \partial\Omega_2$ with possible jumps on the surface $\partial\Omega_2$. We also assume that the functions $\mathbf{g} : [0,T] \times \Omega_2 \to \mathbb{R}^3$, $\mathbf{U}_0, \mathbf{U}_1 : \Omega_2 \to \mathbb{R}^3$, and $\phi : [0,T] \to \mathbb{R}$ are given and have sufficient smoothness. As concerns the functional Φ, we assume that it is linear and depends only on the spatial variables. For example, we may assume that the functional Φ has the form

$$\Phi[\mathbf{E}] = \int_\Omega \mathbf{K}(x) \cdot \mathbf{E}(x)\, dx, \qquad (5.4.23)$$

where $\mathbf{K} : \Omega \to \mathbb{R}^3$ is a given sufficiently smooth vector-function.

Remark 5.4.2. It is the additional information (5.4.22) only that connects the solutions to the Lamé equations and the Maxwell equations. If instead of the additional information (5.4.22) we specify a certain information about the vector-function $\mathbf{U}$, for example, of the form

$$\Phi[\mathbf{U}] = \phi(t), \quad t \in [0,T], \qquad (5.4.24)$$

then the IP 5.4.1 splits into the inverse problem of determining the function $f(t)$ from equations (5.4.12)–(5.4.15) and (5.4.24) and then the direct problem (5.4.16)–(5.4.21).

Before proceeding to solution of the IP 5.4.1 we investigate in detail the smoothness properties of the solution to the direct problem (5.4.12)–(5.4.21). The results of this investigation will be used essentially in solution of the IP 5.4.1.

Solution of the direct problem (5.4.12)–(5.4.15)

In solving the direct problem (5.4.12)–(5.4.15) we assume that the function f is known. It is worth noting that the term $\partial \widetilde{\mathbf{U}}/\partial t \times \mathbf{H}$ occurring in equation (5.4.16) creates certain difficulties during solution of the direct problem (5.4.12)–(5.4.21). In fact, we cannot confine ourselves to consideration of the weak solution $(\mathbf{U}, \mathbf{E}, \mathbf{H})$ of the problem since in this case the product $\partial \widetilde{\mathbf{U}}/\partial t \times \mathbf{H}$ may fail to be an element of the space $L^2(\Omega_2; \mathbb{R}^3)$. In order to avoid imposing too severe constraints on the function, we should require

Chapter 5. Coupled Maxwell and Lamé systems

at least that both the multipliers $\partial \tilde{U}/\partial t$ and H be elements of the space $L^4(\Omega_2; \mathbb{R}^3)$. For this, we assume that the density ρ, the Lamé coefficients λ and $\varkappa$, the free term fg, and the initial data U_0 and U_1 of the problem (5.4.12)–(5.4.15) satisfy the following conditions:

$$\rho \in H^2(\Omega_2; \mathbb{R}), \qquad \varkappa, \lambda \in W^{2,\infty}(\Omega_2; \mathbb{R}), \tag{5.4.25}$$

$$\min\bigl(\rho(x), \lambda(x), \varkappa(x)\bigr) \geq \rho_0 > 0, \quad \forall x \in \overline{\Omega_2}, \tag{5.4.26}$$

$$f \in L^p((0,T); \mathbb{R}), \quad g \in L^{p'}\bigl((0,T); H^2(\Omega_2; \mathbb{R}^3)\bigr) \cap H_0^1(\Omega_2; \mathbb{R}^3))$$

$$(1/p + 1/p' = 1), \tag{5.4.27}$$

$$U_0 \in H^3(\Omega_2; \mathbb{R}^3) \cap H_0^1(\Omega_2; \mathbb{R}^3), \qquad \text{Div } T \in H_0^1(\Omega_2; \mathbb{R}^3), \tag{5.4.28}$$

$$U_1 \in H^2(\Omega_2; \mathbb{R}^3) \cap H_0^1(\Omega_2; \mathbb{R}^3). \tag{5.4.29}$$

Theorem 5.4.1. *Let ρ, λ, $\varkappa$, f, g, U_0, and U_1 satisfy conditions (5.4.25)–(5.4.29). Then there exists a unique solution $U(f)$ to the problem (5.4.12)–(5.4.15). This solution satisfies the conditions*

$$U(f) \in C\bigl([0,T]; H^3(\Omega_2; \mathbb{R}^3) \cap H_0^1(\Omega_2; \mathbb{R}^3)\bigr) \cap C^1\bigl([0,T]; H^2(\Omega_2; \mathbb{R}^3)\bigr)$$

$$\cap W^{2,1}\bigl((0,T); H^1(\Omega_2; \mathbb{R}^3)\bigr), \tag{5.4.30}$$

$$\left(\|U(f)(t)\|_{3,2}^2 + \left\|\frac{\partial}{\partial t}U(f)(t)\right\|_{2,2}^2 + \left\|\frac{\partial^2}{\partial t^2}U(f)(t)\right\|_{1,2}^2\right)^{1/2}$$

$$\leq C_1\bigl(\rho_0^{-1}, \|\rho\|_{2,2}, \|\varkappa\|_{2,\infty}, \|\lambda\|_{2,\infty}\bigr)$$

$$\times \left[\bigl(\|U_0\|_{3,2}^2 + \|U_1\|_{2,2}^2\bigr)^{1/2} + \|g\|_{t,0,p',2,2} \cdot \|f\|_{t,0,p}\right],$$

$$\forall t \in [0,T], \quad \forall f \in L^p((0,T); \mathbb{R}), \tag{5.4.31}$$

$$\left(\|U(f_2)(t) - U(f_1)(t)\|_{3,2}^2 + \left\|\frac{\partial}{\partial t}U(f_2)(t) - \frac{\partial}{\partial t}U(f_1)(t)\right\|_{2,2}^2 \right.$$

$$\left. + \left\|\frac{\partial^2}{\partial t^2}U(f_2)(t) - \frac{\partial^2}{\partial t^2}U(f_1)(t)\right\|_{1,2}^2\right)^{1/2}$$

$$\leq C_1\bigl(\rho_0^{-1}, \|\rho\|_{2,2}, \|\varkappa\|_{2,\infty}, \|\lambda\|_{2,\infty}\bigr) \|g\|_{t,0,p',2,2} \cdot \|f_2 - f_1\|_{t,0,p},$$

$$\forall t \in [0,T], \quad \forall f_1, f_2 \in L^p((0,T); \mathbb{R}). \tag{5.4.32}$$

Here $\|\cdot\|_{j,2}$, $\|\cdot\|_{t,0,q,j,2}$, and $\|\cdot\|_{t,0,p}$ are the norms in the spaces $H^j(\Omega_2; \mathbb{R}^3)$, $L^q((0,t); H^j(\Omega_2; \mathbb{R}^3))$, and $L^p((0,t); \mathbb{R})$, respectively, and C_1 is a nonnegative function continuous and nondecreasing in each of its arguments.

To prove the main result on solvability of the IP 5.4.1 we shall also need some properties of the solution to the direct problem (5.4.16)–(5.4.21).

We make the following assumptions about the coefficients and initial data of the problem:

$$\varepsilon, \mu, \sigma \in W^{1,\infty}(\Omega_1; \mathbb{R}) \cap W^{1,\infty}(\Omega_2; \mathbb{R}),$$
$$\min\{\varepsilon(x), \mu(x)\} \geq \gamma^{-1} > 0, \quad \forall x \in \Omega_1 \cup \Omega_2; \quad (5.4.33)$$

$$\boldsymbol{E}_0 \in H(\mathrm{rot}; \Omega), \quad \boldsymbol{H}_0 \in H(\mathrm{rot}; \Omega) \cap H^1(\Omega_1; \mathbb{R}^3) \cap H^1(\Omega_2; \mathbb{R}^3),$$
$$\mu \boldsymbol{H}_0 \in H(\mathrm{div}; \Omega); \quad (5.4.34)$$

$$\boldsymbol{n} \times \boldsymbol{E}_0 = 0, \quad x \in \partial\Omega; \quad \mathrm{div}\,\mu \boldsymbol{H}_0 = 0, \quad x \in \Omega;$$
$$\boldsymbol{n} \cdot \boldsymbol{H}_0 = 0, \quad x \in \partial\Omega. \quad (5.4.35)$$

Theorem 5.4.2. *Let the vector-functions* g, $\boldsymbol{U}_0$, $\boldsymbol{U}_1$, $\boldsymbol{E}_0$, *and* $\boldsymbol{H}_0$ *satisfy conditions* (5.4.27)–(5.4.29), (5.4.34), *and* (5.4.35). *Then for every function* $f \in L^p((0,T); \mathbb{R})$ *the problem* (5.4.16)–(5.4.21) *has a unique solution* $(\boldsymbol{E}, \boldsymbol{H}) = (\boldsymbol{E}(f), \boldsymbol{H}(f))$ *satisfying the conditions*

$$\boldsymbol{E}(f) \in C\big([0,T]; H(\mathrm{rot}; \Omega)\big) \cap C^1\big([0,T]; L^2(\Omega; \mathbb{R}^3)\big), \quad (5.4.36)$$
$$\boldsymbol{H}(f) \in C\big([0,T]; H(\mathrm{rot}; \Omega) \cap H^1(\Omega_2; \mathbb{R}^3)\big) \cap C^1\big([0,T]; L^2(\Omega; \mathbb{R}^3)\big), \quad (5.4.37)$$

$$\left\|\frac{\partial}{\partial t}\boldsymbol{E}(f)(t)\right\|_{0,2,\Omega} + \left\|\frac{\partial}{\partial t}\boldsymbol{H}(f)(t)\right\|_{0,2,\Omega} + \|\mathrm{rot}\,\boldsymbol{E}(f)(t)\|_{0,2,\Omega}$$
$$+ \|\boldsymbol{E}(f)(t)\|_{0,2,\Omega} + \|\boldsymbol{H}(f)(t)\|_{1,2,\Omega_1} + \|\boldsymbol{H}(f)(t)\|_{1,2,\Omega_2}$$
$$\leq C_2(T) + T C_3(T, \|f\|_{T,0,p}), \quad \forall t \in (0,T), \quad (5.4.38)$$

where $\|\cdot\|_{j,2,\Omega_k}$ *denotes the norm in* $H^j(\Omega_k; \mathbb{R}^3)$ *and* C_2 *and* C_3 *are positive nondecreasing continuous functions depending also on the norms of the data of the problem. Moreover, for every pair of functions* $f_1, f_2 \in L^p((0,T); \mathbb{R})$ *the following estimate is true:*

$$\left\|\frac{\partial}{\partial t}\boldsymbol{E}(f_2)(t) - \frac{\partial}{\partial t}\boldsymbol{E}(f_1)(t)\right\|_{0,2,\Omega} + \left\|\frac{\partial}{\partial t}\boldsymbol{H}(f_2)(t) - \frac{\partial}{\partial t}\boldsymbol{H}(f_1)(t)\right\|_{0,2,\Omega}$$
$$+ \|\mathrm{rot}\,\boldsymbol{E}(f_2)(t) - \mathrm{rot}\,\boldsymbol{E}(f_1)(t)\|_{0,2,\Omega} + \|\boldsymbol{E}(f_2)(t) - \boldsymbol{E}(f_1)(t)\|_{0,2,\Omega}$$
$$+ \|\boldsymbol{H}(f_2)(t) - \boldsymbol{H}(f_1)(t)\|_{1,2,\Omega_1} + \|\boldsymbol{H}(f_2)(t) - \boldsymbol{H}(f_1)(t)\|_{1,2,\Omega_2}$$
$$\leq C_4(T, \|f_1\|_{T,0,p}) \int_0^t h(f_2)(t-s) \|f_2 - f_1\|_{s,0,p}\, ds, \quad (5.4.39)$$

where

$$h(f)(t) = \exp\left[t\left(\gamma \|\sigma\|_{0,\infty,\Omega} + C_5(T)\|f\|_{T,0,p} \cdot \|g\|_{T,0,p',2,2}\right)\right] \quad (5.4.40)$$

and C_4 and C_5 are positive nondecreasing continuous functions depending also on the norms of the data of the problem.

The proofs of Theorems 5.4.1 and 5.4.2 are rather bulky and are therefore omitted. For their complete proofs the reader is referred to the original paper by Lorenzi and Priimenko (1996).

Solution of the Inverse Problem 5.4.1

We can now prove the solvability of the inverse problem. We assume that the function f in the free term of equation (5.4.12) is unknown and the function g is known. By Theorems 5.4.1 and 5.4.2, our IP 5.4.1 is equivalent to the following problem:

Inverse Problem 5.4.2 (IP 5.4.2). *Find a function $f \in L^p((0,T);\mathbb{R})$ such that*
$$\Phi[\boldsymbol{E}(f)(t,\cdot)] = \phi(t), \quad \forall t \in [0,T]. \tag{5.4.41}$$

To prove the solvability of the IP 5.4.2 we require the fulfilment of the following regularity conditions for the data ϕ, the function g, and the kernel $\boldsymbol{K}$ of the functional Φ:

$$\phi \in W^{2,p}((0,T_0);\mathbb{R}), \quad \boldsymbol{g} \in L^{p'}((0,T_0); H^2(\Omega_2;\mathbb{R}^3) \cap H_0^1(\Omega_2;\mathbb{R}^3)),$$
$$T_0 > 0; \tag{5.4.42}$$
$$\|\boldsymbol{g}(t,\cdot)\|_{2,2,\Omega_2} \leq C_6(T_1)\, t^{-\alpha} \tag{5.4.43}$$

for almost all $t \in (0,T_1)$ and some $\alpha \in (0,1/p')$ and $T_1 \in (0,T_0)$; and

$$\varepsilon^{-1}\boldsymbol{K} \in H(\mathrm{rot};\Omega), \quad \boldsymbol{n} \times \boldsymbol{K} = 0, \quad \forall x \in \partial\Omega. \tag{5.4.44}$$

Remark 5.4.3. The first assumption in (5.4.44) is equivalent to the requirement that $\boldsymbol{K} \in H(\mathrm{rot};\Omega_1) \cap H(\mathrm{rot};\Omega_2)$ and $\boldsymbol{n} \times [\varepsilon^{-1}\boldsymbol{K}]_{\partial\Omega_2} = 0$.

Using the results of (Sheen, 1992) and formulas (5.4.23) and (5.4.44) we easily obtain the relation

$$\Phi[\varepsilon^{-1}\mathrm{rot}\,\boldsymbol{H}(t,\cdot)] = \int_\Omega \boldsymbol{H}(t,x) \cdot \mathrm{rot}(\varepsilon^{-1}\boldsymbol{K})(x)\,\mathrm{d}x := \Phi_1[\boldsymbol{H}(t,\cdot)]. \tag{5.4.45}$$

Replacing $\boldsymbol{U}$ in equations (5.4.12) and (5.4.16) by $\boldsymbol{U}(f)$, applying the linear functional Φ to both sides of equation (5.4.16), and using the information (5.4.41), we obtain

$$\phi'(t) = \Phi_1[\boldsymbol{H}(f)(t,\cdot)] - \Phi\left[\frac{\sigma}{\varepsilon}\boldsymbol{E}(f)(t,\cdot)\right]$$
$$- \Phi_2\left[\frac{\sigma\mu}{\varepsilon}\frac{\partial}{\partial t}\boldsymbol{U}(f)(t,\cdot) \times \boldsymbol{H}(f)(t,\cdot)\right], \quad \forall t \in [0,T], \quad (5.4.46)$$

where

$$\Phi_2[\boldsymbol{E}] = \int_{\Omega_2} \boldsymbol{K}(x) \cdot \boldsymbol{E}(x)\,\mathrm{d}x. \qquad (5.4.47)$$

Equations (5.4.41) and (5.4.46) imply that the data $(\boldsymbol{U}_1, \boldsymbol{E}_0, \boldsymbol{H}_0, \phi)$ should satisfy the following consistency conditions:

$$\phi(0) = \Phi[\boldsymbol{E}_0], \qquad \phi'(0) = \Phi_1[\boldsymbol{H}_0] - \Phi\left[\frac{\sigma}{\varepsilon}\boldsymbol{E}_0\right] - \Phi_2\left[\frac{\sigma\mu}{\varepsilon}\boldsymbol{U}_1 \times \boldsymbol{H}_0\right]. \quad (5.4.48)$$

Conversely, if the function $f \in L^p((0,T);\mathbb{R})$ solves equation (5.4.46), then, using formulas (5.4.45) and (5.4.47), we can conclude that the function f solves the equation

$$\frac{\partial}{\partial t}\{\Phi[\boldsymbol{E}(f)(t,\cdot)] - \phi(t)\} = 0, \quad \forall t \in [0,T]. \qquad (5.4.49)$$

Using equation (5.4.49) and the first condition from (5.4.48), we can easily show that the function f is a solution to equation (5.4.41). Differentiating equation (5.4.46) once again with respect to the variable t and taking into account relation (5.4.12), we finally obtain

$$\phi''(t) = \Phi_1\left[\frac{\partial}{\partial t}\boldsymbol{H}(f)(t,\cdot)\right] - \Phi\left[\frac{\sigma}{\varepsilon}\frac{\partial}{\partial t}\boldsymbol{E}(f)(t,\cdot)\right]$$
$$- \Phi_2\left[\frac{\sigma\mu}{\varepsilon}\frac{\partial}{\partial t}\boldsymbol{U}(f)(t,\cdot) \times \frac{\partial}{\partial t}\boldsymbol{H}(f)(t,\cdot)\right]$$
$$- \Phi_2\left[\frac{\sigma\mu}{\varepsilon\rho}\operatorname{Div} T(\boldsymbol{U}(f)(t,\cdot)) \times \boldsymbol{H}(f)(t,\cdot)\right]$$
$$- f(t)\,\Phi_2\left[\frac{\sigma\mu}{\varepsilon\rho}\boldsymbol{g}(t,\cdot) \times \boldsymbol{H}(f)(t,\cdot)\right], \quad \text{for a.a. } t \in (0,T). \quad (5.4.50)$$

Note that according to the assumptions (5.4.42)–(5.4.44) and Theorems 5.4.1 and 5.4.2, each term in (5.4.50) has sense.

Conversely, if $f \in L^p((0,T);\mathbb{R})$ is a solution to equation (5.4.50), then f should be a solution to the equation

$$\frac{\partial}{\partial t}\left\{\phi'(t) - \Phi_1[\boldsymbol{H}(f)(t,\cdot)] + \Phi\left[\frac{\sigma}{\varepsilon}\boldsymbol{E}(f)(t,\cdot)\right]\right.$$
$$\left. + \Phi_2\left[\frac{\sigma\mu}{\varepsilon}\frac{\partial}{\partial t}\boldsymbol{U}(f)(t,\cdot) \times \boldsymbol{H}(f)(t,\cdot)\right]\right\} = 0, \quad \text{for a.a. } t \in (0,T). \quad (5.4.51)$$

But as follows from (5.4.51) and from the last condition in (5.4.48) the function f is a solution to equation (5.4.46). Thus, we have shown that the solutions of equations (5.4.41) and (5.4.50) are equivalent.

In order that it were possible to rewrite equation (5.4.50) in the form of an equation with a stationary point for f, we assume that g and H_0 satisfy the condition

$$\left| \int_{\Omega_2} \frac{\sigma(x)\,\mu(x)}{\varepsilon(x)\,\rho(x)} K(x) \cdot [g(t,x) \times H_0(x)]\, dx \right|$$
$$= \left| \Phi_2\!\left[\frac{\sigma\mu}{\varepsilon\rho} g(t,\cdot) \times H_0\right] \right| \geq 2m, \quad \text{for a.a. } t \in (0,T), \quad (5.4.52)$$

where m is a certain positive constant.

Now we can formulate our main result.

Theorem 5.4.3. *Let $\rho, \lambda, \varkappa$ and ε, μ, σ satisfy conditions (5.4.25), (5.4.26), and (5.4.33); and let the functions U_0, U_1, E_0, H_0 and g, ϕ, K satisfy conditions (5.4.27)–(5.4.29), (5.4.34), (5.4.35), (5.4.42)–(5.4.44), and (5.4.48). Then there exists a number $T^* \in (0, \min\{T_0, T_1\})$ for which the IP 5.4.1 has a unique solution (U, E, H, f) satisfying conditions (5.4.27), (5.4.30), (5.4.36), and (5.4.37) for every $T \in (0, T^*)$.*

Remark 5.4.4. We can also show that the solution (U, E, H, f) of the IP 5.4.1 depends continuously on the data $(U_0, U_1, E_0, H_0, g, \phi)$ in the norms of the respective spaces. For this, we need to prove the results on continuous dependence of the solution on the data, just as it was done in the case of solutions of the direct problems (5.4.12)–(5.4.15) and (5.4.16)–(5.4.21). Since the proof is very bulky, we omit this part.

Now we proceed to the proof of Theorem 5.4.3. Introduce the family $X(M, T)$ of complete metric subspaces in $L^p((0, T); \mathbb{R})$, this family depending on two positive constants M and T:

$$X(M, T) = \{f \in L^p((0, T); \mathbb{R}) \mid \|f\|_{T,0,p} \leq M\}. \quad (5.4.53)$$

If we assume that $f \in X(M, T)$ is a solution to the operator equation (5.4.50), then we can write it in the following form convenient for application of the theorem about a stationary point:

$$f(t) = \left(\Phi_2\left[\frac{\sigma\mu}{\varepsilon\rho} \boldsymbol{g}(t,\cdot) \times \boldsymbol{H}(f)(t,\cdot)\right]\right)^{-1}$$

$$\times \left\{-\phi''(t) + \Phi_1\left[\frac{\partial}{\partial t}\boldsymbol{H}(f)(t,\cdot)\right] - \Phi\left[\frac{\sigma}{\varepsilon}\frac{\partial}{\partial t}\boldsymbol{E}(f)(t,\cdot)\right]\right.$$

$$- \Phi_2\left[\frac{\sigma\mu}{\varepsilon}\frac{\partial}{\partial t}\boldsymbol{U}(f)(t,\cdot) \times \frac{\partial}{\partial t}\boldsymbol{H}(f)(t,\cdot)\right]$$

$$\left. - \Phi_2\left[\frac{\sigma\mu}{\varepsilon\rho}\operatorname{Div}T(\boldsymbol{U}(f)(t,\cdot)) \times \boldsymbol{H}(f)(t,\cdot)\right]\right\} := N(f)(t) \quad (5.4.54)$$

for a.a. $t \in (0,T)$.

Our task is to show the local solvability of equation (5.4.54). For this, we set $T_3(M) = \min\{T_0, T_1, T_2(M)\}$, where $T_2(M)$ is the unique positive root of the equation

$$\|\boldsymbol{K}\|_{0,2,\Omega_2}\left[C_2(T_0) + TC_3(T_0,M)\right]C_6\gamma\rho_0^{-1}\|\sigma\|_{0,\infty,\Omega_2}\|\mu\|_{0,\infty,\Omega_2}T^{1-\alpha} = m. \quad (5.4.55)$$

Using the estimate (5.4.38) and condition (5.4.52), we can show that each function $f \in X(M,T)$ satisfies the following basic inequality for a.a. $t \in (0,T) \subset (0,T_3(M))$:

$$\left|\Phi_2\left[\frac{\sigma\mu}{\varepsilon\rho}\boldsymbol{g}(t,\cdot) \times \boldsymbol{H}(f)(t,\cdot)\right]\right|$$

$$\geq \left|\Phi_2\left[\frac{\sigma\mu}{\varepsilon\rho}\boldsymbol{g}(t,\cdot) \times \boldsymbol{H}_0\right]\right| - \left|\Phi_2\left[\frac{\sigma\mu}{\varepsilon\rho}\boldsymbol{g}(t,\cdot) \times \int_0^t \frac{\partial}{\partial t}\boldsymbol{H}(f)(s,\cdot)\,ds\right]\right|$$

$$\geq 2m - \|\boldsymbol{K}\|_{0,2,\Omega_2}\gamma\rho_0^{-1}\|\sigma\|_{0,\infty,\Omega_2}\|\mu\|_{0,\infty,\Omega_2}\|\boldsymbol{g}(t,\cdot)\|_{0,\infty,\Omega_2}$$

$$\times \left\|\frac{\partial}{\partial t}\boldsymbol{H}(f)(t,\cdot)\right\|_{T,0,0,2} \cdot t$$

$$\geq 2m - \|\boldsymbol{K}\|_{0,2,\Omega_2}\left[C_2(T_0) + TC_3(T_0,M)\right]C_6\gamma\rho_0^{-1}\|\sigma\|_{0,\infty,\Omega_2}$$

$$\times \|\mu\|_{0,\infty,\Omega_2}T^{1-\alpha} \geq m. \quad (5.4.56)$$

Taking into account formulas (5.4.30), (5.4.38), and (5.4.56) and the embedding $H^1(\Omega_2;\mathbb{R}) \cdot H^1(\Omega_2;\mathbb{R}) \hookrightarrow L^2(\Omega_2;\mathbb{R})$ (where the dot denotes the

Chapter 5. Coupled Maxwell and Lamé systems

functional product), we can estimate the nonlinear operator N:

$$\|N(f)\|_{T,0,p} \leq \frac{1}{m} \Big\{ \|\phi''\|_{T_0,0,p} + \|\text{rot}(\varepsilon^{-1}\boldsymbol{K})\|_{0,2,\Omega} \Big\| \frac{\partial}{\partial t} \boldsymbol{H}(f)(t,\cdot) \Big\|_{0,2,\Omega}$$

$$+ \gamma \|\sigma\|_{0,\infty,\Omega} \|\boldsymbol{K}\|_{0,2,\Omega} \Big\| \frac{\partial}{\partial t} \boldsymbol{E}(f)(t,\cdot) \Big\|_{0,2,\Omega}$$

$$+ \gamma \|\sigma\|_{0,\infty,\Omega_2} \|\mu\|_{0,\infty,\Omega_2} \|\boldsymbol{K}\|_{0,2,\Omega} \Big(\Big\| \frac{\partial}{\partial t} \boldsymbol{U}(f)(t,\cdot) \Big\|_{0,\infty,\Omega_2} \Big\| \frac{\partial}{\partial t} \boldsymbol{H}(f)(t,\cdot) \Big\|_{0,2,\Omega_2} \Big)$$

$$+ \rho_0^{-1} C_7 \big(\|\lambda\|_{0,\infty,\Omega_2}, \|\varkappa\|_{0,\infty,\Omega_2}, \Omega_2 \big) \|\boldsymbol{U}(f)(t,\cdot)\|_{3,2,\Omega_2} \|\boldsymbol{H}(f)(t,\cdot)\|_{1,2,\Omega_2} \Big) \Big\}$$

$$\leq \frac{1}{m} \Big\{ \|\phi''\|_{T_0,0,p} + \big(1 + \|g\|_{T_0,0,p',2,2} + \|\boldsymbol{K}\|_{0,2,\Omega} + \|\text{rot}(\varepsilon^{-1}\boldsymbol{K})\|_{0,2,\Omega} \big)$$

$$\times \big[C_8(T_0) + T C_9(T_0, M) \big] \Big\}, \quad \forall T \in (0, T_3(M)]. \quad (5.4.57)$$

Let us choose a pair of numbers (M, T^*) from the conditions

$$M = \frac{2}{m} \Big\{ \|\phi''\|_{T_0,0,p} + C_8(T_0) \big(1 + \|g\|_{T_0,0,p',2,2} + \|\boldsymbol{K}\|_{0,2,\Omega}$$

$$+ \|\text{rot}(\varepsilon^{-1}\boldsymbol{K})\|_{0,2,\Omega} \big) \Big\}, \quad (5.4.58)$$

$$\frac{1}{m} \big(1 + \|g\|_{T_0,0,p',2,2} + \|\boldsymbol{K}\|_{0,2,\Omega} + \|\text{rot}(\varepsilon^{-1}\boldsymbol{K})\|_{0,2,\Omega} \big) C_9(T_0, M) T^* \leq M,$$

$$T^* \in (0, T_3(M)]. \quad (5.4.59)$$

Formulas (5.4.57)–(5.4.59) imply that the operator N maps $X(M,T)$ into itself for all $T \in (0, T^*]$.

Now we estimate the difference $N(f_2) - N(f_1)$ for arbitrary functions $f_1, f_2 \in X(M,T)$, $T \in (0, T^*]$. For this, we consider the equation

$$N(f_2)(t) - N(f_1)(t) = \Big(\Phi_2 \Big[\frac{\sigma\mu}{\varepsilon\rho} g(t,\cdot) \times \boldsymbol{H}(f_2)(t,\cdot) \Big] \Big)^{-1}$$

$$\times \Big\{ \Phi_1 \Big[\frac{\partial}{\partial t} \boldsymbol{H}(f_2)(t,\cdot) - \frac{\partial}{\partial t} \boldsymbol{H}(f_1)(t,\cdot) \Big] - \Phi \Big[\frac{\sigma}{\varepsilon} \Big(\frac{\partial}{\partial t} \boldsymbol{E}(f_2)(t,\cdot) - \frac{\partial}{\partial t} \boldsymbol{E}(f_1)(t,\cdot) \Big) \Big]$$

$$- \Phi_2 \Big[\frac{\sigma\mu}{\varepsilon} \Big(\frac{\partial}{\partial t} \boldsymbol{U}(f_2)(t,\cdot) - \frac{\partial}{\partial t} \boldsymbol{U}(f_1)(t,\cdot) \Big) \times \frac{\partial}{\partial t} \boldsymbol{H}(f_2)(t,\cdot) \Big]$$

$$- \Phi_2 \Big[\frac{\sigma\mu}{\varepsilon} \frac{\partial}{\partial t} \boldsymbol{U}(f_1)(t,\cdot) \times \Big(\frac{\partial}{\partial t} \boldsymbol{H}(f_2)(t,\cdot) - \frac{\partial}{\partial t} \boldsymbol{H}(f_1)(t,\cdot) \Big) \Big]$$

$$- \Phi_2 \Big[\frac{\sigma\mu}{\varepsilon\rho} \text{Div}\, T \big(\boldsymbol{U}(f_2)(t,\cdot) - \boldsymbol{U}(f_1)(t,\cdot) \big) \times \boldsymbol{H}(f_2)(t,\cdot) \Big]$$

$$- \Phi_2 \Big[\frac{\sigma\mu}{\varepsilon\rho} \text{Div}\, T \big(\boldsymbol{U}(f_1)(t,\cdot) \big) \times \big[\boldsymbol{H}(f_2)(t,\cdot) - \boldsymbol{H}(f_1)(t,\cdot) \big] \Big] \Big\}$$

$$-\left(\Phi_2\left[\frac{\sigma\mu}{\varepsilon\rho}g(t,\cdot)\times \boldsymbol{H}(f_2)(t,\cdot)\right]\right)^{-1}$$
$$\times\Phi_2\left[\frac{\sigma\mu}{\varepsilon\rho}g(t,\cdot)\times\left[\boldsymbol{H}(f_2)(t,\cdot)-\boldsymbol{H}(f_1)(t,\cdot)\right]\right]N(f_1)(t),$$
$$\text{for a.a. } t\in(0,T). \quad (5.4.60)$$

Using formulas (5.4.31), (5.4.32), (5.4.38), (5.4.39), (5.4.57), and (5.4.60), we can easily derive the estimate

$$|N(f_2)(t)-N(f_1)(t)|$$
$$\leq C_{10}(T_0,M,m^{-1})\left(1+\|g\|_{T_0,0,p',2,2}+\|\boldsymbol{K}\|_{0,2,\Omega_2}+\|\text{rot}(\varepsilon^{-1}\boldsymbol{K})\|_{0,2,\Omega}\right)$$
$$\times\left(\int_0^t h_1(t-s)\|f_2-f_1\|_{s,0,p}\,ds+\|f_2-f_1\|_{t,0,p}\right),\quad \forall t\in[0,T], \quad (5.4.61)$$

where
$$h_1(t)=\exp\left\{t\|\sigma\|_{0,\infty,\Omega}+C_6(T_0)M\|g\|_{T_0,0,p',2,2}\right\}.$$

The estimate (5.4.61) implies the inequality

$$\|N(f_2)-N(f_1)\|_{t,0,p}$$
$$\leq C_{10}(T_0,M,m^{-1})\left(1+\|g\|_{T_0,0,p',2,2}+\|\boldsymbol{K}\|_{0,2,\Omega_2}+\|\text{rot}(\varepsilon^{-1}\boldsymbol{K})\|_{0,2,\Omega}\right)$$
$$\times[1+T_0 h_1(T_0)]\int_0^t \|f_2-f_1\|_{s,0,p}\,ds$$
$$:= C_{11}(T_0,M,m^{-1})\int_0^t \|f_2-f_1\|_{s,0,p}\,ds,\quad \forall t\in[0,T]. \quad (5.4.62)$$

Inequality (5.4.62) allows us to establish the following estimate for the iterations N^n of the operator N:

$$\|N^n(f_2)-N^n(f_1)\|_{t,0,p}\leq \frac{C_{11}^n(T_0,M,m^{-1})}{(n-1)!}\int_0^t (t-s)^{n-1}\|f_2-f_1\|_{s,0,p}\,ds$$
$$\text{for a.a. } t\in(0,T),\quad \forall n\in\mathbb{N}. \quad (5.4.63)$$

We now apply the generalized contraction mapping principle (Schwartz, 1967, p. 103). This guarantees that equation (5.4.50) has a unique solution $f\in X(M,T)$ for every $T\in(0,T^*]$, which proves the theorem.

Chapter 6.

Numerical solution of inverse problems: some examples

6.1. SHORT REVIEW OF NUMERICAL APPROACHES TO SOLVING INVERSE PROBLEMS

Here we would like to give a short overview of numerical approaches to solve inverse problems (see also Chapter 1). We are going to mention the most popular methods; for details we refer the reader to the cited literature.

The so-called *ray method* was proposed for solving the inverse kinematic problem of seismics at the beginning of the twentieth century. The first statements of such problems (as was discussed in Chapter 1) were studied by Herglotz and Wiechert (1905). The ray methods for reconstruction of parameters of media through acoustic/elastic wave data were well-developed and widely used for seismic prospecting (Cherveny, Molotkov, and Psenchik, 1977). These methods were mainly used to study seismic waves in rather simple media such as spherically symmetric or vertically inhomogeneous structures. The ray method cannot compete with other methods in the precision of results. On the other hand, the ray method can be applied to some complex media where other methods can hardly be used at the moment (Babich and Buldirev, 1972).

Systematic study of dynamic inverse problems (which involves into consideration the time evolution of measured fields) was essentially developed comparatively recently, say, starting from the end of the sixties of twenties century.

The so-called *Gel'fand – Levitan method* or the *Inverse Scattering method* is extensively used for solving the inverse problems of electrodynamics and inverse problems for elastic media, in line transmission theory, etc. This is an "exact" method based on the analytic representation of the solution to inverse problem. Such approach is based on the idea to use the results of quantum scattering theory which make it possible to reduce a nonlinear inverse problem to a one-parameter family of linear integral Fredholm equations of second kind. The first results in the area were obtained by Gel'fand and Levitan (1955), Krein (1951, 1954) and Marchenko (1955).

Later on, the Gel'fand – Levitan integral equation was used for inversion of wave fields by Alekseev (1962), Ware and Aki (1969). The equation of motion was reduced to the 1D Schrödinger equation. As was proved by Ware and Aki (1969), the discrete version of the Gel'fand – Levitan equation happens to be equivalent to the inversion method proposed by Baranov and Kunetz (1960). In the papers by Symes (1979, 1981) a nonlinear version of the Gel'fand – Levitan equation has been formulated. Another inversion scheme based on the Gel'fand – Levitan equation was intoduced by Santosa (1982).

The idea of the *finite-difference inversion method* was proposed by Alekseev (1967) as a general way of determination of coefficients in hyperbolic equations. The main idea of this method is as follows. An inverse problem is replaced by its finite-difference version. As a result, a system of nonlinear algebraic equations arises. A solution to this system gives an approximate solution to the original inverse problem. This method is rather natural from the physical point of view since it is based on the method of characteristics (recall that the main information about the features of the direct solution and the medium is usually concentrated along the characteristics).

As was established by Alekseev and Dobrinskii (1975), there exists a connection between the finite-difference inversion method and the algorithm suggested by Baranov and Kunetz (1960).

The aforementioned methods demonstrate some instability or/and high computational cost. That is why such methods were not often used for solution to inverse problems.

Now we consider in more details some of those numerical methods which are most widely used in practice for solving *the wave equation in a layered medium*. We follow the paper of Ursin and Berteussen (1986). These methods do not take into account the multiple wave reflections from the interfaces between layers. The elastic medium is supposed to consist of horizontal homogeneous layers with equal "travelling time" $\Delta \tau$ (Goupillaud model). It

Chapter 6. Numerical solution

is required to determine the propagation velocities and densities of all the layers (the acoustic impedance) from the reflection response measured at the free surface.

The first numerical method of inversion for plane-wave propagation in a vertically-inhomogeneous medium was proposed by Baranov and Kunetz (1960). It is the so-called *"classical" method.* The seismic response is assumed to be generated by a perfect spike (a Dirac's mass), and there is no measurement noise (perfectly deconvolved seismic data). The elastic medium is assumed to consist of homogeneous layers with equal travel time; and the reflection coefficients at the interfaces between the layers are estimated. When the first k reflection coefficients are known, the effects of those layers are removed from the data by subtracting a synthetic seismogram. Then the $(k+1)$th reflection coefficient can be calculated from the reduced seismogram. The reflection coefficients are converted into the acoustic impedance by a simple recursive formula.

An *iterative inversion method* in the temporal frequency domain has been proposed by Gjevik, Nilsen, and Höyen (1976). Here and below, "temporal frequency" is the frequency appearing in the Fourier transform with respect to time, in contrast to the "spatial frequencies" appearing in the Fourier transform with respect to the space variables.) The method is based on an integral equation (derived from the Riccati equation) which can be used to calculate the reflection response at depth. The iterative method is adapted to a model with continuously varying parameters. In practice, a quantization has to be made, and a model with layers of equal travel time is also used.

The two aforementioned methods realize the so-called *direct inversion schemes.* That is, the parameters of the medium are estimated directly from the reflection seismogram.

The "classical" method and the iterative inversion method have demonstrated similar stability properties. But the iterative method requires considerably more computational resources and execution time in comparison with the classical Baranov–Kunetz method.

The next step in the development of numerical methods was to apply the *detection scheme.* In this case the wave field is divided into downgoing and upgoing waves, which are downward continued at depth. The k-th reflection coefficient is computed as the ratio of the upgoing and downgoing waves at the depth (one-way travel time) $k \cdot \delta\tau$. There exist a number of different modifications of this approach, namely:

Downward continuation. The wavefield is divided into downgoing and upgoing waves which are both estimated at depth (Mendel, 1981; Mendel and Habibi-Ashrafi, 1980).

Surface calculations. Under assumption that the first k layers are known, the effect of those is removed from the registered seismograms by subtracting a synthetic seismogram. From the reduced seismogram, the parameters of layer $k+1$ are then estimated (Goupillaud, 1961).

Layer removal. The effect of known "upper layers" is removed by applying a nonlinear formula in the frequency domain, and then the parameters of the next layer are estimated (Ursin and Berteussen, 1986).

All of the above three methods give better results than the classical method and the iterative frequency-domain method.

All the aforementioned methods require very good, almost perfect, "inversion data". They are very sensitive to noise and deviations in the data from the assumed model. In practice it means that in view of normally high noise level, only the largest contrasts in acoustic impedance can be detected.

Another problem also arises, which is very well known as the *trend component problem*. The point is that all the aforementioned methods become very unstable and do not yield any reasonable solution in case the low-frequency component of the source function spectrum is not available.

In the number of papers (Bamberger, Chavent, Hemon, and Lailly, 1982; Kolb, Collino, and Lailly, 1986; Tarantola, 1986; Tarantola and Valette, 1982), the *optimization approach* to numerical solution of inverse problems was proposed, analysis and numerical tests were also given. The unknown medium parameters are searched for as the minimum point of the data *misfit functional* (cost function) characterizing the deviation, in some suitable norm, of the observed data from the numerical solution of the direct problem.

One of the advantages of this approach is the possibility to take into account the full *a priori* information about the structure of the medium and to use it to improve the process performance.

Nowadays, the optimization approach (using such methods as the steepest descent method, the conjugate gradient method, the quasi-Newton method, and variety of their modifications) is widely used for numerical solution of inverse problems.

To use the optimization technique, such questions as the structure of the set of solutions, the area of convergence of the method, the rate of its convergence, and others must be answered. By now, these questions have been studied only for some particular statements.

Several examples of application of the optimization approach are given in this chapter.

Chapter 6. Numerical solution

Here we would also like to discuss some rather recent numerical approaches of solution to inverse problems, which are still not that well known.

By analogy with the singular value decomposition (SVD) analysis of matrices, the so-called *Truncated SVD method* is used for solving ill-posed inverse problems, for details we refer the reader to the papers by Bertero and Boccacci (1998), Engl, Hanke, and Neubauer (1996), Kirsch (1996).

One modification of the TSVD approach, the r-*solution method*, was proposed by Cheverda and Kostin (1995), Khajdukov, Kostin, and Cheverda (1997).

The r-solution for a compact operator in a Hilbert space is regarded as a restriction of the initial finite-dimensional operator, i.e., the projection onto the span of the first r eigenvectors. The main features of this notion are its stability and existence of numerical algorithms for its reliable computing. In application to the inverse problem under consideration, r is nothing but a parameter. Its choice depends on the structure of the problem (distribution of eigenvalues, accuracy of approximation, level of noise in data, etc.).

The r-solution method can be described schematically as follows:

First, the original nonlinear inverse problem is reduced to a sequence of linear ones.

Second, the SVD analysis is used to select the first r eigenvectors in the case where the Fréchet derivative is a compact operator. In alternative cases some projecting methods are used (besides, in such cases the r-solution itself is fruitful as well).

Third, an approximation of these linearized problems is used to provide better convergence.

Finally, modern applications of *Linear Algebra* algorithms are used to obtain a reliable solution.

Now the *neural network approach* becomes very popular for solving a special class of inverse scattering problems, namely, the acoustic or electromagnetic characterization of objects (defects or structures) embedded in stratified media. The main difficulties are to determine the optimal design parameters for neural networks as well as the optimal characteristics of the inverse scattering experiments from the viewpoint of neural network data processing, for more references see Gorban' and Rossiev (1996), Rychagov and Duchene (1998).

Wavelets have been used in several works (Louis, Maaß, and Rieder, 1997; Mallat, 1998) as basis functions to solve linear inverse problems by Galerkin methods. *Wavelet-based inversion methods* are discussed also by De Mol (1999). The method is expected to give a valuable contribution in increasing the resolution in the reconstruction of non-stationary or spatially inhomogeneous objects with very localized fine-scale or high-frequency details.

In conclusion of this section, it should be noted that we did not pursue the aim of giving the complete review of numerical methods for solving the inverse problems. Our aim was to point out the most popular numerical approaches. At the same time, we gave references to the most frequently cited publications which contain further references.

6.2. NUMERICAL SOLUTION OF A 3D INVERSE KINEMATIC PROBLEM OF SEISMICS

In this section we are mostly following the papers by Alekseev, Lavrentiev, Mukhometov, and Romanov (1969), Alekseev, Lavrentiev, Mukhometov, Nersesov, and Romanov, (1971).

One of the principal questions of planetary geophysics is the question of presence of lateral (horizontal) variations of the physical properties of the substance in the Earth's mantle and what is their spatial structure and depth of stretching.

The interest to this question is due to their possible role of lateral variations in the mechanism of development of the Earth's crust, in particular, in those features of this mechanism which resulted in the horizontal variability of global structures of the Earth's crust (its splitting into continental and oceanic regions) and regional distinctions in the regime of its motion.

Construction of "one-dimensional" spherically-symmetric models of distribution of the mechanical and physical properties of the Earth's substance belongs to the number of fundamental results of geophysics. These models clarified a number of general questions concerning the character and energetics of deep processes. These models take an important place in the system of geological conceptions about the structure and development of the Earth.

However, from the viewpoint of modern information on the structure of geophysical fields, the one-dimensional models of the Earth are conceived as a rather rough approximation to the reality. They appeared in geophysics in the beginning of the last century as a result of averaging approaches to interpretation of the fields observed. We need to note that the procedure of averaging the fields over geographical coordinates was inevitable until not long ago because methods of solution of 3D inverse problems were unavailable or the available information on the structure of geophysical fields was not detailed enough.

At present, both these reasons loose their strength. Nowadays, due to development of numerical methods of mathematics and wide use of modern computers in geophysics, there are real possibilities to solve many 3D inverse problems of geophysics. As concerns the level of details and the accuracy of observations, there was much progress in recent time, too. Due to space satellites a detailed data on the structure of the gravitational and magnetic fields became available. The number of stations in the teleseismic network increased dramatically in the recent decades. There are achievements in other areas of geophysical observations, too.

Thus, the problem of determining the 3D "geophysical" structure of the Earth becomes a high-priority problem of geophysics.

At present, in various fields of geophysics there are facts which allow us to hypothesize the existence of lateral inhomogeneities in the crust and upper mantle of the Earth at depths up to several hundreds of kilometers. These facts include, e.g., regional deviations of the travel-times of seismic waves from the averaged travel-time curves of Jeffreys and Bullen, the asymmetry of the gravitational and electromagnetic fields, etc. The attempts of quantitative description of deep inhomogeneities on the base of observed anomalies of geophysical fields are still in their infancy.

In recent years, many works dealing with interpretation of the anomalies of dispersion characteristics of surface waves and also with interpretation of the global and regional structures of the electromagnetic field appeared. In this section, we describe a method for determining the structural inhomogeneities of the Earth's mantle on the basis of the travel-time curves of bulk seismic P-waves of refracted type.

The physical statement of the problem here considered is as follows. Many earthquakes occur in different regions of the terrestrial globe each year. The data registered at seismic stations allow us to construct, for each earthquake, a surface travel-time curve of bulk P-waves (i.e., a table of travel-times of longitudinal seismic waves from the earthquake to stations). It is known that the seismic rays, along which the disturbances from earthquakes propagate before they go out on the Earth's surface, pass through the deep layers of the Earth. Therefore, the travel-time curve of P-waves contains integral information about the structure of the deep layers of the Earth. The more is the epicentral distance the deeper, generally speaking, a seismic ray penetrates into the Earth. If we have detailed travel-time curves from many earthquakes, we may hope to reconstruct the pattern of distribution of the velocity of seismic waves inside the Earth. Evidently, this is possible only if the seismic rays cover the interior of the Earth sufficiently dense.

Attempting to solve the problem of reconstructing the velocity distribution pattern in the Earth's mantle, we immediately come across a number of difficulties which are due to various reasons. The main of these reasons are apparently the following ones: insufficient knowledge of the structure of the mantle and inaccuracy in determination of the earthquake coordinates. Meanwhile, to determine the velocity structure of the mantle we need to recalculate the travel-time curve and thus remove the effect of the crust. If sufficient information on the structure of the mantle and exact coordinates of the earthquake are not available, we make considerable errors in this recalculation, which leads inevitably to errors in reconstruction of the velocity structure of the mantle. Undoubtedly, more detailed investigation of the Earth's crust can promote substantially the investigations of the Earth's mantle.

6.2.1. Statement of the problem

Proceeding to the mathematical formulation of the problem, we idealize the statement in a certain sense. Namely, we assume that the Earth with its crust stripped is a ball of unit radius and we know the times it takes a perturbation to travel along seismic rays between any two points on the surface of this ball. It is required to find the velocity of propagation of perturbations inside the ball from this information. Let M be an arbitrary point of the unit ball; let $n(M)$ be the quantity inverse to the velocity of the seismic P-wave at this point; and let $\tau(M_0, M)$ be the travel-time from a point M_0 to the point M. It is known that for a fixed point M_0 the function $\tau(M_0, M)$ satisfies the equation

$$|\operatorname{grad} \tau|^2 = n^2(M). \tag{6.2.1}$$

We shall not investigate this problem in its exact formulation. Well-grounded methods for its solution have not been developed yet. Instead, we consider a linearized statement which allows us to obtain an approximate solution of the problem. Namely, we assume that the function $n(M)$ can be represented as

$$n(M) = n_0(M) + n_1(M), \tag{6.2.2}$$

where the function $n_0(M)$ is known and depends only on the distance to the center of the ball and the function $n_1(M)$ is small as compared with $n_0(M)$. We represent the function $\tau(M_0, M)$ in a similar form

$$\tau(M_0, M) = \tau_0(M_0, M) + \tau_1(M_0, M), \tag{6.2.3}$$

where $\tau_0(M_0, M)$ corresponds to the function $n_0(M)$. Substituting the expressions for $n(M)$ and $\tau(M_0, M)$ into (6.2.1) and neglecting the squares of the quantities $n_1(M)$ and $|\text{grad}\,\tau_1|$, we obtain

$$(\text{grad}\,\tau_0, \text{grad}\,\tau_1) = n_0 n_1. \tag{6.2.4}$$

Now we divide the left and right sides of the last equality by $n_0(M)$ and notice that the vector $(n_0)^{-1}\,\text{grad}\,\tau_0$ gives the directing cosines of the ray which connects the points M_0 and M in case $n = n_0(M)$. As a result, we conclude that equality (6.2.4) can be written as

$$\frac{\mathrm{d}\tau_1}{\mathrm{d}s_0} = n_1. \tag{6.2.5}$$

Here $\mathrm{d}s_0$ is the length element of a ray in the medium with the refraction index $n_0(M)$. Now we can write equality (6.2.5) in the form

$$\tau_1(M_0, M) = \int_{\Gamma_0(M_0, M)} n_1\,\mathrm{d}s_0, \tag{6.2.6}$$

where $\Gamma_0(M_0, M)$ is the ray connecting the points M_0 and M in case $n = n_0(M)$. Equality (6.2.6) is the main relation for further studies.

Since the function $\tau(M_0, M)$ is known on the surface of the ball, therefore, the function $\tau_1(M_0, M)$ can also be considered as a known function of two points on the surface of the ball and we come to the following problem: we know the integrals of the function $n_1(M)$ along certain curves; it is required to find the function $n_1(M)$ itself.

The above hypothesis that the function $n(M)$ is representable in the form (6.2.2) is apparently true for the Earth on the whole since the deviations from the averaged travel-time curve of Jeffreys and Bullen that are obtained in practice are very small. For the function $n_0(M)$ we can take the velocity cross-section corresponding to the Jeffreys–Bullen travel-time curve or any other velocity cross-section sufficiently close to it.

Let us study equation (6.2.6). Since the function $n_0(M)$ depends only on the distance to the center of the ball, therefore, the rays $\Gamma_0(M_0, M)$ are planar; they lie in the plane which passes through the points M_0, M, and the centre of the ball. In view of this, solution of equation (6.2.6) splits into a number of planar problems, namely, problems of the type (6.2.6) for a unit circle, i.e., when the points M_0 and M are located on the boundary of the unit circle and we need to find the function $n_1(M)$ inside this circle. Note that this implies the following possible statement of the original problem:

find the velocity cross-section of the Earth in the plane of the global circle if we know the travel-time curve of P-waves along a certain arc lying on the boundary of the large circle.

Let us consider the cross-section of the ball by a plane passing through its center and show how one can construct the function $n_1(M)$ in this cross-section[i]. Introduce the polar coordinates r, φ with the origin at the center of the circle. Then in this cross-section we have $n_1(M) \equiv n_1(r, \varphi)$ and $n_0(M) \equiv n_0(r)$. We shall assume that the function $m(r) = r \cdot n_0(r)$ is twice continuously differentiable and such that $m'(r) \geq m_0 > 0$. It is known that in this case the equation of the seismic ray Γ_0 can be written as

$$m(r) \cdot \sin(l, r) = p, \qquad (6.2.7)$$

where p is a constant along the ray $\Gamma_0(M_0, M)$ and (l, r) is the angle between the tangent to the ray at some its point and the radius vector of this point. Every pair of points M_0 and M lying on the contour of the cross-section can be associated, in a one-to-one manner, with certain values of the parameter p and angle α which specifies the angular coordinate of the vertex of the ray. Thus, $\tau_1(M_0, M) \equiv \tau_1(p, \alpha)$. It is easy to write an explicit equation of the ray (6.2.7) in polar coordinates. It has the form

$$\varphi = \alpha \pm p \int_p^q \frac{\chi(t) \, dt}{\sqrt{t^2 - p^2}}, \qquad (6.2.8)$$

where the variables q and r are connected by the formulas

$$q = m(r), \qquad r = M(q) \qquad (6.2.9)$$

and

$$\chi(t) = \frac{d}{dt} \ln M(t). \qquad (6.2.10)$$

Then equality (6.2.6) for the cross-section in question takes the form

$$\tau_1(p, \alpha) = \int_p^{p_0} n_1(r, \varphi) \frac{M'(q) \, q}{\sqrt{q^2 - p^2}} \, dq, \qquad p_0 = m(1), \qquad (6.2.11)$$

where, in the last formula, φ is found by formula (6.2.8) and r is found by formula (6.2.9). Introduce the notation

---

[i] The method of construction of the function $n_1(M)$ which is used below is a paraphrase, for this particular case of curves, of a more general integral geometry problem considered by Romanov (1967b).

$$n_1(M(q), \varphi) M'(q) q = N(q, \varphi). \qquad (6.2.12)$$

In these notations, equation (6.2.11) becomes

$$\tau_1(p, \alpha) = \int_p^{p_0} N(q, \varphi) \frac{dq}{\sqrt{q^2 - p^2}}. \qquad (6.2.11')$$

Having found the function $N(q, \varphi)$, we can easily find the function $n_1(r, \varphi)$ since the function $M(q)$ is known.

Let $\tau_1^k(p)$ and $N_k(q)$ ($k = 0, \pm 1, \pm 2, \ldots$) be the Fourier coefficients of the functions $\tau_1(p, \alpha)$ and $N(q, \varphi)$ in the systems $e^{ik\alpha}$ and $e^{ik\varphi}$ ($0 \leq \alpha, \varphi \leq 2\pi$). Then equation (6.2.11') implies that these coefficients are connected by the formula

$$\tau_1^k(p) = \int_p^{p_0} N_k(q) \frac{2R_k(p, q)}{\sqrt{q^2 - p^2}} dq, \quad k = 0, \pm 1, \pm 2, \ldots, \qquad (6.2.13)$$

where

$$R_k(p, q) = \cos\left(kp \int_p^q \frac{\chi(t)}{\sqrt{t^2 - p^2}} dt\right). \qquad (6.2.14)$$

Relations (6.2.13) are integral equations with singularity of the type of the generalized Abel equation. Applying to each of them the operator

$$L\tau_1^k(p) = \frac{1}{\pi} \frac{d}{dP} \int_P^{p_0} \frac{p\,\tau_1^k(p)\,dp}{\sqrt{p^2 - P^2}}, \qquad (6.2.15)$$

we find that

$$L\tau_1^k(p) = -N_k(P) + \int_P^{p_0} N_k(q) T_k(P, q) dq, \quad k = 0, \pm 1, \pm 2, \ldots, \qquad (6.2.16)$$

where

$$T_k(P, q) = \frac{1}{\pi} \frac{\partial}{\partial P} \int_P^q \frac{2p}{\sqrt{(p^2 - P^2)(q^2 - p^2)}} R_k(p, q) dp$$

$$= \frac{P}{\pi} \int_0^\pi \left[\frac{\partial}{\partial p} R_k(p, q)\right]_{p = \frac{q^2 + P^2}{2} + \frac{q^2 - P^2}{2} \cos\theta} \cdot (1 - \cos\theta) d\theta. \qquad (6.2.17)$$

It could be deduced from (6.2.17) that the kernels of equations (6.2.16) are continuous functions. Consequently, each of equations (6.2.17) has a unique solution which can be found, e.g., by the method of successive approximations. Thus, under the hypothesis that the function $m(r)$ is monotone, the function $n_1(r, \alpha)$ is determined uniquely in each global cross-section of the unit ball.

In practical realization of this method, we confine ourselves to a finite partial sum of the Fourier series for the function $N(q, \alpha)$. This is necessary because of two reasons. First, the values of the function $\tau_1(M_0, M)$ are known with some errors, and at the same time the kernels of equations (6.2.16) grow infinitely with the growth of k and oscillate; therefore, because of the errors in calculation of $\tau_1^k(p)$, we shall have large errors in determination of $N_k(q)$ for large values of the parameter k, which will result in divergence of the Fourier series. Second, due to poor behavior of the kernels for large k, any method of solution of equations (6.2.16) for sufficiently large k leads to large accumulation of errors because of approximate calculation of the integrals.

In the theoretical examples below, we have used another algorithm which requires less computations. It is based on the equations

$$\tau_1^k(p) = 2 \int_p^1 n_1^k(r) \frac{m(r) \cos\left(k\pi\varphi(r,p)/\alpha_0\right)}{\sqrt{m^2(r) - m^2(p)}} \, dr, \quad k = 0, \pm 1, \pm 2, \ldots, \tag{6.2.18}$$

where $\tau_1^k(p)$ and $n_1^k(r)$ are the Fourier coefficients of the functions $\tau_1(p, \alpha)$ and $n_1(r, \alpha)$, respectively, and the angle α_0 will be described later.

6.2.2. Numerical algorithm

The algorithm of numerical solution of the problem is based on expansion of the function $\tau_1(p, \alpha)$ in a series in terms of a certain system of special functions which are naturally connected with equations (6.2.18). Note that if we expand $n_1^k(r)$ in equations (6.2.18) in terms of a certain system of functions with unknown coefficients, this will generate an expansion of $\tau_1^k(p)$ in terms of some other system of functions with the same coefficients. In our algorithm, we used the polynomial representation of the functions $n_1^k(r)$:

$$n_1^k(r) = \sum_{n=0}^{N} (A_{nk} + iB_{nk}) (1-r)^n, \quad B_{0k} \equiv 0. \tag{6.2.19}$$

In the Fourier series for the function $\tau_1(p,\alpha)$, we keep only M first harmonics. This yields the following approximate expression for $\tau_1(p,\alpha)$ in terms of the coefficients A_{nk} and B_{nk} and some special functions:

$$\tau_1(p,\alpha) = \sum_{n=0}^{N} \left[A_{n0}\psi_{n0}(p) + \sum_{k=1}^{M} \left(A_{nk} \cos \frac{k\pi\alpha}{\alpha_0} + B_{nk} \sin \frac{k\pi\alpha}{\alpha_0} \right) \psi_{nk}(p) \right], \quad (6.2.20)$$

where

$$\psi_{nk}(p) = 2 \int_p^1 (1-r)^n \frac{m(r) \cos(k\pi\varphi(r,p)/\alpha_0)}{\sqrt{m^2(r) - m^2(p)}} \, dr. \quad (6.2.21)$$

If we know the function $\tau_1(p,\alpha)$ for sufficiently many pairs p, α, we can find the coefficients A_{nk} and B_{nk} and, consequently, the function

$$n_1(r,\alpha) \approx \sum_{k=0}^{M} \sum_{n=0}^{N} \left(A_{nk} \cos \frac{k\pi\alpha}{\alpha_0} + B_{nk} \sin \frac{k\pi\alpha}{\alpha_0} \right).$$

In the examples given below, we use the least-squares method for finding the coefficients.

6.2.3. The domain of solution

Using the algorithm described above, we can solve the problem of reconstructing the function $n_1(r,\alpha)$ in the whole unit ball or in the global circle of this ball if we know the travel-times of waves along seismic rays which fill these domains. However, we do not have such information at present time. In the practical example, seismic rays fill the domain bounded from above and below by arcs of circles and from the sides by the rays themselves. The same domain was taken in the theoretical examples considered below. In this case the sought function $n_1(r,\alpha)$ was reconstructed better in the domain bounded from above and below by the same arcs of circles and from the sides by the curves which represent the geometric location of the vertices of all rays originating from the extreme points. In the sequel, we shall denote this domain by G.

In Figure 6.2.1, the boundary of the domain is shown by a solid line and the lateral rays by dashed lines. The dimensions of the domain are taken proportional to the practical example. The equation of the lateral curves of the domain G in polar coordinates is

$$\varphi(1,p) = \pm \int_p^1 \frac{m(t) \, dt}{t\sqrt{m^2(t) - m^2(p)}}. \quad (6.2.22)$$

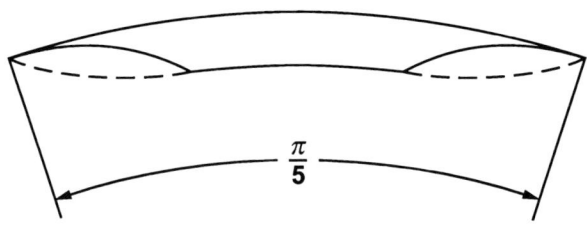

Figure 6.2.1

In this equation, the angle φ is counted from the extreme radii. In the case of constant velocity these curves are arcs of circles of radius 0.5 centered at the midpoints of the lateral radii.

6.2.4. Numerical experiments

To test our method we first solved the direct problem: for a given function $n(r,\varphi) = n_0(r) + n_1(r,\varphi)$ we calculate the travel-times $\tau(p,\alpha)$. Then we solve the inverse problem: from the travel-times $\tau(p,\alpha)$ we reconstruct the function $n_1(r,\varphi)$ (see Subsection 6.2.2) and compare it with the original function.

The following values were used in calculations: $0.94 = p_0 \leq r \leq 1$, $v_0(r) = [n_0(r)]^{-1} = 23.07 - r \cdot 20.33$, and $\alpha_0 = \pi/10$, where α_0 is the half-angle of our domain G; these quantities correspond to the initial data of the practical example. In the above-mentioned direct problem, we took 20 receivers and 20 sources located uniformly over an arc of the circle of radius $r = 1$. The distance between two neighboring points scaled to the Earth's surface corresponds to 200 km. The function $\tau(p,\alpha)$ was calculated in 210 points of the domain G.

In order to find the coefficients A_{nk} and B_{nk} in equation (6.2.20) the functions $\psi_{nk}(p)$ are used. A software application was developed for computation of these functions. Figures 6.2.2–6.2.5 show the functions $\bar{\psi}_{nk}(p) = 10^n \cdot \psi_{nk}(p)$ for various n and k.

Comparison of Figures 6.2.2–6.2.5 shows that for these values of n and k, the function $\psi_{nk}(p)$ becomes monotone for $n > k$.

As was shown in Subsection 6.2.1, the function $n_1(r,\varphi)$ should be small then compared with $n_0(r)$. In our examples, the absolute value of $n_1(r,\varphi)$ was not greater than 10% of $n_0(r)$.

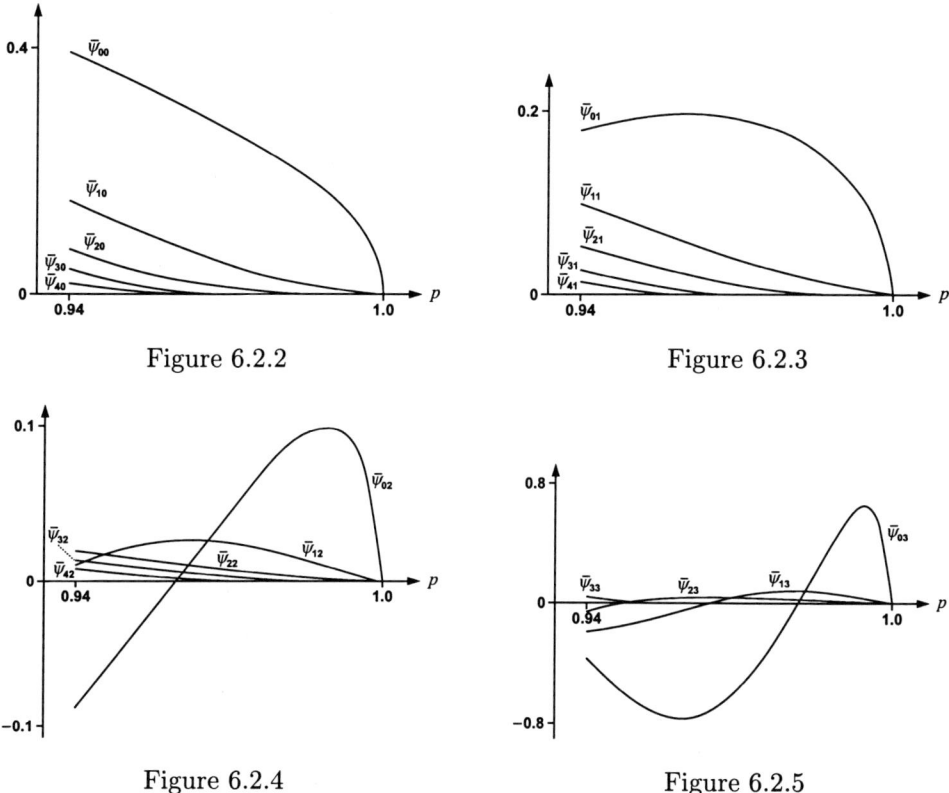

Figure 6.2.2

Figure 6.2.3

Figure 6.2.4

Figure 6.2.5

In conclusion, we give the results of calculations for four models of the medium:

$$n_1(r,\varphi) = 0.004\left(\sin 10\varphi + r\cos 10\varphi + r\cos 20\varphi + \sin 30\varphi\right), \qquad \text{I}$$

$$n_1(r,\varphi) = 0.0024 \sin 10\varphi \cdot \sin \frac{\pi(1-r)}{0.03}, \qquad \text{II}$$

$$n_1(r,\varphi) = 0.0009\left(\cos 10\varphi + \sin 20\varphi\right)\left[1 + \left(\frac{1-r}{0.06}\right)^2 + \sin \frac{\pi(1-r)}{0.03}\right], \qquad \text{III}$$

$$n_1(r,\varphi) = 0.001 \sin 10\varphi \cdot \sin \frac{\pi(1-r)}{0.06}. \qquad \text{IV}$$

The number of the model is indicated in each figure in the upper right corner.

The function $n_1(r,\varphi)$ depends on two arguments, so the plots of this function were calculated along cross-sections $\varphi = \text{const}$ or $r = \text{const}$. In Figures 6.2.6–6.2.11, the dashed lines show the reconstructed function $n_1(r,\varphi)$ and the solid lines show the true $n_1(r,\varphi)$ for comparison. The parameters N and M from (6.2.20), with which the problem was solved, are also indicated.

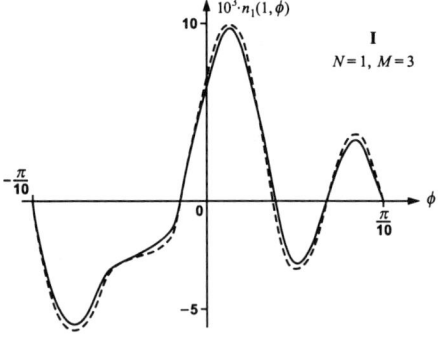

Figure 6.2.6

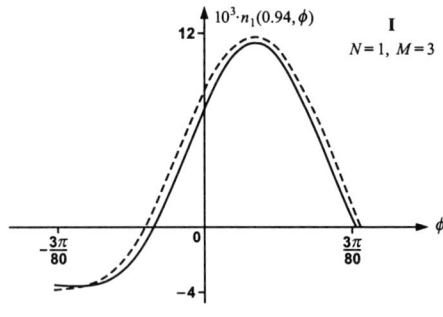

Figure 6.2.7

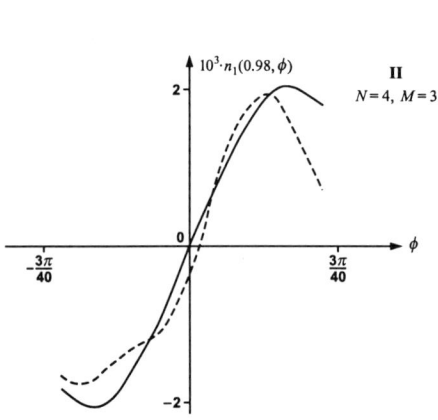

Figure 6.2.8

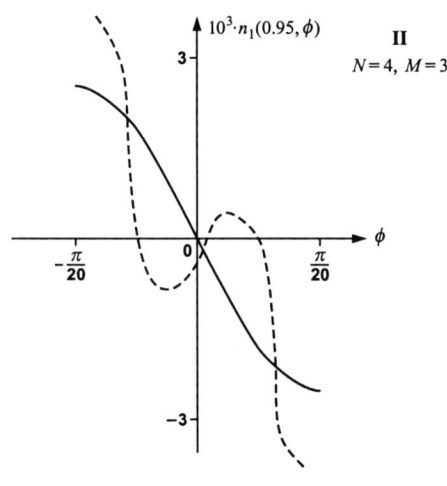

Figure 6.2.9

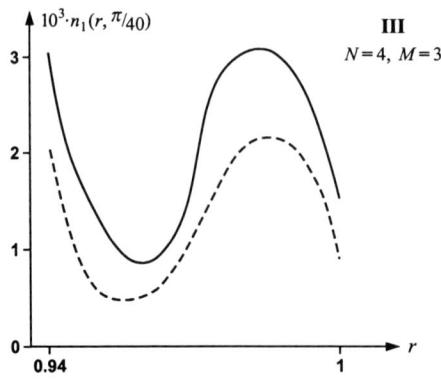

Figure 6.2.10

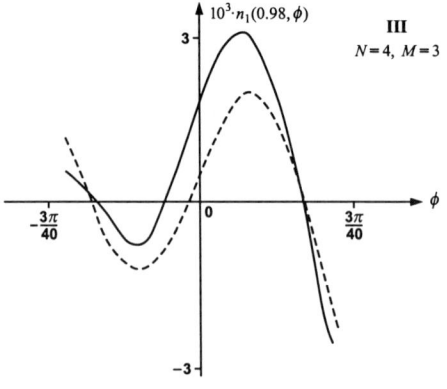

Figure 6.2.11

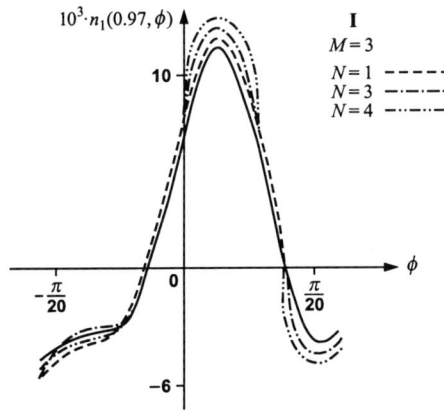

Figure 6.2.12

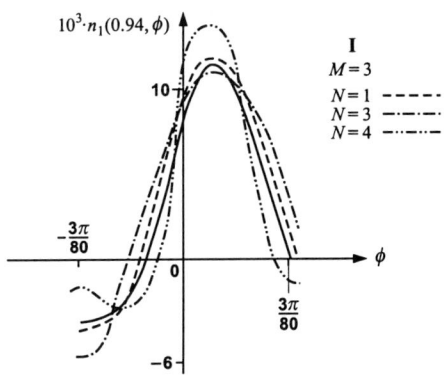

Figure 6.2.13

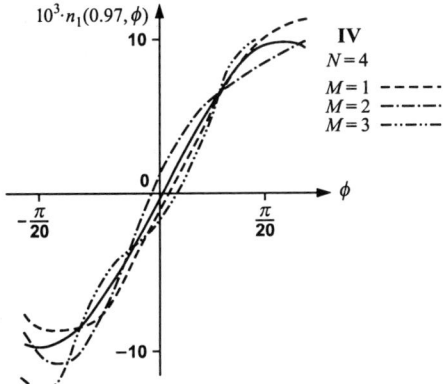

Figure 6.2.14

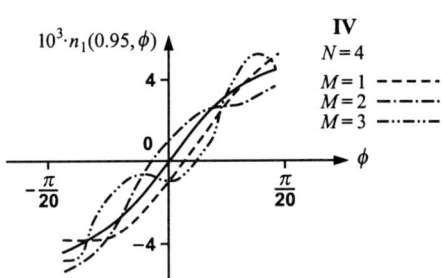

Figure 6.2.15

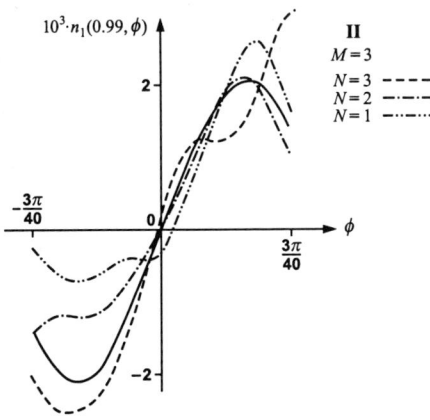

Figure 6.2.16

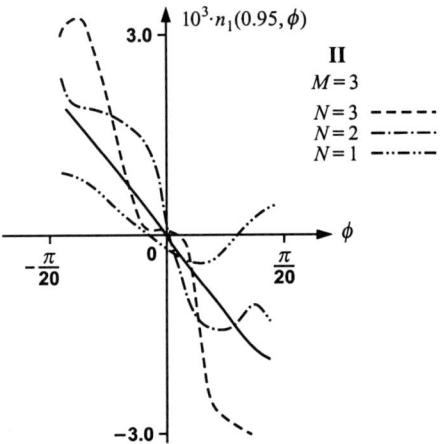

Figure 6.2.17

Our problem is classically ill-posed. However, it is well-posed in the Tikhonov sense (Lavrentiev and Mukhometov, 1969). We carried out a numerical investigation of the stability of the solution depending on the number of the approximating functions (i.e., N and M), the depth r, and the error of the initial data $\tau_1(p, \alpha)$.

When we use the least-squares method, the number of points at which the function $\tau_1(p, \alpha)$ is known should be much greater than the number of unknown parameters A_{nk} and B_{nk} of the sought function $n_1(r, \varphi)$. This condition is fulfilled in all examples concerned.

Figures 6.2.12 and 6.2.13 show the solutions for different N. The best reconstruction of $n_1(r, \varphi)$ occurs for $N = 1$. In the next example (Figures 6.2.14 and 6.2.15), the solutions were obtained for different M. In these two examples, the number of approximating functions was sufficient for the reconstructed function $n_1(r, \varphi)$ to reflect the behavior of the solution. We also studied the example when this is not true ($N = 1$), see Figures 6.2.16 and 6.2.17. In this case, the function $n_1(r, \varphi)$ is reconstructed in a certain average sense.

The effect of the error in the travel-times of waves along seismic rays has been also studied.

Figures 6.2.6–6.2.9 show the function $n_1(r, \varphi)$ for different r. The analysis of these and other examples shows that as the depth increases (i.e., r decreases), the error of the reconstructed function $n_1(r, \varphi)$ become grower. However, this does not change essentially the behavior of $n_1(r, \varphi)$ on this interval of r.

The above investigation of theoretical examples shows that the method for reconstructing the function $n_1(r, \varphi)$ from travel-times of P-waves can be effectively used in practice. Within the framework of this research, I. L. Nersesov (at the Complex Seismological Expedition of the Moscow Institute of Earth Physics) prepared practical material about earthquakes on the arc of the large circle from Pamir to Lake Baikal. Altogether, 43 powerful earthquakes that occurred on this profile in 1961–63 were taken. These earthquakes were registered on stations of regional type. The corrections for the crust were made on the base of the crust thickness map which was compiled by M. E. Artemiev on the base of gravitational data and information about the relief of the land. The average velocity in the crust was taken to be equal to 6.1 km/s.

Chapter 6. Numerical solution

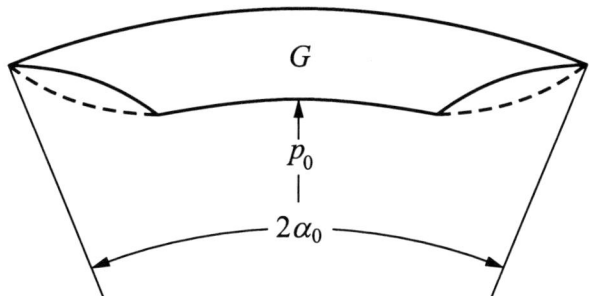

Figure 6.3.1

Substituting this expression in (6.3.7), we obtain

$$\frac{1}{R}\tau_1(p,\alpha) = \sum_{n=0}^{N}\left[A_{n0}\psi_{n0}(p) + \sum_{k=1}^{M}\left(A_{nk}\cos\frac{k\pi\alpha}{\alpha_0} + B_{nk}\sin\frac{k\pi\alpha}{\alpha_0}\right)\psi_{nk}(p)\right], \tag{6.3.12}$$

where

$$\psi_{nk}(p) = 2\int_{p}^{1}\left(\frac{1-r}{1-p_0}\right)^{n}\left[\cos\frac{k\pi\varphi(r,p)}{\alpha_0}\right]\frac{m(r)\,dr}{\sqrt{m^2(r)-m^2(p)}}. \tag{6.3.13}$$

The graphs of the functions $\psi_{nk}(p)$ for various n and k are given in Section 6.2.

If we know the function $\tau_1(p,\alpha)$ at sufficiently many points, we can find the coefficients A_{nk} and B_{nk} and, consequently, using (6.3.11), the function $n_1(r,\alpha)$ too.

To find the coefficients we use the least-squares method. In Section 6.2, we study the numerical stability of the solution $n_1(r,\alpha)$ obtained by this method depending on the number of approximating functions (i.e., N and M), the depth r, and the error of the initial data $\tau_1(p,\alpha)$. In these theoretical examples, we take those values of the parameters which correspond to our profile.

In fact, this investigation showed that with modern methods of registration of the source data (the times $\tau(p,\alpha)$ and the coordinates of earthquakes), the sought function $n_1(r,\alpha)$ can be reconstructed by our method.

As was mentioned in connection with Figure 6.3.1, the domain G corresponds to the domain where the initial data $\tau_1(p,\alpha)$ are specified. Now let us clarify the domain in which the solution $n_1(r,\alpha)$ can be found. It is easy to see from the basic relation (6.3.10) and Figure 6.3.1 that when the point (p,α) is in a sufficiently small neighborhood of the lateral arcs of the boundary of the domain G, the values of the function $n_1(r,\alpha)$ at points lying outside the domain G are used in formula (6.3.10). The whole area of used values of $n_1(r,\alpha)$ consists of the extension of the domain G up to the lateral rays, which are shown by dashed lines in Figure 6.3.1.

However, in investigation of this method on the theoretical examples in Section 6.2, the function $n_1(r,\alpha)$ was reconstructed well enough only in the domain G. For reconstruction of the function $n_1(r,\alpha)$ we can use the method of successive approximations in solution of the Volterra integral equation of second kind for the Fourier harmonics of $n_1(r,\alpha)$ (see Section 6.2). It is easy to see that in this case, again, the solution is obtained only in the domain G. In view of these facts it is clear that solving a practical problem, the reconstructed function $n_1(r,\alpha)$ should be considered only at the points of the domain G.

6.3.3. Processing of seismic data

According to the modern geophysical concepts the Earth's structure is as follows. The upper part, the crust, has thickness 30–50 km; the next is the mantle with thickness about 3000 km; and the last is the core. Each of these parts has its specific structure because they differ much from each other in the character of propagation of seismic waves through them.

It is known that in transition from the crust to the mantle, the velocity of longitudinal seismic waves changes stepwise by approximately 2 km/s. If we use our algorithm immediately, with the Earth in the capacity of our sphere, then such jumps in the velocity could lead to considerable errors. To avoid this, all travel-time curves were referred to a certain sphere of averaged mantle in which the velocity changes smoothly, at least to a certain maximal depth. Moreover, in our example we consider near-surface earthquakes which occurred at depths about 10–20 km, i.e., their epicenters were in the crust. So, after conversion to our sphere of averaged mantle, it turns out that all sources are on its surface. This procedure is described below in details.

Introduce a sphere of radius R_1 which will be called the sphere of averaged mantle. It was mentioned above that the Earth's upper part (crust) differs from the mantle in the velocity of seismic waves; moreover, the thickness of the crust is not constant but depends on the geographical coordinates. Therefore, for solution of the problem of determining the velocity distribution in the Earth's mantle it is convenient to remove the effect of the crust.

Chapter 6. Numerical solution

The sphere of averaged mantle of radius R_1 is a sphere with the radius that is obtained if the average thickness of the crust is subtracted from the Earth's radius. In the sequel, we shall refer to all data about the travel times of seismic waves to the sphere of radius R_1. Here we assume that the velocity of seismic waves in the crust is equal to a certain average value.

Let $A_j(\varphi_j, \theta_j, \bar{h}_j, t_j)$ be the points associated with earthquakes and $B_k(\varphi_k, \theta_k, \bar{h}_k, h_k)$ be the points associated with seismic stations. Here $j = 1, 2, \ldots, L$; $k = 1, 2, \ldots, K$; L is the number of earthquakes A_j; K is the number of stations B_k; θ is the latitude ($-\pi/2 \le \theta \le \pi/2$); φ is the longitude ($0 \le \varphi \le 2\pi$); $\bar{h}_j$ and $\bar{h}_k$ are the distances from the points to the sphere R_1; h_k is the distance from B_k to the mantle; and t_j is the time of the earthquake A_j.

Let t_{jk} be the moment when the earthquake A_j is registered by the station B_k. All these values are assumed to be known. We take only those points from the set $\{B_k\}$ which lie at a distance greater than a certain number H from the point A_j. This constraint removes those rays which pass only through the crust.

Assume that an earthquake at the point A_j occurred at the time t_j and was registered by the seismic stations B_k at times t_{jk}, then

$$t_{jk} = t_j + T(\theta_j, \varphi_j, \bar{h}_j, \theta_k, \varphi_k, \bar{h}_k, h_k), \qquad (6.3.14)$$

where T is the travel time of the seismic wave through the crust and the mantle from A_j to B_k. Represent the quantity T in the form

$$T = \Delta t_{jk} + T'(\theta_j, \varphi_j, \theta_k, \varphi_k), \qquad (6.3.15)$$

where Δt_{jk} is the travel time of the seismic wave from A_j to the sphere of averaged mantle of radius R_1 and from this sphere to the point B_k; and $T'(\theta_j, \varphi_j, \theta_k, \varphi_k)$ is the travel time of the wave inside the sphere of radius R_1.

Write the function $T'(\theta_j, \varphi_j, \theta_k, \varphi_k)$ as a sum of two terms

$$T'(\theta_j, \varphi_j, \theta_k, \varphi_k) = T'_0(\theta_j, \varphi_j, \theta_k, \varphi_k) + T'_1(\theta_j, \varphi_j, \theta_k, \varphi_k), \qquad (6.3.16)$$

where T'_0 is the travel time of the wave in the medium with radial distribution of velocity and T'_1 is the deviation of the travel time due to nonradial character of the medium. Now our main aim is to find the function T'_1.

Consider a single earthquake A_j. Let us denote by ψ_{kj} the central angle between the points A_j and B_k. With θ_j, φ_j, θ_k, and φ_k known, the angle ψ_{kj} is given by the formula

$$\psi_{kj} = 2 \arcsin \sqrt{\frac{1 - \cos\theta_j \cdot \cos\theta_k \cdot \cos(\varphi_j - \varphi_k) - \sin\theta_j \cdot \sin\theta_k}{2}}. \qquad (6.3.17)$$

Let v_c be the average velocity of seismic waves in the crust. Let v_{cm} be the velocity of waves in the mantle on its boundary. With v_c, v_{cm}, R_1, $\bar{h}_k$, h_k, and $\bar{h}_j$ known, we find Δt_{jk} and ψ'_{jk}, the angle between the points at which the ray $A_j B_k$ enters the sphere of radius R_1 and goes out of this sphere.

We shall also assume that we know the angle at which the seismic ray enters (leaves) the mantle, i. e., the angle between the radius and the tangent to the ray (more precisely, to its segment in the mantle) at the point where the ray enters (leaves) the mantle. We shall denote this angle by α_{cm}. Later we shall derive a formula for determining the angle α_{cm}.

To find ψ'_{jk} and Δt_{jk}, we use the notation of Figures 6.3.2 and 6.3.3. In these figures, 1 is the Earth's surface, 2 is the interface between the mantle and the crust, and 3 is the sphere of averaged mantle of radius R_1 centered at the point O.

From the triangles $A_j A'_j O$ and $B_k OD$ we have

$$\frac{\sin \varepsilon}{R_1 + \bar{h}_j} = \frac{\sin \alpha_j}{R_1}, \tag{6.3.18}$$

$$\frac{\sin \delta}{R_1 + \bar{h}_k} = \frac{\sin \alpha_k}{R_1 + \bar{h}_k - h_k}. \tag{6.3.19}$$

From the triangle $B'_k OD$ we have, for the case in Figure 6.3.2a,

$$\frac{\sin \delta}{R_1} = \frac{\sin \alpha_2}{R_1 + \bar{h}_k - h_k} \tag{6.3.20}$$

and, for the case in Figure 6.3.2b,

$$\frac{\sin \alpha_2}{R_1 + \bar{h}_k - h_k} = \frac{\sin \alpha_{cm}}{R_1}. \tag{6.3.21}$$

By the law of refraction of a ray passing across the interface between two media, we have

$$\frac{\sin \varepsilon}{v_c} = \frac{\sin \alpha_{cm}}{v_{cm}}, \tag{6.3.22}$$

$$\frac{\sin \alpha_{cm}}{v_{cm}} = \frac{\sin \delta}{v_c}. \tag{6.3.23}$$

Chapter 6. Numerical solution

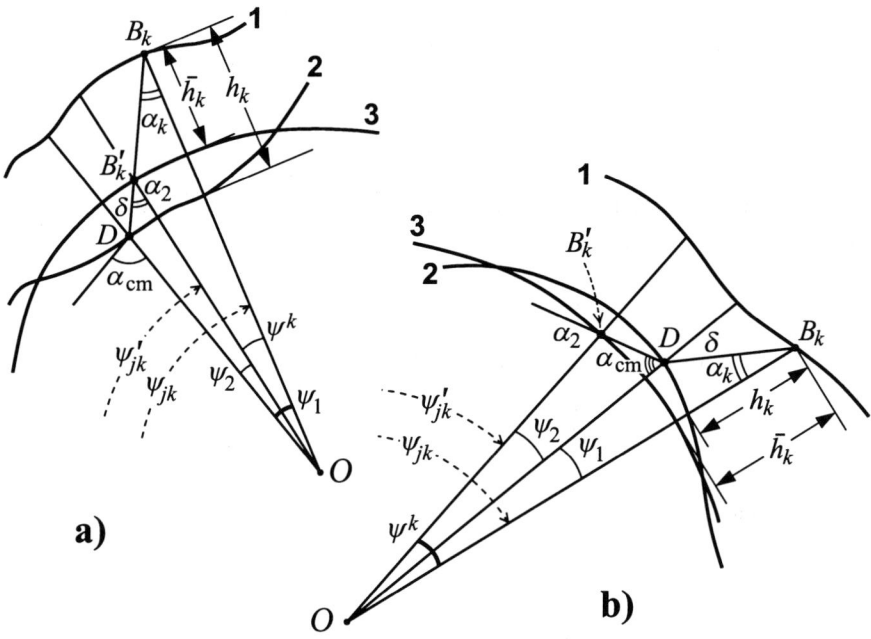

Figure 6.3.2

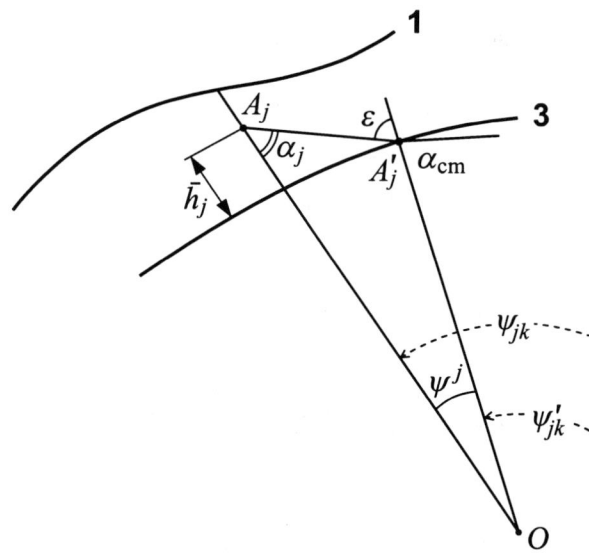

Figure 6.3.3

Formulas (6.3.18)–(6.3.23) imply that

$$\sin\delta = \sin\alpha_{\rm cm} \cdot \frac{v_{\rm c}}{v_{\rm cm}}, \qquad (6.3.24)$$

$$\sin\varepsilon = \sin\alpha_{\rm cm} \cdot \frac{v_{\rm c}}{v_{\rm cm}}, \qquad (6.3.25)$$

$$\sin\alpha_2 = \frac{v}{v_{\rm cm}} \cdot \frac{R_1 + \bar{h}_k - h_k}{R_1} \sin\alpha_{\rm cm}, \qquad (6.3.26)$$

where

$$v = \begin{cases} v_{\rm c}, & \text{if } h_k > \bar{h}_k \text{ (Figure 6.3.2a)}, \\ v_{\rm cm}, & \text{if } h_k < \bar{h}_k \text{ (Figure 5.3.2b)}, \end{cases}$$

$$\sin\alpha_k = \sin\delta \cdot \frac{R_1 + \bar{h}_k - h_k}{R_1 + \bar{h}_k}, \qquad (6.3.27)$$

$$\sin\alpha_j = \sin\varepsilon \cdot \frac{R_1}{R_1 + \bar{h}_j}. \qquad (6.3.28)$$

From Figs. 2 and 3 we easily get

$$\psi'_{jk} = \psi_{jk} - \psi^j - \psi^k, \qquad (6.3.29)$$

where

$$\psi^j = \varepsilon - \alpha_j, \qquad \psi^k = \psi_1 + \psi_2,$$

$$\psi_1 = \delta - \alpha_k, \qquad \psi_2 = \begin{cases} \alpha_2 - \delta, & \text{if } h_k > \bar{h}_k \text{ (Figure 5.3.2a)}, \\ \alpha_2 - \alpha_{\rm cm}, & \text{if } h_k < \bar{h}_k \text{ (Figure 5.3.2b)}, \end{cases}$$

or

$$\psi_2 = \alpha_2 - \arcsin\frac{R_1 \sin\alpha_2}{R_1 + \bar{h}_k - h_k}.$$

Assuming that the velocity of seismic waves in the crust is constant and equal to $v_{\rm c}$, from Figures 6.3.2 and 6.3.3 we obtain the following expressions for Δt_{jk}:

$$\Delta t_{jk} = \begin{cases} \dfrac{B_k B'_k + A_j A'_j}{v_{\rm c}}, & \text{if } \bar{h}_k < h_k \text{ (Figure 6.3.2a)}, \\[2mm] \dfrac{DB_k + A_j A'_j}{v_{\rm c}} + \dfrac{B'_k D}{v_{\rm cm}}, & \text{if } \bar{h}_k > h_k \text{ (Figure 6.3.2b)}, \end{cases} \qquad (6.3.30)$$

where

$$B_k B'_k = \frac{\sin \psi^k}{\sin \alpha_k} R_1, \quad \text{if} \quad \bar{h}_k < h_k;$$

$$A_j A'_j = \frac{\sin \psi^j}{\sin \alpha_j} R_1;$$

$$DB_k = \frac{\sin \psi_1}{\sin \alpha_k}(R_1 + \bar{h}_k - h_k), \quad DB'_k = \frac{\sin \psi_2}{\sin \alpha_{cm}} R_1, \quad \text{if} \quad \bar{h}_k > h_k.$$

Thus, in the expressions (6.3.29) for ψ'_{jk} and (6.3.30) for Δt_{jk}, only the angle α_{cm} remains unknown.

Let us find the angle ψ_{jk} under the assumption that the velocity of seismic waves in the ball of radius R_1 is equal to $v_0(r)$, where r is counted in fractions of radius (i.e., we refer to the unit ball). We obtain

$$\psi'_{jk}(p) = 2\varphi(p, 1), \tag{6.3.31}$$

where

$$\varphi(p, 1) = \int_p^1 \frac{m_0(t)\, dt}{t\sqrt{m_0^2(t) - m_0^2(p)}}. \tag{6.3.32}$$

As is known from geophysics, the exit angle $\alpha(p)$ in this medium is

$$\alpha(p) = \arcsin \frac{m_0(p)}{m_0(1)}. \tag{6.3.33}$$

As an approximation, we assume that

$$\alpha_{cm} = \alpha(p). \tag{6.3.34}$$

Substituting (6.3.31) and (6.3.34) in (6.3.29), we obtain the following equation for p:

$$\psi_{jk} = 2\varphi(p, 1) + \psi^j(p) + \psi^k(p). \tag{6.3.35}$$

This equation was solved by partition of the interval $[p_{\min}, 1]$, where $p_{\min}$ is the minimum p for the domain G; we assume that $p_{\min}$ is known.

Having found p from equation (6.3.35), we can easily find ψ'_{jk} by formulas (6.3.31) and (6.3.32), the quantity Δt_{jk} by (6.3.30), and the time $T'_0(\theta_j, \varphi_j, \theta_k, \varphi_k)$ by formula (6.3.9). Having all these quantities known, we find the function $T'_1(\theta_j, \varphi_j, \theta_k, \varphi_k)$ from (6.3.14)–(6.3.16).

For an earthquake at the point A_j and a station at the point B_k, we have points A'_{jk} and B'_{kj}, respectively, on the sphere of radius R_1. Now, for this pair of points we know the quantity T_1' that corresponds to $\tau_1(p,\alpha)$ from (6.3.10).

Let R be the average radius of the Earth. On the sphere of radius R we consider a strip whose central line is an arc of the large circle specified by two endpoints $X(\varphi_1,\theta_1,R)$ and $Y(\varphi_2,\theta_2,R)$ (in geographical coordinates). The width of the strip is specified by the central angle 2γ. The equation of the plane passing through three points $X(x_1,y_1,z_1)$, $Y(x_2,y_2,z_2)$, and $O(0,0,0)$ is

$$\begin{vmatrix} x & y & z \\ x_1 & y_1 & z_1 \\ x_2 & y_2 & z_2 \end{vmatrix} = 0. \qquad (6.3.36)$$

In geographical coordinates, the points are $X(\varphi_1,\theta_1,R)$, $Y(\varphi_2,\theta_2,R)$, and $O(0,0,0)$; and equation (6.3.36) takes the form

$$\sin\theta\,\cos\theta_1\,\cos\theta_2\,\sin(\varphi_2-\varphi_1) + \cos\theta\,\cos\theta_1\,\sin\theta_2\,\sin(\varphi_1-\varphi)$$
$$- \cos\theta\,\sin\theta_1\,\cos\theta_2\,\sin(\varphi_2-\varphi) = 0. \quad (6.3.37)$$

It is easy to derive an inequality for the strip from this equation.

Let us consider the earthquakes and seismic stations located in the part of the strip bounded by the points X and Y. Denote the central angle between the points X and Y by 2β; in our setting $2\beta < \pi$. It is easy to see that seismic rays pass through that part of the ball of radius R whose points belong to the strip and are between the planes that pass through the lines XO and YO perpendicularly to the central line of the strip. This domain is shown in Figure 6.3.4. The corresponding domain for the sphere of radius R_1 is also shown in the same figure. Since we shall deal only with the sphere of radius R_1, we denote the corresponding points by the same symbols X and Y.

For an earthquake at the point A_j and a station at the point B_k we have the points A'_{jk} and B'_{kj}, respectively, on the sphere of radius R_1 after the crust is removed. For this pair of points, we know T_1' which corresponds to $\tau_1(p,\alpha)$ from (6.3.10). In order to reduce this problem to the one considered in Subsection 6.3.1, we need to transfer the sources A'_{jk} and receivers B'_{kj} to the arc XY of the large circle of the sphere of radius R_1. In this transformation, the angle between these points (with the vertex at the point O), which was earlier denoted by ψ'_{kj}, should be preserved. Below we describe how this can be done.

So, suppose that we have a source at the point $A'_{jk}(\varphi_1,\theta_1,R_1)$ and a receiver at the point $B'_{kj}(\varphi_2,\theta_2,R_1)$. Let $C_{jk}(\varphi,\theta,R_1)$ be the midpoint of

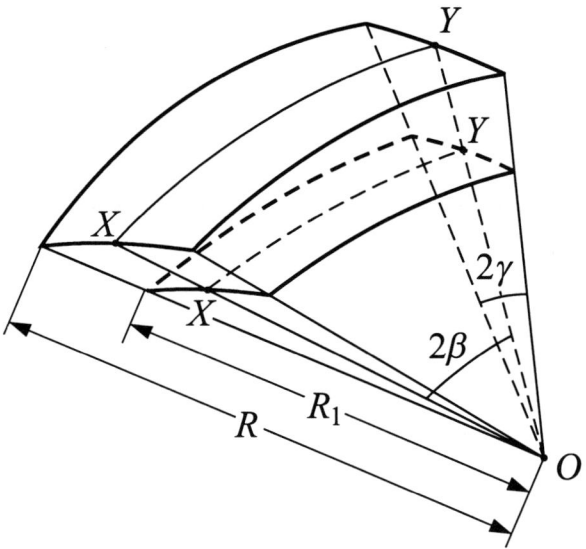

Figure 6.3.4

the arc $A'_{jk}B'_{kj}$ of the large circle. The coordinates of the point C_{jk} are given by the formulas

$$\varphi = \arcsin \frac{\cos\theta_1 \sin\varphi_1 + \cos\theta_2 \sin\varphi_2}{\sqrt{\cos^2\theta_1 + \cos^2\theta_2 + 2\cos\theta_1 \cos\theta_2 \cos(\varphi_1 - \varphi_2)}}, \qquad (6.3.38)$$

$$\theta = \arcsin \frac{\sin\theta_1 + \sin\theta_2}{\sqrt{2}\sqrt{1 + \cos\theta_1 \cos\theta_2 \cos(\varphi_1 - \varphi_2) + \sin\theta_1 \sin\theta_2}}. \qquad (6.3.39)$$

Take a point $C'_{jk}(\varphi, \theta, R)$ on the arc XY such that the arc $C'_{jk}C_{jk}$ belonging to the circle of radius R_1 is perpendicular to the arc XY at the point of their intersection C'_{jk}.

Now take points A''_{jk} and B''_{kj} on the arc XY such that the arcs $C'_{jk}A''_{jk}$ and $C'_{jk}B''_{kj}$ are equal and the angle $A''_{jk}OB''_{kj}$ is equal to the angle ψ'_{jk}. The points A''_{jk} and B''_{kj} constructed in this way are now considered to be the source and receiver, respectively. We make such transfer for each pair of an earthquake A_j and a station B_k. With each such pair we associate the quantity T'_1.

Having processed all data in this way, we can use them in the mathematical problem formulated in Subsection 6.3.1.

In numerical solution of the practical example we modify this method in the following way. For an earthquake at the point A_j, we consider the point of intersection of the sphere of radius R_1 with the ray OA_j and denote this point by A'_j. Further, for this point A'_j we take a point A''_j on the arc XY of the sphere of radius R_1 such that the arc $A'_j A''_j$ of the large circle of the sphere of radius R_1 is perpendicular to the arc XY at the point A''_j. Now we have only one source A''_j instead of all sources A''_{jk}.

Now, on the arc XY, we lay off the arcs equal to the angles ψ'_{jk} from the point A''_j in the proper directions; this yields the points of receivers B''_{kj}. Now that we have a fixed source A''_j we assign to each receiver B''_{kj} the quantity T'_1 which was already found.

In the general case, the available data are such that the points B''_{kj} are distributed nonuniformly. So, in order to reduce the error in solution of the problem, we can find approximately the values of the function T'_1 at the points which would be distributed uniformly over the arc XY and take these values as the initial data.

For this, we represent the function T'_1 as a truncated Fourier series. Introduce the polar coordinates. Let the pole be at the point O and the polar axis $\bar{a}$ pass through the middle of the arc XY. Let ψ_j and ψ_{kj} be the central angles between the axis $\bar{a}$ and the points A''_j and B''_{kj}, respectively. Then the quantity T'_1 can be considered as a function of the angles ψ_j and ψ_{kj}. Expand the function $T'_1(\psi_j, \psi_{kj})$ in the Fourier series on the interval $[-\beta, \beta]$ and write the first N terms

$$T'_1(\psi_j, \psi_{kj}) = \frac{a_0(\psi_j)}{2} + \sum_{n=1}^{N} \left[a_n(\psi_j) \cos \frac{n\pi \psi_{kj}}{\beta} + b_n(\psi_j) \sin \frac{n\pi \psi_{kj}}{\beta} \right]. \quad (6.3.40)$$

For each earthquake A_j, using the values of T'_1 known at individual points, we find the coefficients $a_n(\psi_j)$ and $b_n(\psi_j)$ by the least-squares method. Now we can easily find T'_1 for ψ_j at every point of the interval $[-\beta, \beta]$.

6.3.4. Analysis of seismic data

Before using the data processed in this way in our problem, we carry out the following analysis. For each source A''_j, we plot the obtained functions (6.3.40). Among the earthquakes used, there were earthquakes with close epicenters. Then the corresponding functions $T'_1(\psi_j, \psi_{kj})$ should be similar in a certain sense and this might serve, to certain extent, as a criterion of quality of our data.

Substantial distinctions in the behavior of these functions can be apparently attributed to the errors in seismic data (the times of registration of signals at seismic stations and the coordinates of earthquakes) and, possibly to smaller extent, to averaging of our profile over width. Figure 6.3.5 show the plots of the functions T'_1 for earthquakes with close epicenters. The positions of sources are indicated by arrows on the axes for the angle φ.

In Figure 6.3.5a, one can see similar behavior of the functions T'_1. However, the functions in Figure 6.3.5b have nothing in common.

In this case, the analysis of the function T'_1 has a certain advantage over consideration of travel-time curves for earthquakes. Thus, if we had plotted the travel-time curves, these curves would differ a little from each other because the depths at which the earthquakes occur are close to each other. The plots of the function T'_1 for earthquakes which are far from each other differ appreciably. The values of T'_1 are of the order 10^{-3} s.

6.3.5. Waveguides

In 1951, Gutenberg suggested that the Earth's mantle has layers with lower velocity or *waveguides*, as they are usually called, see (Gutenberg, 1959) for details. With such waveguides, the inverse kinematic problem with sources located above waveguides (in particular, on the surface of the medium) cannot be solved uniquely, see (Gerver and Markushevich, 1965). However, as follows from Subsection 6.3.1 and Section 6.2, we solve the inverse kinematic problem not in its exact statement but we seek for a solution $\tilde{n}_1$ of the linearized problem.

As follows from the book by Romanov (1967b), under certain constraints on τ_1 the function $\tilde{n}_1$ is determined uniquely. In this connection, a numerical experiment of finding $\tilde{n}_1$ for a medium with a waveguide and discrete assignment of τ_1 is of interest. In this case the velocity $v_0(r)$ was taken the same as in the practical example; and the layer with lower speed was placed at a depth close to that where, according to geophysicists' hypothesis, a waveguide might exist.

More precisely, the function $n_1(r, \alpha)$ was taken in the form

$$n_1(r, \alpha) = \begin{cases} 0, & 0.98 < r \leq 1, \\ 0.004 \sin^2 \dfrac{0.95 - r}{0.03} \pi, & 0.95 < r \leq 0.98, \\ 0, & 0.94 < r \leq 0.95. \end{cases} \quad (6.3.41)$$

The methodology of this numerical experiment is similar to that in Section 6.2. In Figure 6.3.6, the exact function $n_1(r, \alpha)$ is shown by a solid

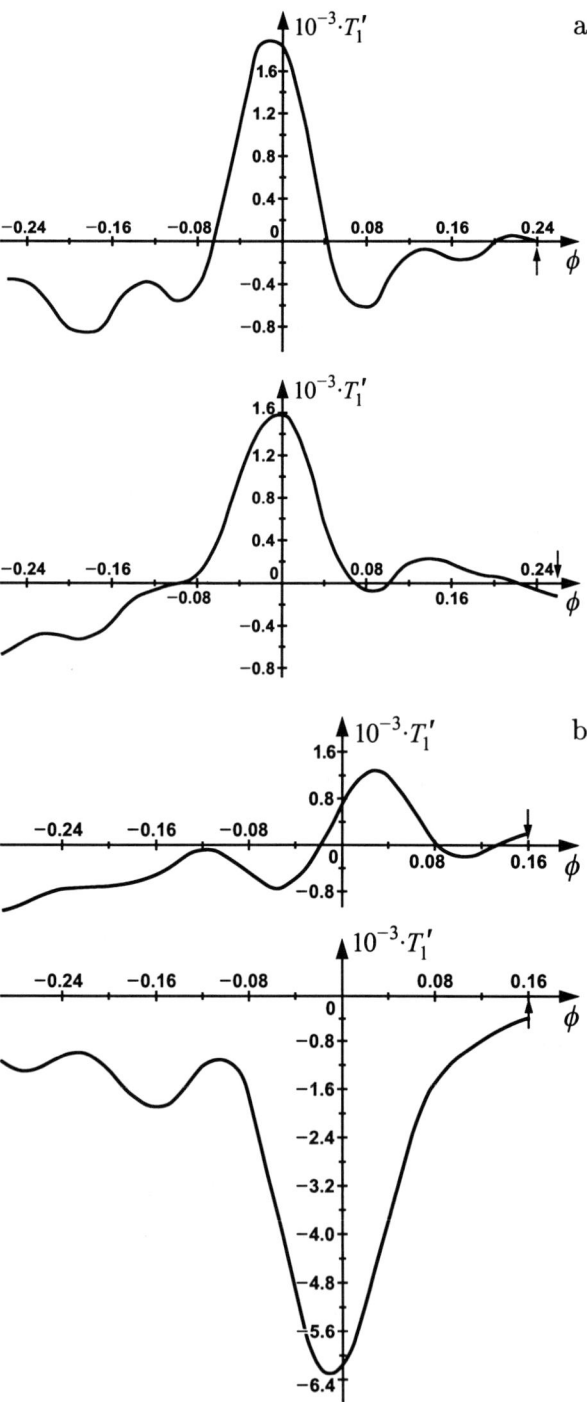

Figure 6.3.5

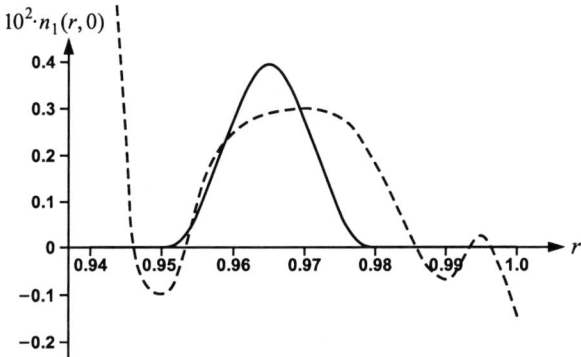

Figure 6.3.6

line, and the reconstructed function is shown by a dashed line. For reconstruction of the function $n_1(r, \alpha)$, a polynomial of degree 5 in variable r was taken. One can see in this figure that we actually obtain a waveguide in the interval $0.95 \leq r \leq 0.98$ and the values of $n_1(r, \alpha)$ on the larger portion of this interval are close to the true values.

This result allows us to conclude that if we obtain domains with lower velocity as a result of solution of practical examples by this method, this can actually reflect the presence of a waveguide in these domains and even give the values of velocity close to the true ones.

6.3.6. Numerical experiments

To use our method in a practical example, we exploit the data about earthquakes on the strip of the large circle from Pamir to Lake Baikal (this material was prepared by I. L. Nersesov in the Complex Seismological Expedition of the Moscow Institute of Earth Physics). The length of this strip was about 4000 km and its width was about 1000 km. The average velocity of P-waves in the crust was taken to be equal to 6 km/s, and at the interface between the crust and mantle 7.74 km/s.

The radius of the averaged mantle has been taken equal to 6338 km. In our problem we consider only those seismic rays which penetrate into the Earth to the depth no greater than 400 km. This depth corresponds to the value $p_{\min} = 0.94$ on the unit sphere of the averaged mantle. On the interval $0.94 \leq p \leq 1$ the one-dimensional law of distribution of velocity is described sufficiently well by the formula

$$v_0(p) = 28.07 - 20.33 \cdot p. \tag{6.3.42}$$

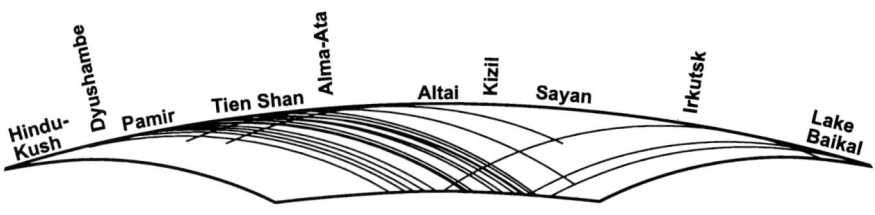

Figure 6.3.7

The corrections for the crust were made on the base of the crust thickness map which was compiled by M. E. Artemiev on the base of gravitational data and information about the relief of the land. Altogether, 22 powerful earthquakes that occurred on the profile in 1961–63 were taken. These earthquakes were registered on stations of regional type.

The distribution of our discrete data in the domain G is shown in Figure 6.3.7.

For each arc, its end point lying on the circle $r = 1$ corresponds to the location of an earthquake on our profile. The arc itself represents the locus of the vertices of all seismic rays which originate from this endpoint.

For clearness, we plot curves in this figure instead of separate points on them. In reality, each earthquake is registered by a finite number of seismic stations; therefore, we have only a finite number of seismic rays.

As is seen from Figure 6.3.7, the distribution of our data in the domain G is far from the uniform one. The available data are especially insufficient in the region of Altai. Therefore, the velocity will be reconstructed in the best way in those parts of the domain G where we have more data. However, reducing the number of parameters of the function which is reconstructed, we can obtain a solution on the average (in a certain sense) on the whole profile.

As follows from (6.3.11), the function $n_1(r, \alpha)$ is sought in the form of a trigonometric polynomial of degree M in the variable α; each coefficient of this polynomial is an algebraic polynomial of degree N.

Figure 6.3.8 shows the plot of the reconstructed function $v(r, \varphi)$ on the whole profile. Here the level lines of the function $v(r, \varphi)$ are plotted since it is a function of two variables. The values of $v(r, \varphi)$ for each level line are shown on the lateral arcs of the boundary of the domain G. This distribution of velocity was obtained with $N = 4$ and $M = 1$.

It should be noted that at a depth about 120 km almost along the whole profile we obtain a layer with lower velocity or, at least, considerable de-

Chapter 6. Numerical solution 227

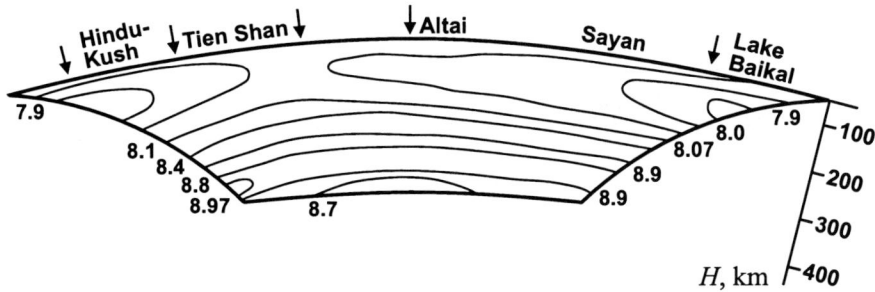

Figure 6.3.8

crease of the rate of growth of the velocity with depth. Because of insufficient amount of data the values of the velocity obtained in a vicinity of $p_{\min} = 0.94$, which corresponds to a depth about 400 km, should be treated carefully.

It is known that the least-squares method is unstable with respect to increase of the number of parameters N and M, starting from certain values of these parameters, with one and the same data. This instability apparently increases even more because of the fact that our problem is ill-posed in the classical sense, as was established by Lavrentiev and Mukhometov (1969). Therefore, to ensure higher reliability of the results obtained, we carry out calculations with other N and M for comparison. For $N = 3$ and $M = 1$, the distribution of velocity $v(r, \varphi)$ obtained was close to that in the case $N = 4$ and $M = 1$.

In Figure 6.3.9, the plots of velocities obtained with $M = 0$ are given. Curves 2 and 3 show the reconstructed velocity in cases it was approximated by polynomials of degree 3 and 4, respectively. For comparison, curve 1 shows the velocity distribution according to Jeffreys and Bullen.

These plots show that at a depth about 120 km we have a layer with lower velocity on curve 2 and considerable slowing-down of the growth of the velocity with depth on curve 3. We also note that the reconstructed velocities on these plots are close to the velocities in Figure 6.3.8.

Then the velocity was reconstructed with $N = 3$ and $M = 2$. However, the solution obtained with these parameters evidently contradict the physical sense. Admissible values of the velocity were obtained only in the lateral parts of the domain G. The values of the velocity obtained in these parts of G are plotted in Figure 6.3.10. In this case we also have either a layer with lower velocity at a depth about 100 km or considerable slowing-down of the growth of the velocity with depth.

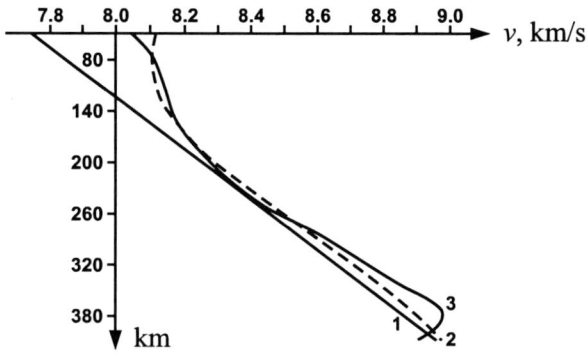

Figure 6.3.9

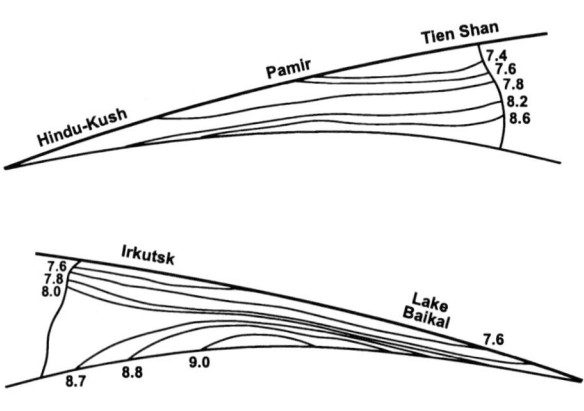

Figure 6.3.10

The values of the velocity on these plots are also close to those in Figure 6.3.8.

Such instability of the solution in the case $N = 3$ and $M = 2$ is a consequence of insufficiency of data in the middle part of the domain G.

However, we obtain admissible values of the velocity in those regions of the domain G which are filled relatively dense with our data.

The analysis of the mathematical method applied, which was carried out in this section and in Section 6.2, and the last analysis of the numerical results of the practical example allow us to hope that the reconstructed distribution of the velocity of P-waves on our profile (Figure 6.3.8) is close to the actual one.

In conclusion, we express our gratitude to I. L. Nersesov and M. E. Artemiev for the seismic data and material about the structure of the crust on the profile considered in our practical example, which were used in numerical calculations.

6.4. NUMERICAL SOLUTION OF INVERSE PROBLEMS OF ELECTROMAGNETOELASTICITY

Here we consider several possible statements of inverse problems connected with Electro-Magnetic (EM) waves propagating through elastic media. In order to introduce the governing equations, we need some preliminary discussion.

The movement of a conductive elastic medium in the electromagnetic field is described by two sets of equations: that of elasticity and that of electrodynamics. Similarly to magnetic hydrodynamics, these equations are connected due to the presence of additional terms that account for effects related to the motion of the elastic conductive medium in the electromagnetic field. The first attempts to apply the theory of electro-magneto-elasticity to the study of the process of wave propagation in elastic conductive media were made by Chadwick (1956), Dunkin and Eringen (1963), Keilis-Borok and Monin (1959), Knopoff (1955). Knopoff has studied the influence of electromagnetic fields on the propagation of elastic waves and arrived at the conclusion that in the class of geophysical problems the effect of electromagnetic phenomena on the process of propagation of elastic waves is negligible, at least in the case of comparatively small electromagnetic disturbances.

We assume that the model under consideration satisfies the basic hypotheses of continuum mechanics: continuity, Euclidity, and absoluteness of time. The first hypothesis means that an uninterrupted continuum is considered, the second one implies the possibility to introduce a Cartesian frame of reference for all points, and according to the third hypothesis relativistic effects are not taken into account. Moreover, the model is inapplicable in the case of strong magnetic fields. We also assume that electro-magneto-elastic waves arise due to the action of mechanical perturbations, and that one can neglect the effect of electromagnetic waves on the process of propagation of elastic oscillations and also neglect the displacement currents as compared with conduction currents. Finally, we shall consider the case of small perturbations.

Now we can write the governing equations. The assumption that we consider the fields of small perturbations allows us to consider the linearized statement of the problem when the displacement vector, the vector of the electric field intensity, and the vector of the magnetic field intensity can be represented in the form

$$(\mathbf{0}, \mathbf{0}, \mathbf{H}^0) + (\mathbf{U}, \mathbf{E}, \mathbf{H}), \qquad (6.4.1)$$

where $(\mathbf{0}, \mathbf{0}, \mathbf{H}^0)$ is the value related to the unperturbed state of the medium

(H^0 is a constant vector); and the vectors U, E, and H correspond to small perturbations of the elastic and electromagnetic fields. Besides, in view of our assumptions, one can reckon that the process of elastic wave propagation is described by the usual system of differential equations of the theory of elasticity:

$$\rho \frac{\partial^2 U}{\partial t^2} = \text{Div}\, T, \qquad (6.4.2)$$

where

$$\text{Div}\, T = \Big(\sum_{j=1}^{3} \frac{\partial}{\partial x_j} T_{ij}\Big)_{i=1}^{3},$$

For the stress tensor T we have the following defining relation:

$$T = \lambda \,\text{tr}\, S \cdot I + 2\varkappa S, \qquad (6.4.3)$$

where S is the strain tensor defined by the formula

$$S_{ij} = \frac{1}{2}\Big(\frac{\partial U_i}{\partial x_j} + \frac{\partial U_j}{\partial x_i}\Big), \qquad i,j = 1,2,3,$$

and I is the unit matrix of order 3×3. Here ρ, λ, and $\varkappa$ denote the density of the medium and the Lamé coefficients, respectively, and δ_{ij} are the Kronecker symbols.

The process of electromagnetic wave propagation through an elastic conductive medium is described in our case by the following system:

$$\text{rot}\, H = J, \qquad \frac{\partial B}{\partial t} = -\text{rot}\, E, \qquad \text{div}\, B = 0, \qquad (6.4.4)$$

where, in virtue of our assumptions, the constitutive relations are written as

$$B = \mu(H^0 + H), \qquad J = \sigma\Big(E + \mu \frac{\partial U}{\partial t} \times H^0\Big). \qquad (6.4.5)$$

Here μ is the magnetic permeability and σ is the conductivity of the medium.

Now we proceed with the statement of the direct problem for differential equations (6.4.2)–(6.4.5). Consider the rectangular Cartesian frame of reference $x = (x_1, x_2, x_3)$. Let the plane $x_3 = 0$ be the interface of two media of the types "air" ($x_3 < 0$) and "conductive ground" ($x_3 > 0$). Electromagnetic and elastic characteristics of the ground are described by piecewise constant functions with planes of break being parallel to the plane $x_3 = 0$. We assume that elastic oscillations arise under the action of a force source concentrated at the origin

$$T_{k3}\big|_{x_3=0} = \delta_{k3} f(t)\, \delta(x_1, x_2), \qquad k = 1, 2, 3, \qquad (6.4.6)$$

where $\delta(\cdot)$ stands for the Dirac's mass.

As concerns the force source and initial data, we assume that the function $f(t)$ and the electromagnetoelastic field are absent before the moment $t = 0$, i. e.,

$$(\boldsymbol{U}, \boldsymbol{E}, \boldsymbol{H})\big|_{t<0} \equiv 0, \quad f\big|_{t<0} \equiv 0. \tag{6.4.7}$$

To single out the unique solution to the direct problem, it is necessary to add the radiation condition at infinity:

$$|\boldsymbol{E}| \to 0, \quad |\boldsymbol{H}| \to 0 \quad \text{for} \quad |x| \to \infty. \tag{6.4.8}$$

Moreover, on the planes where the coefficients of the problem have breaks we assume the standard interface conditions

$$[E_m] = [H_k] = [U_k] = [T_{m3}] = 0, \quad k = 1, 2, \quad m = 1, 2, 3. \tag{6.4.9}$$

Thus, the direct problem consists in finding the vector functions $\boldsymbol{U}$, $\boldsymbol{E}$, and $\boldsymbol{H}$ which satisfy equations (6.4.2)–(6.4.9), provided the elastic and electromagnetic characteristics of the medium and the constant vector $\boldsymbol{H}^0$ characterizing the magnetic field of the Earth are known.

Our main task consists in showing the possibility of applying the optimization approach to the simultaneous determination of electromagnetic and elastic characteristics of the medium and the function $f(t)$ from the system (6.4.2)–(6.4.9) based on some additional information about the components of the vector functions $\boldsymbol{U}$ and $\boldsymbol{E}$. We shall study a special case of this problem which, however, will reflect many principal points of the more general case.

The problem of numerical determining the elastic and electromagnetic characteristics of a medium, taking into account the interaction of two fields, was considered by Avdeev, Goryunov, and Priimenko (1996a, 1997), Avdeev, Priimenko, Goryunov, and Zvyagin (1997a, 1997b), Avdeev, Goryunov, Soboleva, and Priimenko (1999). We also mention the paper by Klimenko (1995) where a numerical algorithm to determine the coefficient of the electromagnetoelastic coupling was proposed.

As concerns the form of the sensing signal (i. e., the function $f(t)$), in most cases of real geophysical investigations it is either unknown or is given only approximately, while its accurate estimate is necessary for practical solution of many inverse problems. Note that problems of simultaneous reconstruction of the structure of the medium and the form of the sensing signal were studied theoretically by Blagovestchensky (1970) and Reznitskaya (1972). The most complete bibliography on the questions of numerical solution can be found in the paper by Carrion, Sacramento, and Pestana (1990).

6.4.1. The first inverse problem

Now let us state the first inverse problem, see (Avdeev, Goryunov, and Priimenko, 1996a, 1997). Let z denote the variable x_3. Consider the functions

$$v(z) = \left(\frac{\lambda + 2\varkappa}{\rho}\right)^{1/2}, \qquad c(z) = \left(\frac{1}{\sigma\mu}\right)^{1/2},$$

where $v(z)$ is the velocity of longitudinal waves and $c(z)$ is the velocity of the diffusion process of electromagnetic waves.

We shall say that the functions $v(z)$, $c(z)$, and $f(t)$ belong to the class $\mathfrak{M}$ if there exist positive constants v_m, c_m, f_m, z_m, z'_m, and t_m such that

$$v(z) = \begin{cases} v_m, & z \in (z'_{m-1}, z'_m), \quad m = \overline{1, k+1}, \\ v_{k+1}, & z > z'_{k+1}, \end{cases} \qquad (6.4.10)$$

$$c(z) = \begin{cases} c_m, & z \in (z_{m-1}, z_m), \quad m = \overline{1, n+1}, \\ c_{n+1}, & z > z_{n+1}, \end{cases} \qquad (6.4.11)$$

$$f(t) = \begin{cases} f_m, & t \in (t_{m-1}, t_m), \quad m = \overline{1, l+1}, \\ 0, & t > t_{m+1}, \end{cases} \qquad (6.4.12)$$

where $z_0 = z'_0 = t_0 = 0$ and $n, k, l \in \mathbb{N}$.

Henceforth we shall always assume that the functions $v(z)$, $c(z)$, and $f(t)$ belong to the class $\mathfrak{M}$.

Consider the functions

$$u(z,t) = \operatorname{Re} F_{x_1 x_2}(U_3)\big|_{\nu_1 = \nu_2 = 0}, \qquad (6.4.13)$$

$$e(z,t) = \operatorname{Re} F_{x_1 x_2}(E_1)\big|_{\nu_1 = \nu_2 = 0}, \qquad (6.4.14)$$

where $F_{x_1 x_2}(\cdot)$ stands for the generalized Fourier transform with respect to the variables x_1, x_2; and (ν_1, ν_2) are the dual variables. Starting from equations (6.4.2)–(6.4.9), we can write the system of relations for the functions u and e in the domain $z \geq 0$

$$\frac{\partial^2 u}{\partial t^2} = v^2(z)\frac{\partial^2 u}{\partial z^2}, \qquad (z,t) \in \Omega' \times \mathbb{R}, \qquad (6.4.15)$$

$$u\big|_{t<0} \equiv 0, \qquad (6.4.16)$$

$$\frac{\partial u}{\partial z}\bigg|_{z=0} = F(t), \qquad (6.4.17)$$

$$[u]_{z=z'_m} = \left[\frac{\partial u}{\partial z}\right]_{z=z'_m} = 0, \qquad m = \overline{1, k+1}, \qquad (6.4.18)$$

Chapter 6. Numerical solution

$$\frac{\partial e}{\partial t} = c^2(z)\frac{\partial^2 e}{\partial z^2} + \mu h^0 \frac{\partial^2 u}{\partial t^2}, \qquad (z,t) \in \Omega \times \mathbb{R}, \qquad (6.4.19)$$

$$e\big|_{t<0} \equiv 0, \qquad \lim_{z \to \infty} e = 0, \qquad (6.4.20)$$

$$\frac{\partial e}{\partial z}\bigg|_{z=0} = 0, \qquad (6.4.21)$$

$$[e]_{z=z_m} = \left[\frac{\partial e}{\partial z}\right]_{z=z_m} = 0, \qquad m = \overline{1, n+1}, \qquad (6.4.22)$$

where

$$\Omega' = \mathbb{R}_+ \setminus \{z = z'_m, \ m = \overline{1, k+1}\},$$
$$\Omega = \mathbb{R}_+ \setminus \{z = z_m, \ m = \overline{1, n+1}\},$$

$F(t) = (\lambda(0) + 2\varkappa(0))^{-1} f(t)$, and h^0 is the constant characterizing the magnetic field of the Earth.

Now we formulate the inverse problem.

Inverse Problem 6.4.1 (IP 6.4.1). *Find the functions $v(z)$, $c(z)$, $f(t) \in \mathfrak{M}$ (i.e., the set of numbers v_m, c_m, and f_m) if the following additional information on the solutions to the problems (6.4.15)–(6.4.18) and (6.4.19)–(6.4.22) is known:*

$$u\big|_{z=0} = u_0(t), \qquad t \in \mathbb{R}_+ \qquad (6.4.23)$$

$$e\big|_{z=0} = e_0(t), \qquad t \in \mathbb{R}_+, \qquad (6.4.24)$$

and the numbers μ and h^0 are known, too.

Remark 6.4.1. Without any loss of generality, we shall assume that $\mu = \mu_0$, where μ_0 is the magnetic permeability of vacuum.

To solve IP 6.4.1 numerically, an optimization approach, based on minimizing the objective misfit functionals, has been used.

At the first stage the initial-boundary value problem (6.4.15)–(6.4.18) describing the propagation of elastic waves in a vertically inhomogeneous medium was considered.

In this model, the medium is assumed to be a stack of homogeneous layers over a homogeneous half-space. A numerical-analytic method proposed by Fatianov (1990), Fatianov and Mikhailenko (1988) allows one to obtain the exact solution in this case and, what is especially important, to arrange the process of constructing the solution to the inverse problem in the most efficient way.

Concerning the system (6.4.15)–(6.4.18), we consider the inverse problem of reconstructing the functions $v(z)$, $f(t) \in \mathfrak{M}$ from the additional information (6.4.23).

Applying the Fourier transform with respect to the variable t, we rewrite the original statement (6.4.15)–(6.4.18), (6.4.23) in the following form:

$$\frac{d^2}{dz^2} u(z,\omega) + \nu^2 u(z,\omega) = 0, \qquad z \in \Omega', \qquad (6.4.25)$$

$$\left. \frac{du(z,\omega)}{dz} \right|_{z=0} = F(\omega), \qquad (6.4.26)$$

$$[u(z,\omega)]_{z=z'_m} = \left[\frac{du(z,\omega)}{dz} \right]_{z=z'_m} = 0, \qquad m = \overline{1, k+1}, \qquad (6.4.27)$$

where $\nu^2 = \omega^2 v^{-2}(z)$ and

$$F(\omega) = \int_0^{+\infty} F(t) \exp(-i\omega t)\, dt.$$

To single out the unique solution, we assume that the principle of the limit absorption is satisfied, i.e.,

$$u(z,\omega) = \lim_{\varepsilon \to +0} u(z, \omega - i\varepsilon), \qquad (6.4.28)$$

where

$$\lim_{z \to +\infty} u(z, \omega - i\varepsilon) = 0. \qquad (6.4.29)$$

The additional information (6.4.23) can be represented as

$$u(z,\omega)\big|_{z=0} = u_0(\omega). \qquad (6.4.30)$$

We shall seek for the solution to the inverse problem as the minimum point of the functional

$$\Phi_1[n(z), F(\omega)] = \int_{\omega_1}^{\omega_2} \big|u_0(\omega) - B_1[n(z), F(\omega)](\omega)\big|^2 d\omega, \qquad (6.4.31)$$

where (ω_1, ω_2) is the range of temporal frequencies defined by the spectral contents $F(\omega)$ of the sensing signal, and $B_1[n(z), F(\omega)]$ is a nonlinear operator mapping the functions $n(z) = v^{-2}(z)$ and $F(\omega)$ into the trace of the solution to the direct problem (6.4.25)–(6.4.29) at the point $z = 0$.

One can prove the Fréchet differentiability of the functional (6.4.31) with respect to its arguments $n(z)$ and $F(\omega)$ and then obtain the following expressions for its gradients:

$$\nabla_{n(z)} \Phi_1[n(z), F(\omega)](\xi) = -2\operatorname{Re} \int_{\omega_1}^{\omega_2} (\omega + i\varepsilon)^2 F(\omega)$$
$$\times \left[u_0(\omega) - B_1[n(z), F(\omega)](\omega)\right] \bar{\mathcal{G}}_1(\xi, \omega) \, d\omega, \quad (6.4.32)$$

$$\nabla_{F(\omega)} \Phi_1[n(z), F(\omega)](\omega)$$
$$= -2\operatorname{Re} \left[u_0(\omega) - B_1[n(z), F(\omega)](\omega)\right] \bar{\mathcal{G}}_1(\xi, \omega)$$
$$- 2\mathrm{i}\operatorname{Im} \left[u_0(\omega) - B_1[n(z), F(\omega)](\omega)\right] \bar{\mathcal{G}}_1(\xi, \omega), \quad (6.4.33)$$

where $\mathcal{G}_1(\xi, \omega)$ is the solution to the problem (6.4.25)–(6.4.29) with $F(\omega) \equiv 1$ and the bar over the symbol of the function denotes the complex conjugation.

Assume that there exists a point (n_s, F_s) at which the gradients of the functional vanish. Then from (6.4.32) and (6.4.33) one can easily obtain the following expression:

$$F_s(\omega) = \frac{\bar{\mathcal{G}}_1(0, \omega) \, u_0(\omega)}{|\mathcal{G}_1^2(0, \omega)|^2}, \quad (6.4.34)$$

where $\mathcal{G}_1(0, \omega)$ is the trace of the solution to the problem (6.4.25)–(6.4.29) with $F(\omega) \equiv 1$ and $n(z) = n_s(z)$ calculated at the point $z = 0$.

It was proposed by Cheverda and Voronina (1994) to use a formula similar to (6.4.34) for computing the impulse $F_k(\omega)$ on the k-th iteration when solving the inverse problem of VSP (vertical seismic profiling). An application of this algorithm to solution of the inverse dynamic problem of seismics with unknown source in the case where the whole wave field is measured on the free surface $z = 0$ was described in the papers by Avdeev (1995), Avdeev and Goryunov (1996a, 1996b).

Using expressions (6.4.32) and (6.4.34), we can apply the optimization methods of steepest descent of the first order to search for the minimum point of the functional (6.4.31), i.e., to reconstruct the unknown functions $v(z)$ and $F(t)$. Succeeding in reconstruction of these functions, then, having solved the direct problem (6.4.25)–(6.4.29), we can determine the spectrum of the wave field $u(z, \omega)$ in the whole of the half-space under study, i.e., find the right-hand side in the differential equation for the electric field in the problem (6.4.19)–(6.4.22).

On the second stage the initial-boundary value problem (6.4.19)–(6.4.22) is considered, which, in terms of the Fourier images with respect to the variable t, can be written as

$$\frac{d^2}{dz^2}e(z,\omega) + \eta^2(z)e(z,\omega) = -i\omega\mu_0 h^0 \eta^2(z) u(z,\omega), \quad z \in \Omega, \quad (6.4.35)$$

$$\left.\frac{de(z,\omega)}{dz}\right|_{z=0} = 0, \quad (6.4.36)$$

$$[e(z,\omega)]_{z=z_m} = \left[\frac{de(z,\omega)}{dz}\right]_{z=z_m} = 0, \quad m = \overline{1, n+1}, \quad (6.4.37)$$

where $\eta^2(z) = i\omega c^{-2}(z)$.

The additional information (6.4.24) is rewritten as

$$\left. e(z,\omega)\right|_{z=0} = e_0(\omega). \quad (6.4.38)$$

We shall look for the solution to the inverse problem (6.4.35)–(6.4.38) as the minimum point of the cost functional

$$\Phi_2[\sigma(z)] = \int_{\omega_1}^{\omega_2} \left| e_0(\omega) - B_2[\sigma(z)](\omega) \right|^2 d\omega, \quad (6.4.39)$$

where $B_2[\sigma(z)]$ is a nonlinear operator mapping the function $\sigma(z)$ (the "test" value of conductivity) into the trace of the solution to the direct problem (6.4.35)–(6.4.37) at $z = 0$.

The gradient of the cost functional (6.4.39) with respect to the conductivity is written as follows:

$$\nabla_\sigma \Phi_2[\sigma(z)](\xi) = A_1(\xi) + A_2(\xi), \quad (6.4.40)$$

where

$$A_1(\xi) = 2\mu_0^2 H \operatorname{Re} \int_{\omega_1}^{\omega_2} \omega^2 \left[e_0(\omega) - B_2[\sigma(z)](\omega) \right] \bar{\mathcal{G}}_2(\xi,\omega)\, \bar{u}(\xi,\omega)\, d\omega, \quad (6.4.41)$$

$$A_2(\xi) = 2\mu_0^2 H \operatorname{Im} \int_{\omega_1}^{\omega_2} \omega^3 \left[e_0(\omega) - B_2[\sigma(z)](\omega) \right]$$

$$\times \bar{\mathcal{G}}_2(\xi,\omega) \int_0^{+\infty} \sigma(\tau) \bar{\mathcal{G}}_2(\tau,\omega)\, \bar{u}(\tau,\omega)\, d\tau\, d\omega, \quad (6.4.42)$$

and $\mathcal{G}_2(\xi,\omega)$ is the Green function for the problem (6.4.35)–(6.4.37).

Using formulas (6.4.40)–(6.4.42), we can apply the optimization methods of the first order for the search for the minimum point of the functional (6.4.39), i.e., for reconstruction of the unknown conductivity $\sigma(z)$.

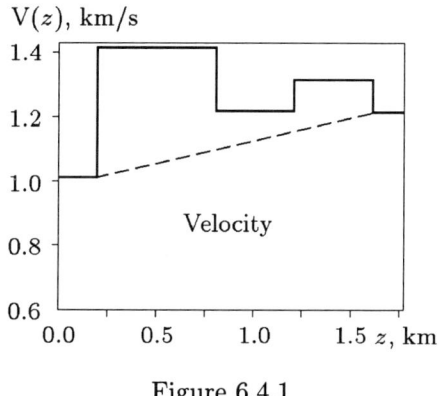

Figure 6.4.1

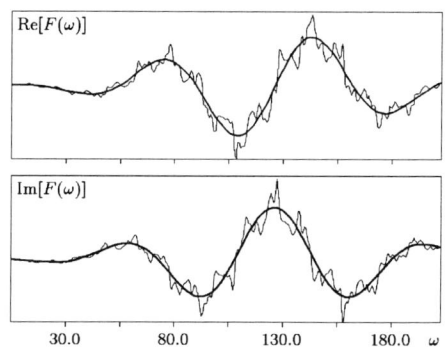

Figure 6.4.2

To carry out numerical experiments a rather complex model of a vertically inhomogeneous medium was chosen. This model incorporated sharp changes of the values of parameters. The reconstruction of the medium was carried out up to the depth of 1.75 km. The medium below this depth was assumed to be homogeneous. All the medium from the surface to the depth 1.75 km was partitioned into 9 layers of equal width.

As a sensing signal, the impulse with a "bell-shaped envelope" was chosen (the dominating frequency $f = 20$ Hz):

$$F(\omega) = \left[\exp\left(-\left(\frac{\omega - 2\pi f}{\pi f}\right)^2\right) + \exp\left(-\left(\frac{\omega + 2\pi f}{\pi f}\right)^2\right)\right] \exp\left(-\frac{i \cdot 1.75\omega}{f}\right). \tag{6.4.43}$$

Computations were made for temporal frequencies from 5 to 40 Hz.

To compute the whole wave field $u(z, \omega)$ and the electrical field intensity $e(z, \omega)$, the numerical-analytical method was used.

To calculate the impulse $F_j(\omega)$ on the j-th iteration, we use the condition of vanishing of the gradient of the functional (6.4.31) with respect to the function F_j on the current velocity $v_j(z)$, i.e., expression (6.4.34).

In Figure 6.4.1, the velocity model of the medium (solid line) and the initial approximation (dashed line) are shown. Figure 6.4.2 represents the spectrum $F(\omega)$ of the input signal impulse $F(t)$ (thick line) and its first approximation (thin line) obtained by formula (6.4.34). In Figure 6.4.3, the function $F(t)$ (thick line) and its first approximation (thin line) are shown. In the result of 35 iterations by the method of conjugate gradients we succeed in reconstructing with good accuracy both the velocity distribution for this medium and the functions $F(\omega)$ and $F(t)$. The results of calculations are plotted in Figures 6.4.4–6.4.6.

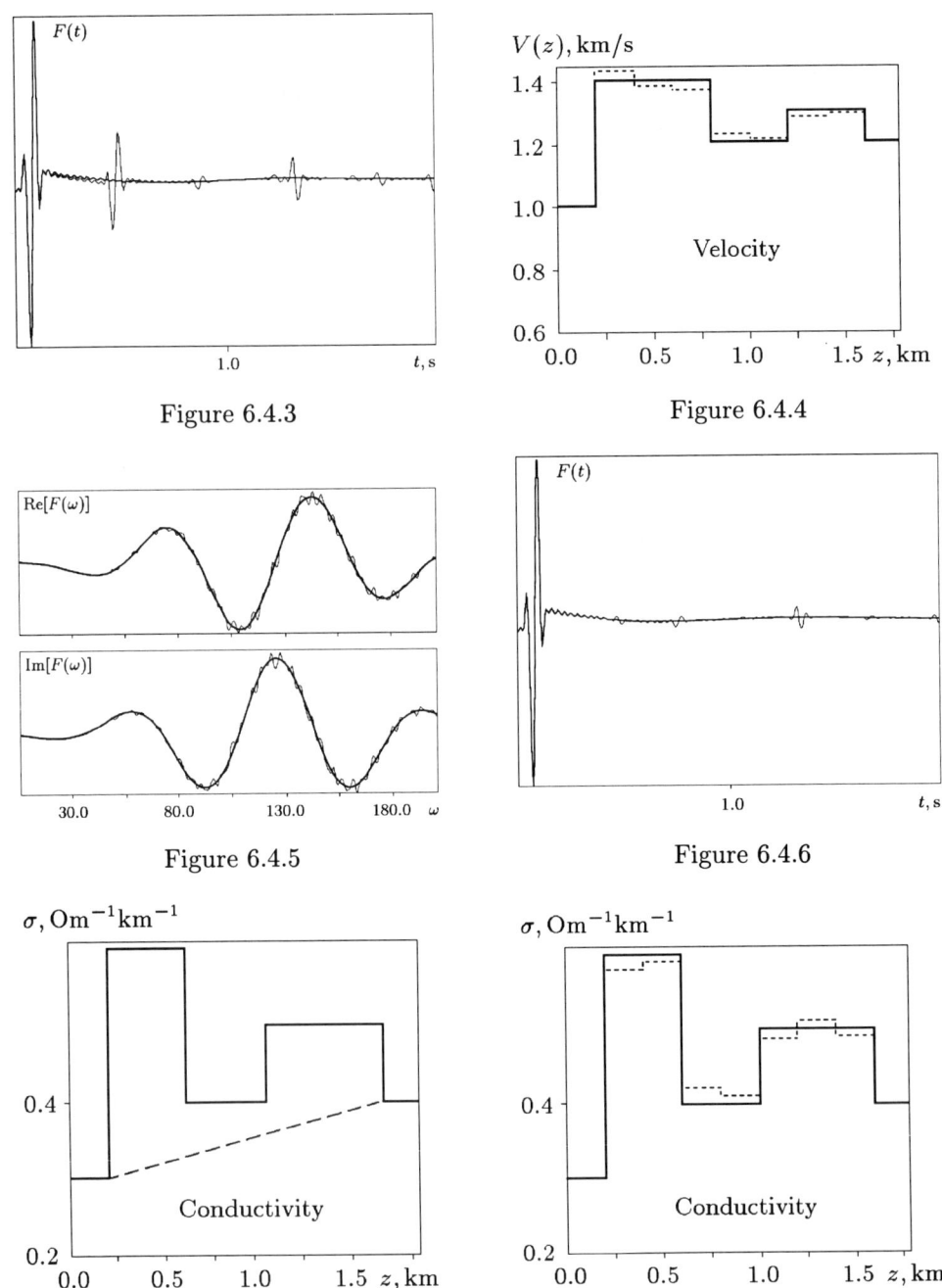

Figure 6.4.3

Figure 6.4.4

Figure 6.4.5

Figure 6.4.6

Figure 6.4.7

Figure 6.4.8

Here and in the sequel, pointing out the number of iterations made, we mean a practically complete stop of the iteration process on the stage concerned. The quality of the approximations obtained was estimated by the closeness of the values of the corresponding functional to zero.

On the next stage, using the reconstructed functions $v(z)$ and $F(t)$, we calculate the spectrum of the wave field $u(z,\omega)$ in the whole of the half-space under study, i.e., the right-hand side in the problem (6.4.35)–(6.4.37) was determined.

In Figure 6.4.7, the "true" function $\sigma(z)$ (solid line, see also the same solid line in Figure 6.4.8) and its initial approximation (dashed line) are shown. The final approximation computed by 68 iterations of the conjugate gradient method is plotted in Figure 6.4.8 (dashed line).

6.4.2. The second inverse problem

Let z denotes the variable x_3. Consider the functions

$$v_p(z) = \sqrt{\frac{\lambda(z) + 2\varkappa(z)}{\rho(z)}}, \qquad v_s(z) = \sqrt{\frac{\varkappa(z)}{\rho(z)}}, \qquad c(z) = \sqrt{\frac{1}{\sigma(z)\mu(z)}},$$

where $v_p(z)$ is the velocity of longitudinal waves, $v_s(z)$ is the velocity of transverse waves, $c(z)$ is the velocity of the diffusion process of electromagnetic waves, ρ is the density of the medium, λ and $\varkappa$ are the Lamé parameters, σ is the electrical conductivity of the medium, and μ is the magnetic permeability of the medium, all these parameters being dependent on one spatial variable z.

Now we formulate the inverse problem that will be studied below, for details see Avdeev, Priimenko, Goryunov, and Zvyagin (1997a, 1997b), Avdeev, Goryunov, Soboleva, and Priimenko (1998).

Inverse Problem 6.4.2 (IP 6.4.2). *Find the functions $v_p(z)$, $v_s(z)$, and $c(z)$ if the following additional information on the solution to the direct problem (6.4.2)–(6.4.9) is known:*

$$\boldsymbol{U}\big|_{z=0} = \boldsymbol{U}_0(x_1, x_2, t), \quad \boldsymbol{H}\big|_{z=0} = \boldsymbol{H}_0(x_1, x_2, t), \quad (x_1, x_2, t) \in \mathbb{R}^2 \times \mathbb{R}_+.$$

We assume that μ and $\boldsymbol{H}^0$ are also known.

Remark 6.4.2. Without any loss of generality, we shall assume that $\mu = \mu_0$, where μ_0 is the magnetic permeability of the air, and that ρ is a known constant.

To solve the direct problem the following algorithms have been used by Mikhailenko and Soboleva (1995).

Applying the Fourier transforms with respect to the variables x_1, x_2, and t (marked by $\sim$ over corresponding functions), we can rewrite the original system of equations (6.4.2)–(6.4.5) in the terms of $\tilde{U}_k, \tilde{H}_k$, $k = 1, 2, 3$, only:

$$\tilde{U}_{1,zz} - \left(\frac{v_p^2}{v_s^2}\nu_1^2 + \nu_2^2 - \frac{\omega^2}{v_s^2}\right)\tilde{U}_1 - \nu_1\nu_2\frac{v_p^2 - v_s^2}{v_s^2}\tilde{U}_2 + i\nu_1\frac{v_p^2 - v_s^2}{V_s^2}\tilde{U}_{3,z} = 0, \quad (6.4.44)$$

$$\tilde{U}_{2,zz} - \nu_1\nu_2\frac{v_p^2 - v_s^2}{v_s^2}\tilde{U}_1 - \left(\nu_1^2 + \frac{v_p^2}{v_s^2}\nu_2^2 - \frac{\omega^2}{v_s^2}\right)\tilde{U}_2 + i\nu_2\frac{v_p^2 - v_s^2}{v_s^2}\tilde{U}_{3,z} = 0, \quad (6.4.45)$$

$$\tilde{U}_{3,zz} + \frac{v_p^2 - v_s^2}{v_s^2}i(\nu_1\tilde{U}_{1,z} + \nu_2\tilde{U}_{2,z}) - \left(\frac{v_s^2}{v_p^2}(\nu_1^2 + \nu_2^2) - \frac{\omega^2}{v_p^2}\right)\tilde{U}_3 = 0, \quad (6.4.46)$$

$$\tilde{H}_{1,zz} + r^2\tilde{H}_1 = \omega c^{-2}\big(\nu_2(H_1^0\tilde{U}_2 - H_2^0\tilde{U}_1) + i(H_3^0\tilde{U}_{1,z} - H_1^0\tilde{U}_{3,z})\big), \quad (6.4.47)$$

$$\tilde{H}_{2,zz} + r^2\tilde{H}_2 = \omega c^{-2}\big(\nu_1(H_2^0\tilde{U}_1 - H_1^0\tilde{U}_2) + i(H_3^0\tilde{U}_{2,z} - H_2^0\tilde{U}_{3,z})\big), \quad (6.4.48)$$

$$\tilde{H}_{3,zz} + r^2\tilde{H}_3 = \omega c^{-2}\big(\nu_1(H_3^0\tilde{U}_1 - H_1^0\tilde{U}_3) + \nu_2(H_3^0\tilde{U}_2 - H_2^0\tilde{U}_3)\big). \quad (6.4.49)$$

The solution to the system (6.4.47)–(6.4.49) is sought in the form

$$\tilde{H}_k = C_{1k}^j e^{\tau_j z} + C_{2k}^j e^{-\tau_j z} + \varphi_k^j, \quad k = 1, 2, 3, \quad (6.4.50)$$

where j is the number of the layer.

If the solutions φ_ν^j to the system are known, the constants C_{lk}^j, $l = 1, 2$, can be determined with the help of the well-known (for such problems) recurrent formulas through the boundary conditions and the interface conditions for layers.

It is non-trivial to find a particular solution of the system (6.4.44)–(6.4.46); the difficulties of construction of an analytical, in every layer, solution to equations of elasticity by matrix methods are well known (Mikhailenko and Soboleva, 1995). To solve equations (6.4.44)–(6.4.46) some modification of the factorization method is used.

Numerical solution of the direct problem allows us to consider some dynamic features of seismomagnetic waves. Each kind of seismic waves generates an electromagnetic wave associated with it and propagating with the same velocity. The electromagnetic wave generated by a seismic wave of a given kind is called the seismomagnetic wave of the same kind (e. g., Rayleigh seismomagnetic wave, longitudinal seismomagnetic wave, transverse seismomagnetic wave, etc.). As compared to the longitudinal wave, the seismomagnetic wave is transverse, just as any other electromagnetic wave is. However, the longitudinal seismomagnetic wave propagates with the velocity close to that of the longitudinal seismic wave. Basic dynamic features of seismomagnetic waves for homogeneous elastic media were considered by Mikhailenko and Soboleva (1995). Further, we convert the components of both the seismic and seismomagnetic fields from x_1, x_2, z to the spherical coordinate system for stratified elastic media. We transformed all the components into dimensionless form. We normalize all the components of the seismomagnetic field and the seismic field to have a unit maximum amplitude in both cases; for this, we divided them by proper numbers.

Figure 6.4.9 shows the radial and tangential components of the displacement of the elastic wave at the point $r_0 = 3\lambda$, where λ is the dominant P-wavelength in the elastic medium, and the radial and tangential components of the seismomagnetic field at the same point for different angles $\hat{\theta}$. Here $\hat{\theta}$ is the angle between the strength vector of the Earth's magnetic field $\boldsymbol{H}^0$ and the vertical axis z. The elastic model is used with an explosive type point source located near the point $z = 0$; in this case the transverse components of all waves are equal to zero. The parameters of the model are the following: $v_{p1} = 1000$ m/s, $v_{p2} = 2000$ m/s, $v_{si} = v_{pi}/1.73$, $i = 1, 2$, the depth of the layer $h = \lambda$, the strength of the geomagnetic field $H^0 = 40$ A/m, and the conductivity $\sigma = 0.01$ S/m. Figure 6.4.9 shows that the phase and the first arrivals of geomagnetic variations coincide with the similar characteristics of the seismic waves. The first wave is the longitudinal wave P, the second wave is the Rayleigh wave, and the third wave is the wave reflected from the boundary of the layer. The radial and tangential components of the P- and the Rayleigh seismomagnetic waves have the same circular polarization. The amplitude of the P-seismomagnetic wave decreases with the increase of the angle $\hat{\theta}$, while the amplitude of the Rayleigh wave increases.

Figure 6.4.10 shows the radial, tangential, and transverse components of the elastic displacements and seismomagnetic waves. The elastic model and its parameters are the same as in Figure 6.4.9. A point source located near $z = 0$ is the source of force of horizontal type. In this case we have nonzero radial, tangential, and transverse components of all waves. The amplitude of the P-seismomagnetic wave decreases with the increase of the angle $\hat{\theta}$, while the amplitude of the Rayleigh wave increases.

To solve the inverse problem numerically an optimization approach was used based on minimizing the misfit functionals of the observed data and the data computed in solving "test" direct problems.

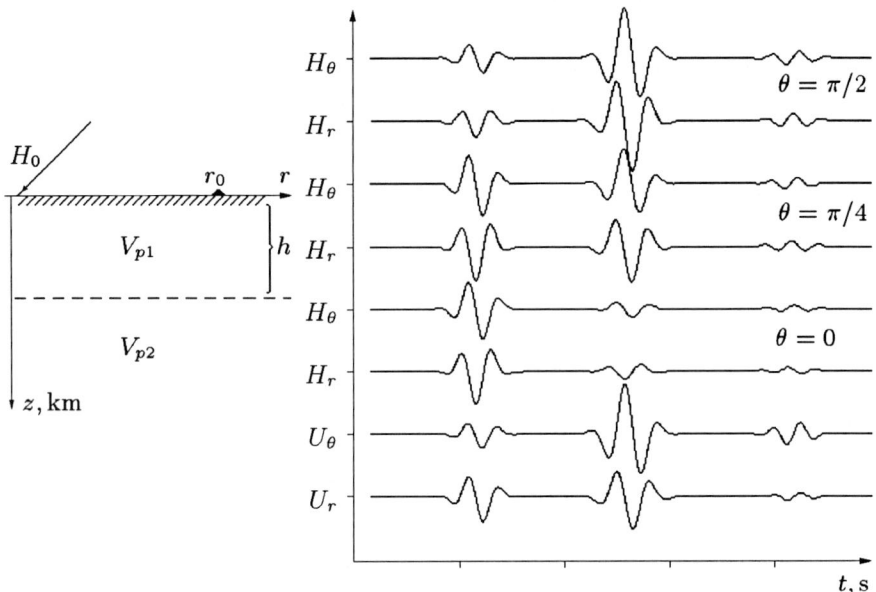

Figure 6.4.9

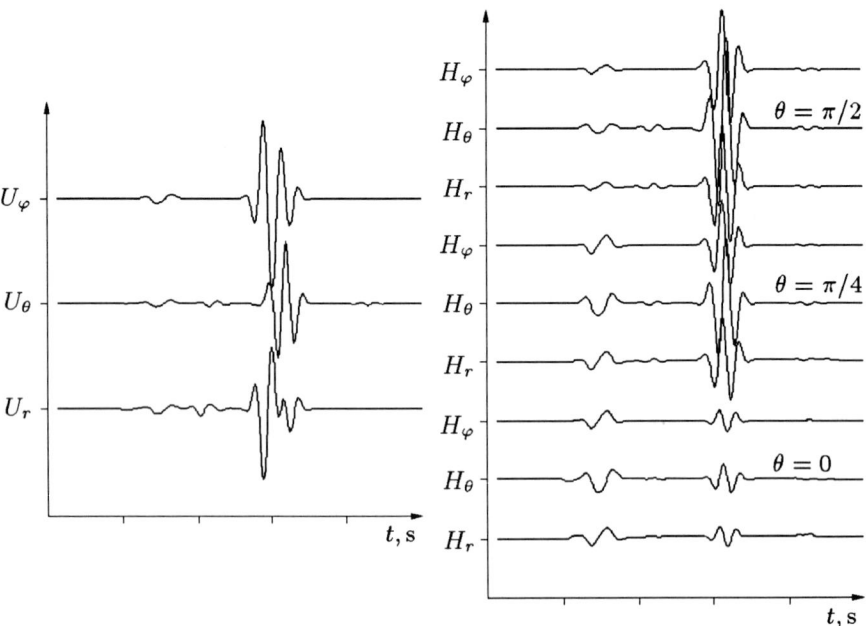

Figure 6.4.10

Chapter 6. Numerical solution

We shall look for the functions $v_p(z)$ and $v_s(z)$ as the minimum point of the functional

$$\Phi_1[v_p, v_s] = \int_{\omega_1}^{\omega_2} \int_{\nu_{1,1}}^{\nu_{1,2}} \int_{\nu_{2,1}}^{\nu_{2,2}} \left|\widetilde{\boldsymbol{U}}_0(\nu_1, \nu_2, \omega) - B_1[v_p, v_s](\nu_1, \nu_2, \omega)\right|^2 d\nu_1 \, d\nu_2 \, d\omega,$$
(6.4.51)

where (ω_1, ω_2) is the range of temporal frequencies defined by the spectral contents $f(\omega)$ of the sensing signal, $(\nu_{1,1}, \nu_{1,2})$ and $(\nu_{2,1}, \nu_{2,2})$ are the ranges of spatial frequencies, and $B_1[v_p, v_s]$ is the nonlinear operator mapping the functions $v_p(z)$ and $v_s(z)$ into the solution of the appropriate direct problem at the point $z = 0$.

If we succeed in reconstructing the functions $v_p(z)$ and $v_s(z)$, then, having solved the direct problem, we can determine the spectrum of the wave vector $\widetilde{\boldsymbol{U}}(\nu_1, \nu_2, z, \omega)$ in the whole of the half-space under study, i.e., we can find the right-hand side in the system of differential equations for the magnetic fields.

Then we seek for the conductivity function $\sigma(z)$ as the minimum point of the cost functional

$$\Phi_2[c(z)] = \int_{\omega_1}^{\omega_2} \int_{\nu_{1,1}}^{\nu_{1,2}} \int_{\nu_{2,1}}^{\nu_{2,2}} \left|\widetilde{\boldsymbol{H}}_0(\nu_1, \nu_2, \omega) - B_2[c(z)](\nu_1, \nu_2, \omega)\right|^2 d\nu_1 \, d\nu_2 \, d\omega,$$
(6.4.52)

where $B_2[c(z)]$ is the nonlinear operator mapping the function $c(z)$ (the "test" value of the conductivity) into the solution of the appropriate direct problem at the point $z = 0$.

To arrange the interactive process of the search for the minimum points of the objective functionals we used the quasi-Newton method.

The reconstruction of the medium was carried out up to the depth of 0.5 km. The medium below this depth was assumed to be homogeneous. The whole medium from the surface to the depth 0.5 km was partitioned into layers of equal width.

As a sensing signal the impulse with a "bell-shaped envelope" was chosen with the dominating frequency $f = 25$ Hz. All computations were made for temporal frequencies from 5 to 50 Hz.

The real model distributions for the functions $v_p(z)$, $v_s(z)$, and $\sigma(z)$ are shown in Figure 6.4.11 by solid lines. The initial approximations for the functions $v_p(z)$, $v_s(z)$, and $\sigma(z)$ are shown by dashed lines.

The results of reconstruction are presented in Figure 6.4.12. These results were obtained for the functions $v_p(z)$ and $v_s(z)$ (in 59 iterations) and the function $\sigma(z)$ (in 38 iterations).

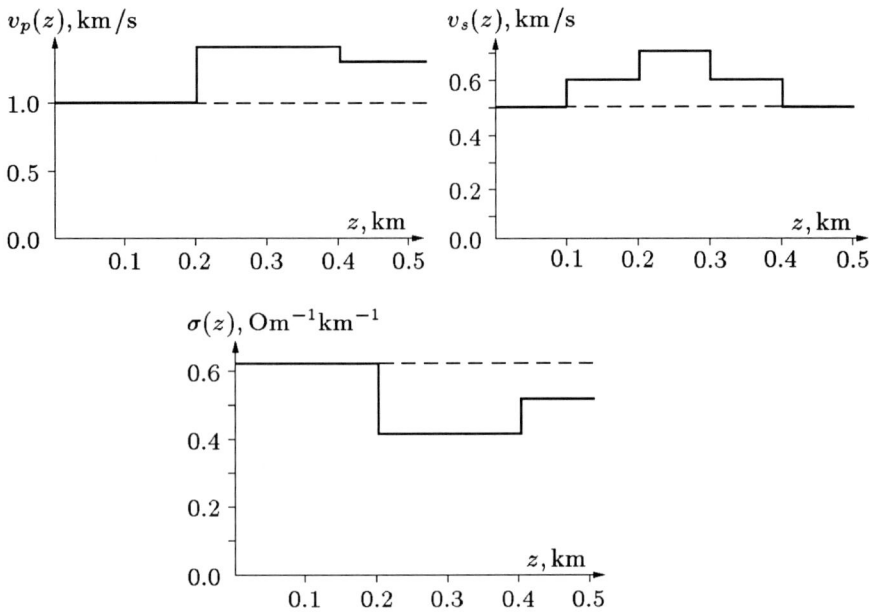

Figure 6.4.11

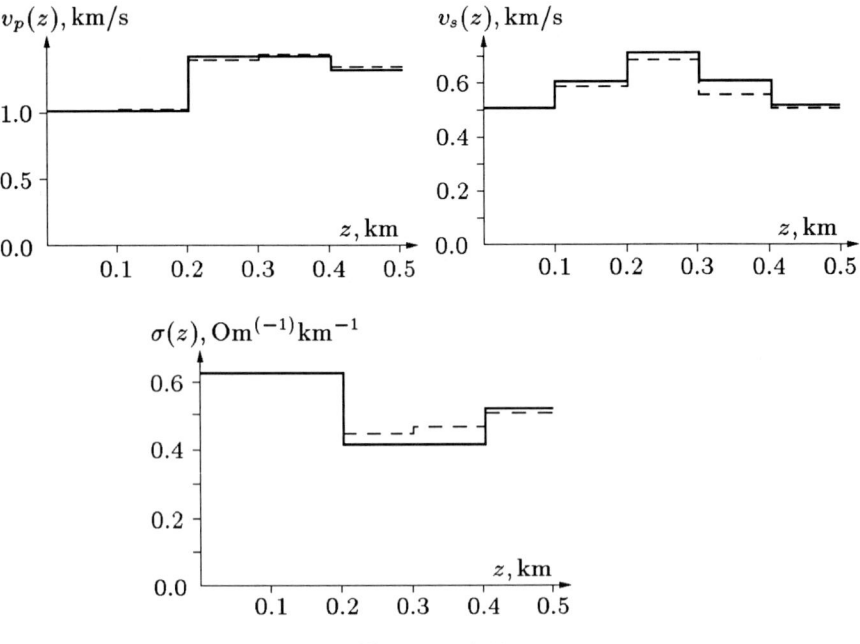

Figure 6.4.12

6.5. SIMULATION OF THE LONG-TERM COASTAL PROFILE EVOLUTION

In this section, we give the introduction to *coastal profile modeling*. Our understanding is that until now the professionals in the theory and applications of inverse problems are not acquainted with this practically important area.

6.5.1. General types of models and scales

Erosion and accretion on beaches as a result of modifications in the coastal environment have been a source of growing concern among coastal engineers. Generally, these modifications occur as structures (e. g., groins, seawalls) built in some cases to protect harbors against severe wave action or to alleviate the effects of erosion on the adjacent shore property. In some cases they arise as a result of man's need to exploit some natural resource (e. g., artificial sand islands used as drilling platforms for oil exploration). In all cases a newly introduced structure disturbs the dynamic equilibrium of sediment transport and thus alters the coastal morphology. In order to minimize possible undesirable effects due to this alteration it is necessary to study the resulting changes in coastal morphology.

Generally, there are two basic approaches in coastal engineering to investigation of this problem, namely:

1) hydraulic models;

2) mathematical models.

In the case of *hydraulic model* studies, a small-scale model of the problem under study is constructed. From this study, forecasts are made about the behavior of the prototype. Although certain difficulties arise due to scale effects in these studies, the results obtained are valuable in understanding the various mechanisms which interact in the coastal environment. However, such physical models are quite expensive and they require a long time if several possible options are to be tested.

With increasing availability of high-speed computers coupled with improved understanding of coastal processes, *mathematical simulation* is becoming a major tool in coastal engineering.

It should be noted that processes under consideration can occur in different temporal and spatial scales, namely:

1) small-scale coastal behavior;

2) meso-scale coastal behavior (the actual development of coastal sections over a period of months to a few decades);
3) large-scale coastal behavior (the development of the entire coast on a time scale of decades to a few centuries);
4) meta-scale coastal behavior (the geological evolution of the coastal plane on a temporal scale of centuries to millennia).

All these scales are divided into *long-term* and *short-term* processes (of order of several years).

Short-term processes are explored well enough on hydraulic (hydrodynamic) models, while long-term modeling of coastal behavior is still in its infancy.

Even if we would have computing power enough to run small-scale models for a sufficiently long period of time, this might be far from the best approach to long-term modeling. Processes which are negligible at smaller scale can have significant long-term effects and conversely. So it is important to study long-term processes regardless of short-term ones.

6.5.2. Overview of basic mathematical models

As a result of the complexity of the coastal processes, the engineers usually proceed by simplifying the processes into a number of modules which can be formulated individually and then assembled together in a quasi-steady process to form the required model. For example, the coastal processes involved in morphology development have been simplified into a 3-stage process of wave transformation, sediment transport, and erosion/accretion (morphology development), see De Vriend, Capobianco, Chesher, T., de Swart, Latteux, and Stive (1993).

The largest and most consistent body of work concerns the coastline models (Hanson, 1987) and the equilibrium state of the system, for example, the equilibrium coastal profiles (Dean, 1991).

One of the key elements in long-term modeling is reduction of information. For example, we need data reduction which enables us to separate the relevant information from the "noise" and to reduce this into a manageable number of parameters. For a long time, these two sources of information were separated, namely, a study was approached either from the empirical viewpoint or from the process-based modeling.

The reduction of information involves, essentially, four levels which concern the input, the physical system or its model, the output, and the interpretation or generalization. These levels are reflected in various approaches to long-term modeling:

1) input reduction based on the idea that we can describe long-term residual effects (e.g., transport fields) with the help of models based on the description of small-scale processes;
2) model reduction based on the idea that, by using more or less formal analysis and integration methods, the model can be reformulated at the scale of interest without describing the details of smaller-scale effects;
3) behavior-oriented modeling which attempts to describe the phenomena without going into the underlying processes.

All of these approaches are discussed by De Vriend, Capobianco, et al. (1993).

The problem of quantization is important in coastal profile modeling. Usually the area under study is discretized into rectangular grids using fixed alongshore and cross-shore grids, in which the depth is allowed to vary in each grid according to the condition of continuity for sediment transport. To ensure that depths are continuous along grid lines, the changes in depth along these lines are found by averaging between adjacent grids. In an alternative method, the area of studies is divided into a fixed number of cross-shore strips. Each strip can be analyzed as a single computational cell or to be divided further into multiple cells. For calculation of morphology, the section of the profile within each cell is assumed to accrete or erode uniformly (i.e., the profile translates) so that the total accretion/erosion equals the net sediment transport into the cell. The movement of the contours between adjacent cells is found by linear interpolation.

In the above quantization, if each strip is analyzed as one cell, the whole profile is assumed to move uniformly and the model is called a *1-line model* (Larson, Hanson, and Kraus, 1987). In this case, the cross-shore length of the strip should be extended only to the critical depth (depth at which the sediment transport is negligible). In the sequel, such depth will be referred to as *the depth of closure*.

Alternatively, if each strip is divided into N cells, the profile is divided into N segments, and each profile segment is used to translate the net sediment transport into each cell as accretion/erosion. In this case, the model becomes an *N-line model* (Bakker, 1968). In order to calculate the morphology changes in each cell, it is necessary to calculate the sediment transport at the ends of each cell.

6.5.3. The diffusion model

In the paper by De Vriend, Capobianco, et al. (1993) a model of diffusion type was proposed to describe the evolution of the coastal profile. The diffusion coefficient in the governing equation (its physical dimension is length squared divided by time) corresponds to the time scale of shoreline change following a disturbance (wave action).

The aforementioned diffusion model can be described by the following basic equation:

$$\frac{\partial(\delta X)}{\partial t} = D^2(z)\frac{\partial^2(\delta X)}{\partial z^2} + g\left(t, z, \delta X, \frac{\partial(\delta X)}{\partial z}\right). \tag{6.5.1}$$

Here $\delta X(z,t)$ represents the change of the cross-shore position (i.e., change in the depth at the distance z from the shore line) of the coastal profile and $D(z)$ is the diffusion coefficient.

Following the paper by De Vriend, Capobianco, et al. (1993), we give some explanations for special cases of the term $g(t, z, \delta X, \partial(\delta X)/\partial z)$. If $g = S(z,t)$ (an external source function), it is possible to introduce the effects of random forcing, along-shore transport gradients, and human interference such as nourishment and sand mining.

The linear choice $g = B(z)\,\partial(\delta X)/\partial z$ or $g = B(z)\delta X$ is also interesting in view of applications. In these models, the coefficient $B(z)$ represents the speed of along-shore sand wave movement. We assume that $g = B(z)\delta X$ in (6.5.1). So, we shall consider the following

Inverse Problem 6.5.1. *Given the function $\delta X_0(t)$ (the change of the cross-shore position at the point $z = 0$), find the coefficients $D(z)$ and $B(z)$ such that the solution $\delta X(z,t)$ to the problem*

$$\frac{\partial(\delta X)}{\partial t} = D^2(z)\frac{\partial^2(\delta X)}{\partial z^2} + B(z)\,\delta X, \tag{6.5.2}$$

$$\delta X\big|_{t=0} = 0, \tag{6.5.3}$$

$$\frac{\partial(\delta X)}{\partial z}\bigg|_{z=0} = \varphi_0(t), \qquad \delta X\big|_{z=H} = 0 \tag{6.5.4}$$

satisfies the equation (surface measurements)

$$\delta X\big|_{z=0} = \delta X_0(t). \tag{6.5.5}$$

Note that the parameter H can be thought of as an estimate of the depth of closure, i.e., the location where the diffusive and transport phenomena virtually end.

Numerical reconstruction of two coefficients of the equation

As the first step to solve the aforementioned Inverse Problem numerically, we apply the Fourier transform. Thus, the original dynamic problem is replaced by a Helmholtz equation with a complex-valued coefficient. The point is that recovering the space-dependent coefficients $D(z)$ and $B(z)$ in the frequency domain, we do not need to go back to the time domain. Here we have used the term *frequency domain* for retaining a formal analogy with inverse problems for the wave equation, in which case the frequency domain concerning Fourier images of solutions has a clear physical meaning, see, e. g., (Aki, Richards, 1980; Avdeev, Lavrentiev, Jr., and Priimenko, 1999).

For numerical processing we apply to the problem (6.5.2)–(6.5.4) the formal Fourier transform under assumption that the coefficients are smooth: $D(z), B(z) \in C^2(0, H)$, and $\omega \in [\omega_1, \omega_2]$.

So, we consider the problem

$$\frac{d^2 V}{dz^2} + \frac{B(z) - i\omega}{D^2(z)} V = 0, \qquad (6.5.6)$$

$$\left.\frac{dV}{dz}\right|_{z=0} = F(\omega), \qquad V\big|_{z=H} = 0, \qquad (6.5.7)$$

where $V(z, \omega) = \int_0^\infty e^{-i\omega t} \delta X \, dt$ and $F(\omega) = \int_0^\infty e^{-i\omega t} \varphi_0(t) \, dt$.

The inverse problem we are interested in consists in reconstructing both functions $D(z)$ and $B(z)$ from the additional information

$$V_0(\omega) = V(0, \omega), \qquad \omega_1 \leq \omega \leq \omega_2, \qquad (6.5.8)$$

where $[\omega_1, \omega_2]$ is the interval of available frequencies.

We solve the inverse problem (6.5.6)–(6.5.8) by minimizing the cost functional

$$\Phi[D, B] = \int_{\omega_1}^{\omega_2} |V_0(\omega) - K[D, B](\omega)|^2 \, d\omega$$
$$+ \beta \sup_z |D - D^{\text{est}}| + \gamma \sup_z |B - B^{\text{est}}|, \qquad (6.5.9)$$

where the operator $K[D, B](\omega)$ maps the current "test" values of $D(z)$ and $B(z)$ into the trace of the solution of the boundary value problem (6.5.6), (6.5.7) at $z = 0$. Here β and γ are some weighted regularization parameters, $0 \leq \beta \leq 1$, $0 \leq \gamma \leq 1$, and D^{est} and B^{est} are the estimated values of D and B, respectively, that can be obtained from physical measurements.

In the modification of the numerical algorithm which is used here, we operate with the gradient of the "uniform" cost functional (6.5.9), i.e., the functional with $\beta = 0$ and $\gamma = 0$. In this case, after rather technical calculations (omitted here since, e.g., a similar approach can be found in McGillivray and Oldenbourgh (1990), we obtain the following formulas for the gradients of the cost functional with respect to D and B:

$$(\nabla_D \Phi[D, B])(z) = -4\text{Re} \int_{\omega_1}^{\omega_2} \frac{B(z) - i\omega}{D^3(z)} \bar{F}(\omega) \cdot [V_0(\omega) - F(\omega) G(z, 0; \omega)]$$
$$\times \bar{G}(z, z; \omega) \, \bar{G}(z, 0; \omega) \, d\omega, \qquad (6.5.10)$$

$$(\nabla_B \Phi[D, B])(z) = 2\text{Re} \int_{\omega_1}^{\omega_2} D^{-2}(z) \bar{F}(\omega) \cdot [V_0(\omega) - F(\omega) G(z, 0; \omega)]$$
$$\times \bar{G}(z, z; \omega) \, \bar{G}(z, 0; \omega) \, d\omega. \qquad (6.5.11)$$

Here $G(z, \zeta; \omega)$ is the Green function of the problem (6.5.6), (6.5.7) and the bar denotes the complex conjugation.

Here we do not consider a rather nontrivial theoretical question of uniqueness as well as existence of the *global* minimum point of the cost functional (6.5.9). Instead, we refer the reader to the paper by Alekseev, Avdeev, et al. (1993) in which similar questions are discussed for a more simple statement of the inverse problem.

To minimize the cost functional (6.5.9), the conjugate gradients method was used, see (Karmanov, 1986).

Numerical experiments for synthetic data

Computer codes were prepared in C++; and Mathematica 3.0 was used for verification and visualization. Only resources of a personal computer are needed. On each iteration, the function $V(z, \omega)$ was computed by the so-called numerical-analytical method described by Fatianov (1990), Fatianov and Mikhailenko (1988).

We approximated the coefficients $D(z)$ and $B(z)$ by piecewise constant functions, i.e., 6 layers with equal width 200 meters. According to the results of auxiliary numerical tests, the regularization parameters β and γ were chosen equal to 0.3 and 0.2, respectively. The results of simultaneous reconstruction of both coefficients $D(z)$ and $B(z)$ (obtained after 98 iterations) are shown in Figure 6.5.1. True functions are shown by solid lines, the initial guess by dotted lines, and the result of reconstruction by dashed lines.

Chapter 6. Numerical solution

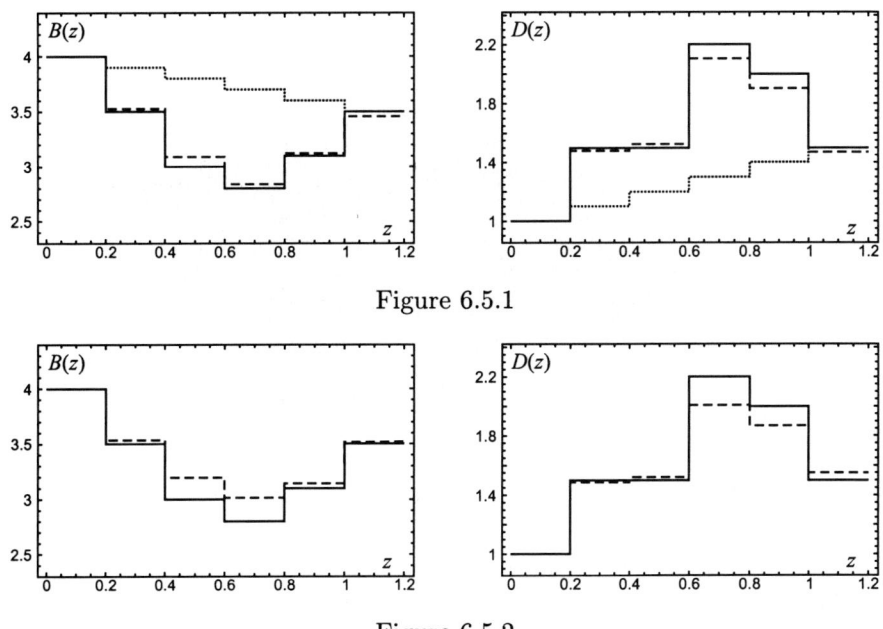

Figure 6.5.1

Figure 6.5.2

In order to test the robustness of the algorithm, the inversion data were artificially corrupted by adding a randomly distributed white noise. A normally distributed random noise with average fluctuations equal to 5% of the data amplitudes was added to the inversion data. The result of simultaneous identification of *two* coefficients $D(z)$ and $B(z)$ from corrupted data is shown in Figure 6.5.2. The initial approximations for $D(z)$ and $B(z)$ were the same as in the previous test, see Figure 6.5.1.

To conclude the section we would like to stress that the proposed version of the numerical inversion algorithm has demonstrated quite reasonable performance and accuracy. Thus, a basis for real data processing has been established.

Validating the diffusion model: numerical experiments for real data

Real data representing the long-term evolution of the cross-shore position of coastal profiles were collected over a period of 10 years from 1981 to 1991 at Duck, North Carolina, and are presented in Figure 6.5.3. Real data $\delta X_{\rm r}(z,t)$ (the subscript "r" stands for "real") for the cross-shore position consist of a 100×250 array. This means 100 observation points in the space variable z, each being an average over 80 meters, and 250 observation times, each being an average over 15 days. In addition, the source function $f(t)$s was measured

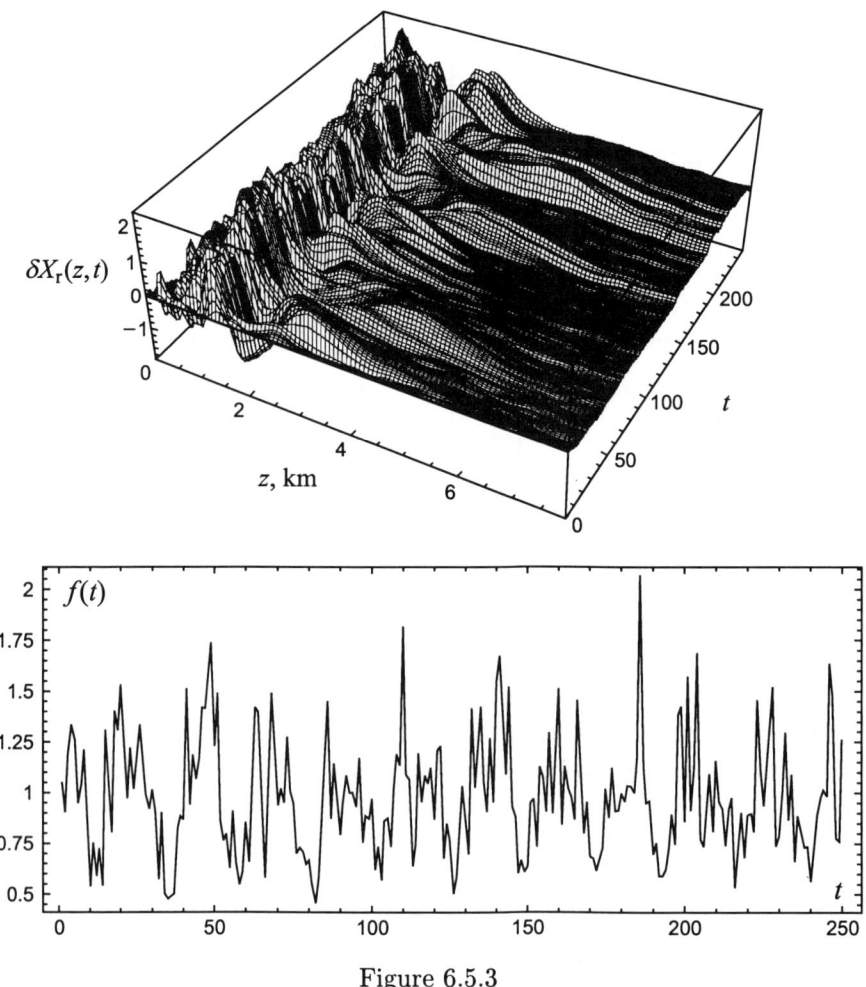

Figure 6.5.3

at the same observation times. This function represents the average height of waves over the observation periods and is also shown in Figure 6.5.3.

It was supposed that the coastal profile evolution is described by the model equations (6.5.2)–(6.5.5), where, in equation (6.5.2), the function $f(t)$ is added to the right-hand side.

First, the data were numerically recalculated in terms of the Fourier transform $\mathcal{F}$ in order to obtain the data $V_\mathrm{r}(z,\omega) = \mathcal{F}[\delta X_\mathrm{r}(z,t)]$ and $\psi(\omega) = \mathcal{F}[f(t)]$. More precisely, to solve the problem we use the Fourier-type representation

$$V(z,\omega) = \int_0^T \delta X(z,t) \mathrm{e}^{-\mathrm{i}\omega t}\,\mathrm{d}t, \qquad (6.5.12)$$

where T is the whole period of observation; with the inverse transform given by the formula

$$\delta X(z,t) = \frac{1}{T} \sum_n V(z,\omega_n) e^{i\omega_n t}, \qquad (6.5.13)$$

where $\omega_n = n\pi/T$, $n = 1, 2, \ldots, N$.

Then the cost functional was computed by the formula

$$\Phi = \int_{\omega_1}^{\omega_N} \int_0^H |V_r - V_{\text{synth}}|^2 \, dz \, d\omega, \qquad (6.5.14)$$

where the data V_{synth} ("synthetic") were obtained as the solution to the direct problem

$$\frac{d^2 V}{dz^2} = -\frac{i\omega - B}{D^2} V + \frac{\psi(\omega)}{D^2}, \qquad (6.5.15)$$

$$\left.\frac{dV}{dz}\right|_{z=0} = F(\omega), \qquad V|_{z=H} = h(\omega). \qquad (6.5.16)$$

The boundary data were obtained as follows:

$$F(\omega) = \frac{V_r(\Delta z, \omega) - V_r(0, \omega)}{\Delta z}, \qquad h(\omega) = V_r(H, \omega) \qquad (\Delta z = 80 \text{ m}).$$

First, the cost functional was studied in the case where both coefficients of the equation are *constants*: $D(z) \equiv \text{const}$ and $B(z) \equiv \text{const}$. Then a contour plot of the cost functional on the plane (B, D) was calculated (Figure 6.5.4). Each point in Figure 6.5.4 corresponds to the value of the functional (6.5.14) for a pair of constants (B, D). The following step-sizes for numerical computations were chosen: $\Delta B = 10$, $\Delta D = 0.02$. The intervals for B and D were taken $(0, 800)$ and $(0, 0.5)$, respectively. The 3D visualization of the cost functional is also shown in Figure 6.5.4, where the point shows the position of the global minimum

We can see that the cost functional has a rather complicated structure even in the simplified case of the governing equation (6.5.2) with *constant* coefficients D and B.

Usual gradient methods provide very poor convergence in such cases and it is practically impossible to find the global minimum.

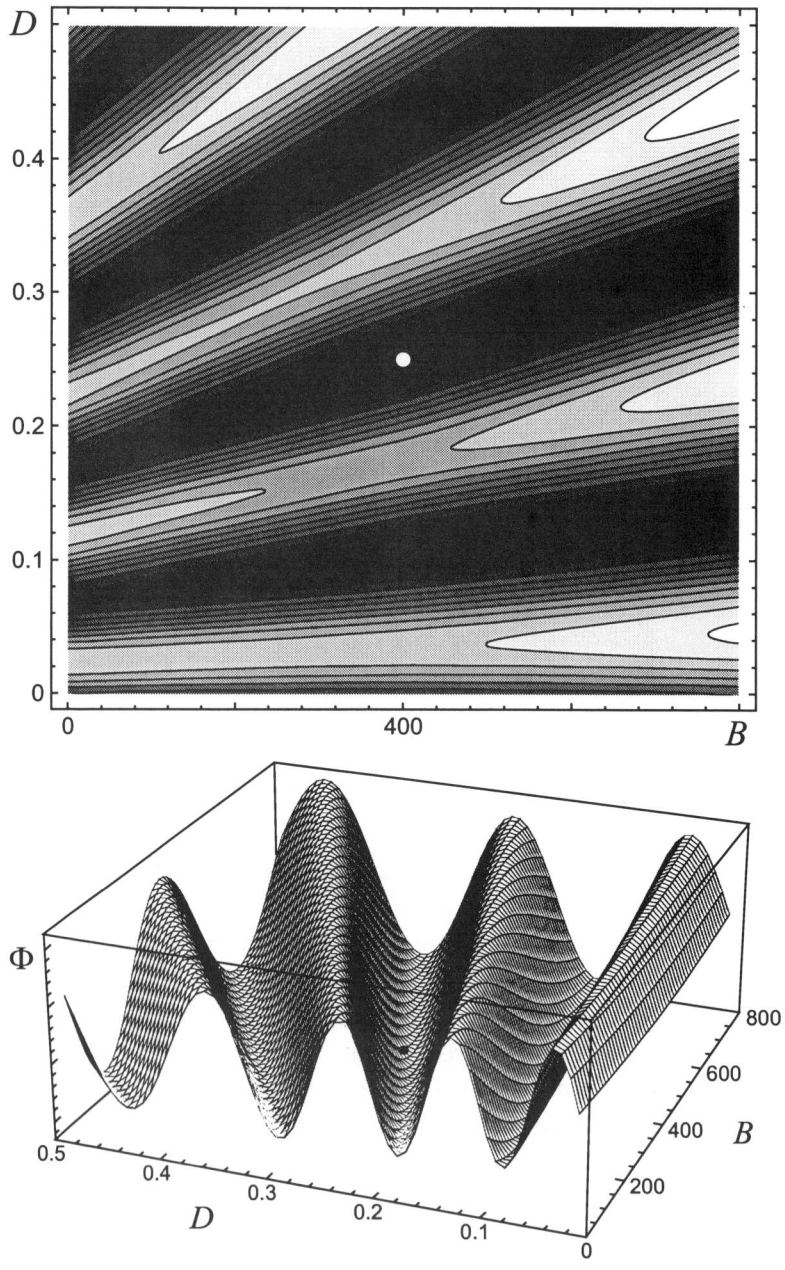

Figure 6.5.4

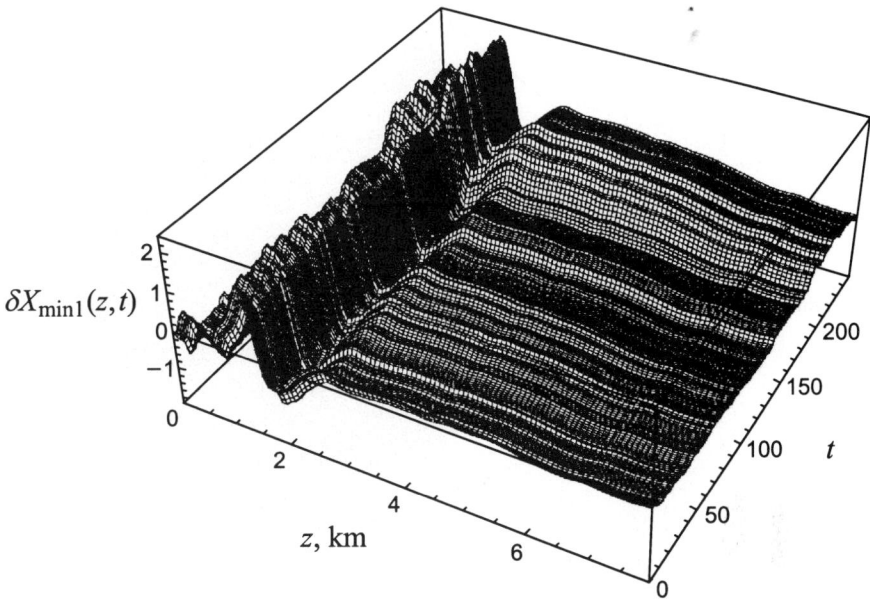

Figure 6.5.5

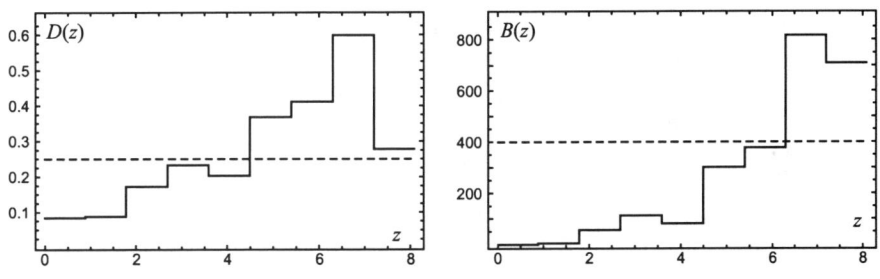

Figure 6.5.6

Therefore the global minimum of the functional (6.5.14) was found numerically by exhaustive search over a rough mesh. We denote the corresponding values of the parameters of the problem by $B_{\min}$ and $D_{\min}$; $B_{\min} = 400$ and $D_{\min} = 0.25$. Next, the solution $V_{\min}(z,\omega)$ to the direct problem (6.5.15), (6.5.16) (with constants $B_{\min}$ and $D_{\min}$ substituted for D and B, respectively) was computed.

Finally, the time-dependent profile, $\delta X_{\min 1}(z,t)$ was determined after inverse Fourier transformation, see (6.5.13).

Results of numerical reconstruction of the cross-shore position (3D plot of the profile $\delta X_{\min 1}(z,t)$) for constant values of D and B are shown in Figure 6.5.5.

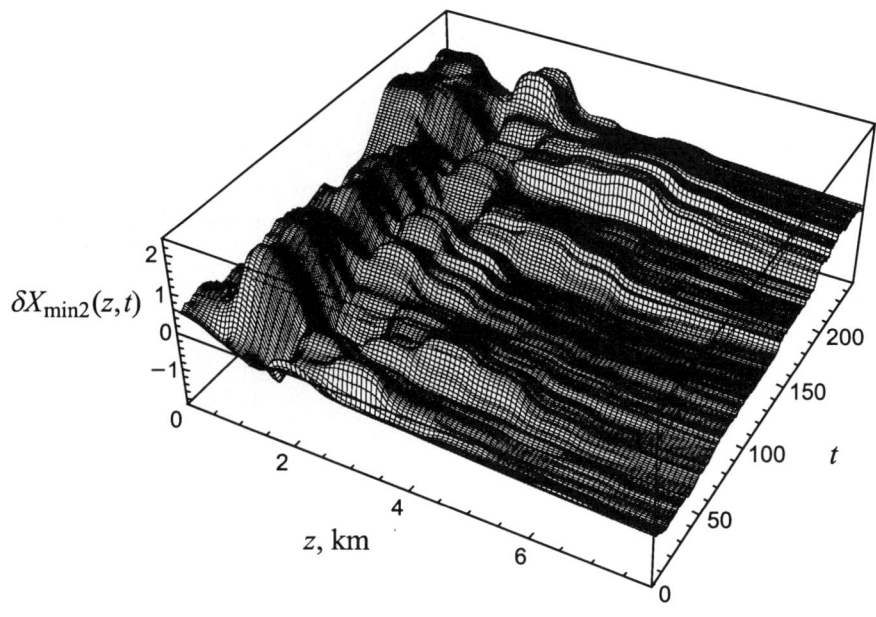

Figure 6.5.7

For comparison with the measured data, we calculated the time integral of the relative error:

$$\delta(z) = \int_0^T \left| \frac{\delta X_{\mathrm{r}} - \delta X_{\mathrm{min1}}}{\delta X_{\mathrm{r}}} \right| \mathrm{d}t. \qquad (6.5.17)$$

The average relative error is 15.8%. We can see that the rough assumption that the parameters of the problem are just constants, $B = B_{\min}$ and $D = D_{\min}$, turns out to be in a rather good qualitative agreement with the measured data, see Figure 6.5.3. It is important that in our case the relative error $\delta(z,t) = |(\delta X_{\mathrm{r}} - \delta X_{\mathrm{min1}})/\delta X_{\mathrm{r}}|$ does not increase with the growth of time.

Then, using the procedure of minimization of the cost functional by the conjugate direction method we reconstructed the coefficients $B(z)$ and $D(z)$ under the assumption that they are piecewise constant functions (see Figure 6.5.6). The range of the spatial variable was divided into 9 layers of equal width. The values $D = D_{\min}$ and $B = B_{\min}$, which were obtained in the previous numerical test, are shown by dashed line.

The corresponding 3D plot of the profile $\delta X_{\mathrm{min2}}(z,t)$ is shown in Figure 6.5.7. Comparison with the measured data was made as above, see (6.5.17). The average relative error is 5.4 %. We can see that piecewise constant coefficients provide a better agreement with the measured data, see Figures 6.5.3, 6.5.5, and 6.5.7.

So, the results obtained can validate the diffusion model which was proposed by De Vriend, Capobianco, et al. (1993) for description of the long-term coastal profile evolution.

We are indebted to William Birkemeier and his coworkers at the US Army Field Research Facility in Duck, North Carolina, for supplying the high-quality data. We also thank Michele Capobianco, Tecnomare, Venice (Italy), for his interest and a number of useful suggestions.

Bibliography

Aki K., Richards P. G. (1980). *Quantitative Seismology: Theory and Methods*. Freeman, San Francisco.

Aleksandryan, R. A. (1950). On one Sobolev problem for a special fourth-order partial differential equation. *Dokl. Akad. Nauk SSSR*, **73** (4), 631–634 (in Russian).

Alekseev, A. S. (1962). Some inverse problems of wave propagation theory. I, II. *Izv. Akad. Nauk SSSR. Ser. Geofiz.*, No. 11, 1514–1531 (in Russian).

Alekseev, A. S. (1967). Inverse dynamic problems of seismics. In: *Some Methods and Algorithms of Interpretation of Geophysical Data*. Nauka, Moscow, 9–84 (in Russian).

Alekseev, A. S. (1992). On statement of combined inverse geophysics problems. *Sib. J. Comput. Math.*, **1** (1), 47–58.

Alekseev, A. S., Avdeev, A. V., Fatianov, A. G., and Cheverda, V. A. (1991). A closed cycle of mathematical modeling of wave propagation processes in vertically inhomogeneous media (direct and inverse problems). *Matem. Modelir.*, **3** (10), 80–94 (in Russian).

Alekseev, A. S., Avdeev, A. V., Fatianov, A. G., and Cheverda, V. A. (1993). Wave processes in vertically-inhomogeneous media: a new strategy for a velocity inversion. *Inverse Problems*, **9** (3), 367–390.

Alekseev, A. S. and Dobrinskii, V. I. (1975). Some questions of practical application of dynamic inverse problems of seismics. In: *Mathematical Problems of Geophysics. Issue 6, Part 2*. Computing Center, Siberian Branch of the USSR Acad. Sci., Novosibirsk, 7–53 (in Russian).

Alekseev, A. S., Lavrentiev, M. M., Mukhometov, R. G., Nersesov, I. L., and Romanov, V. G. (1971). A numerical method for determining the structure of the Earth's upper mantle. In: *Mathematical Problems of Geophysics. Issue 2*. Computing Center, Siberian Branch of the USSR Acad. Sci., Novosibirsk, 143–165 (in Russian).

Alekseev, A. S., Lavrentiev, M. M., Mukhometov, R. G., and Romanov, V. G. (1969). A numerical method for solution of a three-dimensional inverse kinematic problem of seismics. In: *Mathematical Problems of Geophysics. Issue 1*. Computing Center, Siberian Branch of the USSR Acad. Sci., Novosibirsk, 179–201 (in Russian).

Alfvén, H. (1950). *Cosmical Electrodynamics*. Clarendon Press, Oxford.

Anderson, D. L. (1984). Surface wave tomography. *EOS*, **65**, p. 174.

Anderson, D. L. and Dziewonski, A. M. (1984). Seismic tomography. *Sci. Amer.*, **251** (4), 58–66.

Anikonov, Yu. E. (1978). *Some Methods for Studying Multidimensional Inverse Problems for Differential Equations*. Nauka, Novosibirsk (in Russian).

Avdeev, A. V. (1995). Simultaneous determination of sounding signal and wave propagation velocity in a vertically-inhomogeneous medium. In: *Computerized Tomography (Proc. of the Fourth Int. Symposium, Novosibirsk, 1993)*. VSP, Utrecht, 70–77.

Avdeev, A. V. and Goryunov, E. V. (1996a). The inverse problem of acoustics: determination of source wavelet and velocity. *J. Inv. Ill-Posed Problems*, **4** (6), 475–482.

Avdeev, A. V. and Goryunov, E. V. (1996b). Simultaneous determination of source-time function and velocity via full wave field inversion. In: *Bull. of the Novosibirsk Computing Center. Ser. Mathem. Modeling in Geophysics. Issue 2*. NCC Publ., Novosibirsk, 19–27.

Avdeev, A. V., Goryunov, E. V., and Priimenko, V. I. (1996a). *An Inverse Problem of Electromagnetoelasticity with Unknown Source of Elastic Oscillations*. Preprint No. 1074. Computing Center, Siberian Branch of Russian Acad. Sci., Novosibirsk.

Avdeev, A. V., Goryunov, E. V., and Priimenko, V. I. (1996b). A numerical method of solution of one inverse problem of electromagnetoelasticity. In: *Problems of Geoacoustics: Methods and Instruments (Proc. of the 5th*

Session of Russian Acoustic Society, 1996). Moscow State Mining Univ., Moscow, 46–49 (in Russian).

Avdeev, A. V., Goryunov, E. V., and Priimenko, V. I. (1997). An inverse problem of electromagnetoelasticity with unknown source of elastic oscillations. *Matem. Modelir.*, **9** (10), 50–62 (in Russian).

Avdeev, A. V., Goryunov, E. V., and Skazka, V. V. (1995). On the numerical solution of the inverse problems in combined statements. In: *Advanced Mathematics: Computations and Applications. Proc. Intern. Conf. AMCA-95 (Novosibirsk)*. NCC Publ., Novosibirsk, 264–272.

Avdeev, A. V., Goryunov, E. V., Soboleva, O. N., and Priimenko, V. I. (1999). Numerical solution of some direct and inverse problems of electromagnetoelasticity. *J. Inv. Ill-Posed Problems*, **7** (5), 453–462.

Avdeev, A. V., Lavrentiev, M. M. (Jr.), Spigler, R. and Goryunov, E. V. (2003). Diffusion model of the long-term coastal evolution: Numerical validating. *Computational Technologies*, **8** (1), 3–13.

Avdeev, A. V., Lavrentiev, M. M. (Jr.), and Priimenko V. I. (1999). *Inverse Problems and Some Applications*. ICM&MG, Novosibirsk.

Avdeev, A. V., Priimenko, V. I., Goryunov, E. V., and Zvyagin, D. V. (1997a). Direct and inverse problems for the system of differential equations of electromagnetoelasticity. In: *Proc. of the Int. Conf. PRISM'97 (Nijmegen)*. 1–8.

Avdeev, A. V., Priimenko, V. I., Goryunov, E. V., and Zvyagin, D. V. (1997b). Direct and inverse problems of electromagnetoelastity. In: *Proc. of the 5th Int. Congr. of the Brazilian Geophys. Society (Saõ Paulo, Brazil, 1997)*. 658–661.

Avdeev, A. V., Soboleva, O. N., and Priimenko, V. I. (1998). Some numerical approaches to solving inverse problems of electromagnetoelasticity. In: *Advanced Computational Methods in Engineering. Vol. 2*. Shaker Publ. B.V., Maastricht, 673–679.

Babich, V. M. and Buldirev, V. S. (1972). *Asymptotical Methods in Problems of Difraction of Short Waves*. Nauka, Moscow (in Russian).

Bakker, W. T. (1968). The dynamics of a coast with a groyne system. In: *Proc. of the 11th Coastal Engineering Conference*. American Society of Civil Engineers, 492–517.

Balakirev, M. K. and Gilinskii, I. A. (1982). *Waves in Piezocrystals*. Nauka, Novosibirsk (in Russian).

Bamberger, A., Chavent, G., Hemon, Ch., and Lailly, P. (1982). Inversion of normal incidence seismograms. *Geophysics*, **47** (5), 757–770.

Baranov, V. and Kunetz, G. (1960). Synthetic seismograms with multiple reflections: theory and numerical experience. *Geophysical Prospecting*, **8** (2), 315–325 (in French).

Belonosova, A. V. and Alekseev, A. S. (1967). On one statement of the inverse kinematic problem of seismics for a 2D continuously inhomogeneous medium. In: *Some Methods and Algorithms of Interpretation of Geophysical Data*. Nauka, Moscow, 137–154 (in Russian).

Bernstein, I. N. and Gerver, M. L. (1978). On an integral geometry problem for a family of geodesic curves and on an inverse kinematic problem of seismics. *Dokl. Akad. Nauk SSSR*, **243** (2), 302–305 (in Russian).

Bernstein, I. N. and Gerver, M. L. (1980). A condition for distinguishing metrics from travel-time curves. In: *Methods and Algorithms for Interpretation of Seismological Data. (Computational Seismology. Issue 13)*. Nauka, Moscow, 50–73 (in Russian).

Bertero, M. and Boccacci, P. (1998). *Introduction to Inverse Problems in Imaging*. Institute of Physics Publ., Bristol.

Bishop, T. N., Bube, K. P., Cutler, R. T., Langan, R. T., Love, P. L., Resnick, J. R., Shuey, R. T., Spindler, D. A., and Wyld, H. W. (1985). Tomographic determination of velocity and depth in laterally varying media. *Geophysics*, **50** (6), 903–923.

Blagovestchensky, A. S. (1966). On an inverse problem of the theory of waves propagation. In: *Problems of Mathematical Physics. Issue 1*. Leningrad State Univ., Leningrad, 68–81 (in Russian).

Blagovestchensky, A. S. (1970). An inverse problem for the wave equation with unknown source. In: *Problems of Mathematical Physics. Issue 4*. Leningrad State Univ., Leningrad, 27–39 (in Russian).

Blagovestchensky, A. S. (1971). On the local method of solution to nonsteady problem for nonhomogeneous string. Proc. MI AN SSSR, V.115, 28–38 (in Russian).

Bukhgeim, A. L. (1975). Necessary conditions of stability of one class of integro-differential equations. In: *Numerical Methods and Programming*. Computing Center, Siberian Branch of the USSR Acad. Sci., Novosibirsk, 78–86 (in Russian).

Bukhgeim, A. L. (1983a). On an algorithm for solving the inverse kinematic problem of seismics. In: *Numerical Methods in Seismic Investigations*. Nauka, Novosibirsk, 152–155 (in Russian).

Bughgeim, A. L. (1999). *Volterra Equations and Inverse Problems*. VSP, Utrecht.

Bukhgeim, A. L. (2000). *Introduction to the Theory of Inverse Problems*. VSP, Utrecht.

Burdakova, O. A. and Yakhno, V. G. (1989). A one-dimensional inverse problem of electroelasticity. In: *Questions of Correctness of the Problems of Analysis*. Institute of Mathematics, Siberian Branch of the USSR Acad. Sci., Novosibirsk, 33–43 (in Russian).

Cagniard, L. (1953). Basic theory of the magnetotelluric method of geophysical prospecting. *Geophysics*, **18** (3), 605–635.

Carrion, P. M., Sacramento, S. dos S., and Pestana, R. da C. (1990). Source wavelet and its angular spectrum from plane-wave seismograms. *Geophysics*, **55** (8), 1026–1035.

Chadwick, P. (1956). Elastic wave propagation in magnetic field. In: *Proc. of the IXth Intern. Congr. Appl. Mech.* Brussels, 143–153.

Cherveny, V., Molotkov, I. A., and Psenchik, I. (1977). *Ray Methods in Seismology*. Univerzita Karlova, Praha.

Cheverda, V. A. and Kostin, V. I. (1995). R-pseudoinverses for compact operators in Hilbert spaces: existence and stability. *J. Inv. Ill-Posed Problems*, **3** (2), 131–148.

Cheverda, V. A. and Voronina, T. A. (1994). Optimizational approach to data processing in vertical seismic profiling. *J. Inv. Ill-Posed Problems*, **2** (3), 211–226.

Dean, R. G. (1991). Equilibrium beach profiles: characteristics and applications. *J. Coastal Res.*, **7** (1), 53–84.

De Mol, C. (1999). Wavelets for solving inverse problems. In: *Proc. of the 4th Intern. Workshop on Inverse Problems in Electromagnetism and Acoustics (Clermont-Ferrand, France)*.

De Vriend, H. J., Capobianco, M., Chesher, T., de Swart, H. E., Latteux, B., and Stive, M. J. F. (1993). Approaches to long-term modelling of coastal morphology: a review. *Coastal Engineering*, **21** (1/3) (special issue), 225–269.

Dieulesaint, E. and Royer, D. (1974). *Ondes Élastiques dans les Solides. Application au Traitement du Signal*. Masson et Cie, Paris.

Dines, K. A. and Lytle, R. J. (1979). Computerized geophysical tomography. *Proc. IEEE*, **67**, 1065–1073.

Dunkin, J. W. and Eringen, A. C. (1963). On the propagation of waves in an electromagnetic elastic solid. *Int. J. Engineering Sci.*, **1** (4), 461–495.

Dziewonski, A. M. (1984). Mapping the lower mantle: determination of lateral heterogeneity in P velocity up to degree and order 6. *J. Geophys. Res.*, **89** (B7), 5929–5952.

Engl, H. W., Hanke, M., and Neubauer, A. (1996). *Regularization of Inverse Problems*. Kluwer Acad. Publ., Dordrecht.

Eringen, A. C. and Maugin, G. A. (1990). *Electrodynamics of Continua. Vol. I. Foundations and Solid Media. Vol. II. Fluids and Complex Media*. Springer-Verlag, New York—Berlin.

Fatianov, A. G. (1990). A semi-analytical method of solving direct dynamic problems in layered media. *Dokl. Akad. Nauk SSSR*, **310** (2), 323–327 (in Russian).

Fatianov, A. G. and Mikhailenko, B. G. (1988). A method of computation of nonstationary wave fields in inelastic layer-inhomogeneous media. *Dokl. Akad. Nauk SSSR*, **301** (4), 834–839 (in Russian).

Fawcett, J. A. and Clayton, R. W. (1984). Tomographic reconstruction of velocity anomalies. *Bull. Seism. Soc. Amer.*, **74** (6), 2201–2219.

Gel'fand, I. M. and Levitan, B. M. (1955). On the determination of a differential equation from its spectral function. In: *Amer. Math. Soc. Transl. (2). Vol. 1*. Amer. Math. Soc., Providence, R.I., 253–304.

Gerver, M. L. and Markushevich, V. M. (1965). Investigation of ambiguity in determination of a seismic wave velocity by its travel-time curve. *Dokl. Akad. Nauk SSSR*, **163** (6), 1377–1380 (in Russian).

Gjevik, B., Nilsen, A., and Höyen, J. (1976). An attempt at the inversion of reflection data. *Geophys. Prospecting*, **24** (3), 492–505.

Goldin, S. V. (1986). Seismic traveltime inversion. *Series: Investigations in Geophysics*. Tulsa. No. 1.

Gorban', A. N. and Rossiev, D. A. (1996). *Neural Networks on a Personal Computer*. Nauka, Novosibirsk (in Russian).

Goupillaud, P. L. (1961). An approach to inverse filtering of near-surface layer effects from seismic records. *Geophysics*, **26** (6), 754–760.

Gutenberg, B. (1959). *Physics of the Earth's Interior*. Academic Press, New York—London.

Hanson, H. (1987). *GENESIS, a Generalized Shoreline Change Numerical Model for Engineering Use*. Report No. 1007. Univ. of Lund, Dept. of Water Res. Eng.

Herglotz, G. and Wiechert, E. (1905). Über die Elastizität der Erde bei Berücksichtigung ihrer variablen Dichte. *Z. Math. und Phys.*, **52** (3), 275–299.

Herman, G. T. (1980). *Image Reconstruction from Projections. The Fundamentals of Computerized Tomography*. Academic Press, New York.

Humphreys, E., Clayton, R. W., and Hager, B. H. (1984). A tomographic image of mantle structure beneath Southern California. *Geophys. Res. Lett.*, **11** (7), 625–627.

Imomnazarov, Kh. Kh. (1998). On one class of coupled one-dimensional problems for Maxwell's equations and the equations of continual filtration theory. In: *Proc. of the Institute of Computational Mathematics and Mathematical Geophysics (ICM&MG). Ser. Mathem. Modeling in Geophysics. Issue 5*. NCC Publ., ICM&MG (former Computing Center), Siberian Branch of Russian Acad. Sci., Novosibirsk, 61–73 (in Russian).

Isaacson, D. and Isaacson, E. L. (1989). Comment on A. P. Calderón's paper "On an inverse boundary value problem" (1980). *Math. Comp.*, **52** (186), 553–559.

Ivansson, S. (1985). A study of methods for tomographic velocity estimation in the presence of low-velocity zones. *Geophysics*, **50** (6), 969–988.

John, F. (1941). The Dirichlet problem for a hyperbolic equation. *Amer. J. Math.*, **63** (1), 141–154.

Kabanikhin, S. I. and Lorenzi A. (1999). *Identification Problems of Wave Phenomena — Theory and Numerics*. VSP, Utrecht.

Karmanov, V. G. (1986). *Mathematical Programming*. Nauka, Moscow (in Russian).

Kato, Y. and Kikuchi, T. (1950). Phase difference of Earth currents induced by the change of the Earth's magnetic field. I, II. *Sci. Rep. Tohoku Univ. Ser. V. Geophys.*, **2**, 134–145.

Keilis-Borok, V. I. and Monin, A. S. (1959). Magnetoelastic waves and the boundary of the Earth's crust. *Izv. Akad. Nauk SSSR. Ser. Geofiz.*, No. 11, 1529–1541 (in Russian).

Khajdukov, V. G., Kostin, V. I., and Cheverda, V. A. (1997). The r-solution and its applications in linearized waveform inversion for a layered background. In: *Inverse Problems in Wave Propagation*. Springer-Verlag, New York, 277–294.

Kirsch, A. (1996). *An Introduction to the Mathematical Theory of Inverse Problems*. Springer-Verlag, New York.

Klimenko, O. A. (1995). An algorithm for solving a problem in elastoelectrodynamics. In: *Computerized Tomography (Proc. of the Fourth Int. Symposium, Novosibirsk, 1993)*. VSP, Utrecht, 283–287.

Knopoff, L. (1955). The interaction between elastic wave motions and a magnetic field in electrical conductors. *J. Geophys. Res.*, **60** (4), 441–456.

Kolb, P., Collino, F., and Lailly, R. (1986). Prestack inversion of a 1D medium. *Proc. IEEE*, **74** (3), 498–508.

Krein, M. G. (1951). Solution of the inverse Sturm–Liouville problem. *Dokl. Akad. Nauk SSSR*, **76** (1), 21–24 (in Russian).

Krein, M. G. (1954). On a method of effective solution of an inverse boundary value problem. *Dokl. Akad. Nauk SSSR*, **94** (6), 987–990 (in Russian).

Ladyženskaya, O. A., Solonnikov, V. A., and Ural′tseva, N. N. (1968). *Linear and Quasilinear Equations of Parabolic Type.* Amer. Math. Soc., Providence, R.I.

Landau, L. D. and Lifshitz, E. M. (1984). *Course of Theoretical Physics. Vol. 8. Electrodynamics of Continuous Media.* Pergamon Press, Oxford-Elmsford, N.Y.

Larson, M., Hanson, H., and Kraus, N. C. (1987). *Analytical Solutions of the One-Line Model of Shoreline Change.* Report to U.S. Army Corps of Engineers. Coastal Engineering Research Center.

Lavrentiev, M. A. and Shabat, B. V. (1977). *Problems of Hydrodynamics and Their Mathematical Models.* Nauka, Moscow (in Russian).

Lavrentiev, M. M. (1956). On a certain boundary value problem for a hyperbolic system. *Matem. Sb.*, **38 (80)** (4), 451–464 (in Russian).

Lavrentiev, M. M. (1962). *On Some Ill-Posed Problems of Mathematical Physics.* Siberian Branch of the USSR Acad. Sci., Novosibirsk (in Russian).

Lavrentiev, M. M. (1989). A class of problems of integral geometry in the plane. *Sib. Math. J.*, **30** (4), 549–554.

Lavrentiev, M. M. (2001). Mathematical problems of tomography and hyperbolc equations. *Sib. Math. J.*, **42** (5), 1094–1105.

Lavrentiev, M. M. and Mukhometov, R. G. (1969). Investigation of stability of the inverse kinematic problem of seismics. In: *Mathematical Problems of Geophysics. Issue 1.* Computing Center, Siberian Branch of the USSR Acad. Sci., Novosibirsk, 83–91 (in Russian).

Lavrentiev, M. M., Reznitskaya, K. G., and Yakhno, V. G. (1982). *One-dimensional Inverse Problems of Mathematical Physics,* Nauka, Novosibirsk (in Russian).

Lavrentiev, M. M. and Romanov, V. G. (1966). Three linearized inverse problems for hyperbolic equations. *Soviet Math. Dokl.*, **7**, 1650–1652.

Lavrentiev, M. M., Romanov, V. G., and Shishatskii, S. P. (1986). *Ill-Posed Problems of Mathematical Physics and Analysis.* Amer. Math. Soc., Providence, R.I.

Lavrentiev, M. M., Romanov, V. G., and Vasiliev, V. G. (1970). *Multidimensional Inverse Problems for Differential Equations.* Springer-Verlag, Berlin—New York.

Lavrentiev, M. M. and Saveliev, L. Ya. (1995). *Linear operators and ill-posed problems*. Consultants Bureau, New York Plenum Publ. Corp.

Lavrentiev, M. M. and Saveliev, L. Ya. (1999). *Theory of operators and ill-posed problems*. Inst. Math., Novosibirsk (in Russian).

Lavrentiev, M. M. (Jr.) (1992). An inverse problem for the wave Equation. *Sib. Math. Zhurnal*, **33**, (3).

Lavrentiev, M. M. (Jr.) and Priimenko, V. I. (1995). Simultaneous determination of elastic and electromagnetic medium parameters. In: *Computerized Tomography (Proc. of the Fourth Int. Symposium, Novosibirsk, 1993)*. VSP, Utrecht, 302–308.

Lines, L. R., Schultz, A. K., and Treitel, S. (1988). Cooperative inversion of geophysical data. *Geophysics*, **53** (1), 8–20.

Lions, J.-L. and Lattès, R. (1969). *The Method of Quasi-Reversibility. Applications to Partial Differential Equations*. American Elsevier Publ. Co., Inc., New York.

Lorenzi, A. and Priimenko, V. I. (1996). Identification problems related to electro-magneto-elastic interactions. *J. Inv. Ill-Posed Problems*, **4** (2), 115–143.

Lorenzi, A. and Romanov, V. G. (1993). Identification of an electromagnetic coefficient connected with deformation currents. *Inverse Problems*, **9** (2), 301–319.

Louis, A. K., Maaß, P., and Rieder, A. (1997). *Wavelets. Theory and Applications*. John Wiley & Sons, Chichester.

McGillivray P. R. and Oldenbourgh D. W. (1990). Methods for calculating Frechét derivatives and sensitivities for the nonlinear inverse probelms: a comparative study. *Geophysial Prospecting* 38 (5), 499–524.

Mallat, S. (1998). *A Wavelet Tour of Signal Processing*. Academic Press, San Diego.

Marchenko, V. A. (1955). Reconstruction of the potential energy from the phases of scattered waves. *Dokl. Akad. Nauk SSSR*, **104** (5), 695–698 (in Russian).

Matveeva, N. N. (1972). A program for solving the direct and inverse kinematic problems of seismology. In: *Programs for Interpretation of Seismic Observations. Issue 1*. Nauka, Leningrad, 201–229 (in Russian).

Matveeva, N. N. and Alekseev, A. S. (1964). Computer search for velocity profile variants of the upper mantle basing on a set of travel-time curves of deep-focus earthquakes. In: *Problems of Dynamical Theory of Seismic Wave Propagation. Issue 7.* Leningrad State Univ., Leningrad, 130–143 (in Russian).

Maugin, G. A. (1988). *Continuum Mechanics of Electromagnetic Solids.* North-Holland Publ. Co., Amsterdam—New York.

Maurin, K. (1967). *Methods of Hilbert Spaces.* Państwowe Wydawn. Nauk., Warsaw.

McMechan, G. A. (1983). Seismic tomography in boreholes. *Geophys. J. Roy. Astron. Soc.,* **74** (2), 601–612.

Mendel, J. M. (1981). A time-domain approach to the normal-incidence inverse problem. *Geophys. Prospecting,* **29** (5), 742–757.

Mendel, J. M. and Habibi-Ashrafi, F. (1980). A survey of approaches to solving inverse problems for lossless layered media systems. *IEEE Trans. Geosci. Remote Sensing,* **GE-18** (4), 320–330.

Merazhov, I. Z. and Yakhno, V. G. (1995). Direct and inverse problems for systems of electromagnetoelasticity equations. In: *Computerized Tomography (Proc. of the Fourth Int. Symposium, Novosibirsk, 1993).* VSP, Utrecht, 332–335.

Mikhailenko, B. G. and Soboleva, O. N. (1995). Numerical modelling of seismomagnetic effects in an elastic medium. In: *Advanced Mathematics: Computations and Applications. Proc. Intern. Conf. AMCA-95 (Novosibirsk).* NCC Publ., Novosibirsk, 722–730.

Mukhometov, R. G. (1977). The problem of recovering a two-dimensional Riemannian metric and integral geometry. *Soviet Math. Dokl.,* **18** (1), 27–31.

Mukhometov, R. G. (1978). *On the Problem of Reconstructing an Anisotropic Riemannian Metric in the n-Dimensional Domain.* Preprint No. 136. Computing Center, Siberian Branch of the USSR Acad. Sci., Novosibirsk (in Russian).

Mukhometov, R. G. (1981). A problem of reconstructing a Riemannian metric. *Siberian Math. J.,* **22** (3), 420–433.

Mukhometov, R. G. and Romanov, V. G. (1978). On the problem of finding an isotropic Riemannian metric in an n-dimensional space. *Soviet Math. Dokl.,* **19**, 1330–1333.

Nakanishi, I. and Anderson, D. L. (1982). Worldwide distribution of group velocity of mantle Rayleigh waves as determined by spherical harmonic inversion. *Bull. Seism. Soc. Amer.*, **72** (4), 1185–1194.

Nataf, H.-C., Nakanishi, I., and Anderson, D. L. (1984). Anisotropy and shear-velocity heterogeneities in the upper mantle. *Geophys. Res. Lett.*, **11** (2), 109–112.

Natterer, F. (1986). *The Mathematics of Computerized Tomography.* B. G. Teubner, Stuttgart; John Wiley & Sons, Ltd., Chichester.

Neumann-Denzau, G. and Behrens, J. (1984). Inversion of seismic data using tomographical reconstruction technique for investigations of laterally inhomogeneous media. *Geophys. J. Roy. Astr. Soc.*, **79** (1), 305–315.

Nikolaev, A. V. (1988). Seismology: scientific and technological revolution and problems of XXI century. In: *Mathematical Modeling in Geophysics.* Nauka, Novosibirsk, 113–129 (in Russian).

Nolet, G. (1985). Solving or resolving inadequate and noisy tomographic systems. *J. Comput. Phys.*, **61** (3), 463–482.

Nolet, G. (Ed.) (1987). *Seismic Tomography.* Kluwer Acad. Publ., Dordrecht.

Parton, V. Z. and Kudrjavtsev, B. A. (1988). *Electromagnetoelasticity of Piezoelectric and Electrical-Conducting Bodies.* Nauka, Moscow (in Russian).

Pikalov, V. V. and Preobrazhenskii, N. G. (1987). *Reconstructive Tomography in Gas Dynamics and Plasma Physics.* Nauka, Novosibirsk (in Russian).

Poincaré, H. (1899). *Théorie du Potentiel Newtoneon.* Paris.

Pride, S. R. (1994). Governing equations for the coupled electromagnetics and acoustics of porous media. *Phys. Rev. B. Third Series*, **50** (21), 15678–15696.

Priimenko, V. I. (1982). About uniqueness of the definition of electroconductivity tensor in 1D-inhomogeneous medium. In: *Questions of correctness of inverse problems of mathematical physics.* Computing Center, Siberian Branch of the USSR Acad. Sci., Novosibirsk, 120–130 (in Russian).

Priimenko, V. I. (1983). An inverse problem for Maxwell+s system in medium of "crust-air" type. In: *Methods of investigation of ill-posed prob-*

lems of mathematical physics. Computing Center, Siberian Branch of the USSR Acad. Sci., Novosibirsk, 75–83 (in Russian).

Priimenko, V. I. (1986). *An inverse problem for diffusion approximation of Maxwell's system in linear approximation.* Preprint No. 636. Computing Center, Siberian Branch of Russian Acad. Sci., Novosibirsk (in Russian.)

Priimenko, V. I. (1990). *Inverse problems for diffusion approximation of Maxwell's system.* Ph.D. Thesis, Novosibirsk State University, Novosibirsk, Russia (in Russian).

Priimenko, V. I. and Vishnevsky, M. P. (2002). The Cauchy problem for a nonlinear model system of electromagnetoelasticity. In: *56th Seminàrio de Anàlise.* Fluminense Federal University, Niteroi, Brazil, 395–413.

Priimenko, V. I. and Vishnevsky, M. P. (2003a). Direct and inverse problems of electromagnetoelasticity: theoretical and numerical aspects. In: *57th Seminàrio de Anàlise.* Viçosa Federal University, Viçosa, Brazil, 137–213.

Priimenko, V. I. and Vishnevsky, M. P. (2003b). An inverse problem for model electromagnetoelasticity system with complete nonlinear interaction. *J. Inv. Ill-Posed Problems* (in print).

Reznitskaya, K. G. (1972). The unique solvability theorem for a nonlinear one-dimensional inverse dynamic problem of seismics. The Newton–Kantorovich method. In: *Inverse Problems for Differential Equations (Proc. of the All-Union Symp., Novosibirsk, 1971).* Computing Center, Siberian Branch of the USSR Acad. Sci., Novosibirsk, 116–123 (in Russian).

Rikitake, T. (1950). Electromagnetic induction within the Earth and its relation to the electrical state of the Earth's interior. *Bull. Earthquake Res. Inst. Tokyo,* **28**, p. 45.

Romanov, M. E. (1972). The method of characteristics for numerical solution of the inverse kinematic problem of seismics. In: *Mathematical Problems of Geophysics. Issue 3.* Computing Center, Siberian Branch of the USSR Acad. Sci., Novosibirsk, 328–346 (in Russian).

Romanov, M. E. (1975). On numerical solution of problems of integral geometry. In: *Mathematical Problems of Geophysics. Issue 6, Part 1.* Computing Center, Siberian Branch of the USSR Acad. Sci., Novosibirsk, 289–297 (in Russian).

Romanov, M. E. (1983). Numerical methods of solving the inverse kinematic problem for laterally inhomogeneous media. In: *Numerical Methods in Seismic Investigations*. Nauka, Novosibirsk, 90–106 (in Russian).

Romanov, M. E. (1988). Multidimensional inverse kinematic problems of reconstruction of a medium from seismic data. In: *Recent Seismological Investigations in Europe (Proc. of the XIX General Assembly of ESC, Moscow, 1984)*. Nauka, Moscow, 578–581 (in Russian).

Romanov, V. G. (1967a). Reconstruction of a function from its integrals over ellipsoids of revolution with one fixed focus. *Soviet Math. Dokl.*, **8**, 480–483.

Romanov, V. G. (1967b). On reconstruction of a function through integrals over a family of curves. *Sib. Matem. Zh.*, **8** (5), 1206–1208 (in Russian).

Romanov, V. G. (1969). *Some Inverse Problems for Equations of Hyperbolic Type*. Nauka, Novosibirsk (in Russian).

Romanov, V. G. (1973a). On one uniqueness theorem for an integral geometry problem on a set of curves. In: *Mathematical Problems of Geophysics. Issue 4*. Computing Center, Siberian Branch of the USSR Acad. Sci., Novosibirsk, 140–146 (in Russian).

Romanov, V. G. (1973b). On one class of uniqueness of solution of the inverse kinematic problem. In: *Mathematical Problems of Geophysics. Issue 4*. Computing Center, Siberian Branch of the USSR Acad. Sci., Novosibirsk, 147–164 (in Russian).

Romanov, V. G. (1974a). On some classes of uniqueness for the solution of integral geometry problems. *Math. Notes*, **16**, 983–989.

Romanov, V. G. (1974b). On the uniqueness of solution of the inverse kinematic problem in a circle in the class of velocities close to constant. In: *Mathematical Problems of Geophysics. Issue 5, Part 2*. Computing Center, Siberian Branch of the USSR Acad. Sci., Novosibirsk, 108–142 (in Russian).

Romanov, V. G. (1974c). On the uniqueness of the definition of an isotropic Riemannian metric inside a domain in terms of the distances between points of the boundary. *Soviet Math. Dokl.*, **15**, 1341–1344.

Romanov, V. G. (1975). Unique-solution classes for Volterra operator equations of the first kind. *Functional Analysis Appl.*, **9**, 78–79.

Romanov, V. G. (1978a). Integral geometry on the geodesics of an isotropic Riemannian metric. *Soviet Math. Dokl.*, **19**, 847–851.

Romanov, V. G. (1978b). *Inverse Problems for Differential Equations. Inverse Kinematic Problem of Seismics.* Novosibirsk State Univ., Novosibirsk (in Russian).

Romanov, V. G. (1982). Inverse problems of propagation of seismic and electromagnetic waves. In: *Methods for Solving Ill-Posed Problems and Their Applications (Proc. of the All-Union Workshop, Noorus, 1981).* Computing Center, Siberian Branch of the USSR Acad. Sci., Novosibirsk, 111–118 (in Russian).

Romanov, V. G. (1987). The Lamb inverse problem in linear approximation. In: *Numerical methods in seismic study*, Novosibirsk, 170–192 (in Russian).

Romanov, V. G. (1987). *Inverse Problems of Mathematical Physics.* VNU Science Press, Utrecht.

Romanov, V. G. (1995a). Structure of a solution to the Cauchy problem for the system of equations of electrodynamics and elasticity in the case of point sources. *Siberian Math. J.*, **36** (3), 541–561.

Romanov, V. G. (1995b). On an inverse problem for a coupled system of equations of electrodynamics and elasticity. *J. Inv. Ill-Posed Problems*, **3** (4), 321–332.

Romanov, V. G. and Kabanikhin, S. I. (1994). *Inverse Problems for Maxwell's Equations.* VSP, Utrecht.

Romanov, V. G., Kabanikhin, S. I., and Puchnacheva, T. P.(1984). *Inverse Problems of Electrodynamics.* Computing Center, Siberian Branch of the USSR Acad. Sci., Novosibirsk. (in Russian).

Rychagov, M. and Duchene, B. (1998). Binary object identification and reconstruction by using neural network processing of inverse scattering data. In: *Proc. of Int. Conf. "Progress in Electromagnetic Researches" (PIERS'98, Nantes).* P. 1202.

Santosa, F. (1982). Numerical scheme for the inversion of acoustical impedance profile based on the Gelfand–Levitan method. *Geophys. J. Roy. Astron. Soc.*, **70** (1), 229–243.

Schwartz, L. (1967). *Cours d'Analyse.* Hermann, Paris.

Shabat, B. V. (1956). On an analog of the Riemann theorem for linear hyperbolic systems of differential equations. *Uspekhi Matem. Nauk*, **11** (5 (71)), 101–105 (in Russian).

Shabat, B. V. (1970). On certain hyperbolic quasiconformal mappings. In: *Some Problems of Mathematics and Mechanics*. Nauka, Leningrad, 251–266 (in Russian).

Sharafutdinov, V. A. (1994). *Integral Geometry of Tensor Fields*. VSP, Utrecht.

Sheen, D. (1992). A generalized Green's theorem. *Appl. Math. Lett.*, **5** (4), 95–98.

Strakhov, V. N. (1969). The theory of approximate solution of linear ill-posed problems in a Hilbert space and its use in surveying geophysics. I. *Izv. Akad. Nauk SSSR. Ser. Fiz. Zemli*, No. 8, 30–53 (in Russian).

Symes, W. W. (1979). Inverse boundary value problems and a theorem of Gel'fand and Levitan. *J. Math. Anal. Appl.*, **71** (2), 379–402.

Symes, W. W. (1981). Stable solution of the inverse reflection problem for a smoothly stratified elastic medium. *SIAM J. Math. Anal.*, **12** (3), 421–453.

Tamm, I. E. (1976). *Foundations of the Theory of Electricity*. Nauka, Moscow (in Russian).

Tarantola, A. (1986). A strategy for nonlinear elastic inversion of seismic reflection data. *Geophysics*, **51** (10), 1893–1903.

Tarantola, A. and Valette, B. (1982). Generalized nonlinear inverse problems solved using the least-squares criterion. *Rev. Geophys. and Space Phys.*, **20** (2), 219–232.

Tikhonov, A. N. (1943). On stability of inverse problems. *Dokl. Akad. Nauk SSSR*, **39** (5), 195–198 (in Russian).

Tikhonov, A. N. (1946). On the transition to a steady state of the electric current in a homogeneous conductive half-space. *Izv. Akad. Nauk SSSR. Ser. Geograf. Geofiz.*, **10** (3), 213–231 (in Russian).

Tikhonov, A. N. (1950). Determination of electric characteristics of deep layers of the Earth's crust. *Dokl. Akad. Nauk SSSR*, **73** (2), 295–297 (in Russian).

Tikhonov, A. N. (1963). On solution of ill-posed problems and the regularization method. *Dokl. Akad. Nauk SSSR*, **151** (3), 501–504 (in Russian).

Tikhonov, A. N. and Arsenin, V. Ya. (1977). *Solutions of Ill-Posed Problems*. V. H. Winston & Sons, Washington, D.C.; John Wiley & Sons, New York etc.

Ursin, B. and Berteussen, K.-A. (1986). Comparison of some inverse methods for wave propagation in layered media. *Proc. IEEE*, **74** (3), 389–400.

Vasin, V. V. and Ageev, A. L. (1995). *Ill-Posed Problems with A Priori Information*. VSP, Utrecht, 1995.

Ware, J. A. and Aki, K. (1969). Continuous and discrete inverse-scattering problems in a stratified elastic medium. I. Plane waves at normal incidence. *J. Acad. Soc. Amer.*, **45**, 911–921.

Woodhouse, J. H. and Dziewonski, A. M. (1984). Mapping the upper mantle: three-dimensional modeling of Earth structure by inversion of seismic waveforms. *J. Geophys. Res.*, **89** (B7), 5953–5986.

Yakhno, V. G. (1990). *Inverse Problems for Differential Equations of Elasticity*. Nauka, Novosibirsk. (in Russian).

Yakhno, V. G. (1998). Inverse problem for differential equations system of electromagnetoelasticity in linear approximation. In: *Inverse Problems, Tomography, and Image Processing*. Plenum Press, New York, 211–239.

Yakhno, V. G. and Merazhov, I. Z. (2000). Direct problems and a one-dimensional inverse problem of electroelasticity for "slow" waves. *Siberian Advances in Math.*, **10** (1), 87–150.

Zelenyak, T. I. (1970). *Selected Questions of Qualitative Theory of Partial Differential Equations*. Novosibirsk State University, Novosibirsk (in Russian).